Data Democratization with Domo

Bring together every component of your business to make better data-driven decisions using Domo

Jeff Burtenshaw

BIRMINGHAM—MUMBAI

Data Democratization with Domo

Copyright © 2022 Packt Publishing

Publishing Product Manager: Devika Battike

Senior Editor: David Sugarman

Content Development Editor: Manikandan Kurup

Technical Editor: Rahul Limbachiya

Copy Editor: Safis Editing

Project Coordinators: Aparna Ravikumar Nair and Farheen Fathima

Proofreader: Safis Editing

Indexer: Rekha Nair

Production Designer: Prashant Ghare

Marketing Coordinator: Priyanka Mhatre

First published: June 2022
Production reference: 1120522

Published by Packt Publishing Ltd.
Livery Place
35 Livery Street
Birmingham
B3 2PB, UK.

ISBN 978-1-80056-842-6

www.packt.com

Contributors

About the author

Jeff Burtenshaw has over 9 years of experience in leadership roles at Domo. As Domo's first Master Domo, he led a product incubator to identify and test core product features. During this time, he co-authored several patents and generated thousands of apps for the AppStore. As head of Solution Architecture, he created solutions for many Fortune 500 companies. He is passionate about showing the economic impact of BI initiatives.

Currently, as CEO of consultancy Winning Through Value (WtV) Associates, he engages in value assessment initiatives and Domo consultancy.

Before Domo, he held roles in various organizations leading data warehouse buildouts, BI operations, and custom software development teams.

About the reviewer

Andrea Henderson manages data pipeline architecture at Domo, with over a decade of data analysis and engineering experience and a passion for making data flow faster and better. She is passionate about boosting other women in business and raising the next generation of technologists. Outside of work, her nose is always in a book.

Table of Contents

3
Storing Data

4
Sculpting Data

5
Sculpting Data In-Memory

Section 2: Presenting the Message

6
Creating Dashboards

7

Working with Drill Pathways

8

Interacting with Dashboards

9

Interacting with Cards

Section 3: Communicating to Win

10
Telling Relevant Stories

11
Distributing Stories

Section 4: Extending

14

Extending Domo with Domo Apps

15

Using Domo APIs in Python

16

Using Domo Machine Learning

Section 5: Governing

17

Securing Assets

18
Organizing the Team

19
Establishing Standard Procedures

Index

Other Books You May Enjoy

Preface

For the past decade analytical tools have been evolving in the attempt to simplify the tasks of data gathering, sculpting, analyzing, storytelling, communication, distribution, security, and governance. Historically, this exercise has required a multitude of integrated technologies, and organizations had to stitch together solutions across all these technical disciplines. These technological hurdles were the antithesis of data democracy, vesting data power in a privileged few. This data autocracy resulted in data-breadlines with bottlenecked information requests at a time when the volume, variety and velocity of data were exploding. This book will help you learn ow to enable Data Democracy using Domo, a first of a kind integrated cloud platform-as-a-service offering that addresses the needs of both information producers and consumers at cloud scale in record time. Domo's integrated stack of core Business Intelligence technologies and analytical operations brings the power of data to every person in the organization. This data empowerment from the front-line through the corner office and to external constituencies, gives every individual the opportunity to have a voice that matters. These data-literate individual storytellers can make Data Democracy a reality using the Domo platform.

Data Democratization with Domo begins with an overview of the Domo ecosystem. You'll learn how to get data into the cloud with Domo data connectors and Workbench; profile datasets; use Magic ETL to transform data; work with in-memory data sculpting tools (Data Views and Beast Modes); create, edit, and link card visualizations; and create card drill paths using Domo Analyzer. Next, you'll discover options to distribute content with real-time updates using Domo Embed and digital wallboards. As you advance, you'll understand how to use alerts and webhooks to drive automated actions. You'll also build and deploy a custom app to the Domo Appstore and find out how to code Python apps, use Jupyter Notebooks, and insert R custom models. Furthermore, you'll learn how to use Auto ML to automatically evaluate dozens of models for the best fit using SageMaker and produce a predictive model as well as use Python and the Domo Command Line Interface tool to extend Domo. Finally, you'll learn how to govern and secure the entire Domo platform.

By the end of this book, you'll have gained the skills you need to become a successful Domo master.

Who this book is for

This book is for BI developers, ETL developers, and Domo users looking for a comprehensive, end-to-end guide to exploring Domo features for BI. Chief data officers, data strategists, architects, and BI managers interested in a new paradigm for integrated cloud data storage, data transformation, storytelling, content distribution, custom app development, governance, and security will find this book useful. Business analysts seeking new ways to tell relevant stories to shape business performance will also benefit from this book. A basic understanding of Domo will be helpful.

What this book covers

Chapter 1, *Overview of the Domo Ecosystem*: Before Domo, answering business questions was often a laborious, cross-disciplinary process requiring a diverse mix of technical and business resources to accomplish and, far too often, was much too slow and expensive—falling short in the objective to inform decisions. Enter the Domo Ecosystem, a **Platform-as-a-Service (PaaS)** solution consisting of five core integrated parts: Acquire, Present, Communicate, Extend and Govern. This integrated architecture democratizes the process of answering business questions, which means that everything a user needs to use data to answer relevant business questions is available to them without having to wait for specialists.

Chapter 2, *Importing Data*: As a user, often, the first step you may need to do before answering a question is to import data from wherever it resides. We will review all the intake tools, then go through examples of importing data from the following sources: Excel/CSV files, Google Sheets, Email attachments, cloud apps, databases, On-premise systems, Federated queries, webforms. We will also briefly discuss your options for data outtake.

Chapter 3, *Storing Data*: Now that you have created datasets, we will walk through how the data is stored and accessed in the Domo cloud. Domo, by design, enables an iterative and adaptive approach to data storage. The essential storage structure of a dataset is something that would look like a typical Excel spreadsheet. One advantage that this simplicity offers is that dataset schemas are derived from the data as it is ingested: when data is changed at the source, the schema is also automatically updated. That is important because, in traditional data warehousing, it was all about rigid schema building and compliance – which, honestly, was slow and failed to deliver business value at cloud speed. So, although one can enforce highly normalized schemas in Domo, there is simply not a pressing business reason to do so.

In fact, many Domo customers find that a year or so into the implementation, their Domo instance has organically accumulated more tabular data than their legacy data warehouse. This is primarily due to the low-hassle manner in which data is ingested and stored, which was enabled by the fundamental assumption (now proved over years) that the data itself contains all the information needed to create and maintain the raw schema. Storing the data becomes useful when it is also easy to find, profile, and secure. We will walk through searching, browsing, data profiling, and securing datasets as well.

Chapter 4, Sculpting Data: Artistic sculpting requires the ability to see in raw material the potential for a work of art. The sculptor has a range of tools at their disposal to shape raw stone into something more valuable. Data Analysts, like sculptors, can see potential in raw data to answer questions. Analysts also require tools to shape and refine the raw data. The Domo platform has many tools to enable data sculpting. If you are not a technical data manipulation guru, no worries, we will cover Domo's very approachable graphical drag and drop tool that requires no coding, and that any business analyst can use. However, if you are a SQL expert, Domo has also had tools to enhance your scripting skills for sculpting data we will discuss as well.

Chapter 5, Sculpting Data In-Memory: In working with data, sometimes it is very necessary for performance or convenience to conduct operations on the data that are not persistently stored to disk. For performance, an example is combining billions of rows of transactions with millions of rows of customer data. For convenience, an example might be a simple lookup code mapping or applying business logic. We will cover these fine-tuning adjustments enabled by tools that work on in-memory cache and query technology, in Domo's case Adrenaline which is essentially a queryable in-memory cache. These in-memory transforms are magnitudes faster than persistent transforms via ETL and when speed is what you are after these tools are the right choice. Occasionally, the use of Adrenaline dataflows becomes critical in cases when the rows of data being processed reach billions of rows and joins are required and materialization is not practical. If you are familiar with creating non-materialized views on a relational database or using formulas in excel, then this is similar technology.

In this chapter we will cover Beast Mode, Data Views, and Data Blend tools that are part of the Domo MAGIC transforms family (magic because they are so easy to use) that operate in memory cache.

Chapter 6, Creating Dashboards: This chapter covers creating cards and dashboards on the data. Analyzer is the tool used for creating individual Cards that are contained in dashboard pages. For navigation, the dashboard pages are arranged into a logical tree structure up to three layers deep organizing how the content is presented. A Card is simply a chart contained inside of a tile on a dashboard page. Analyzer is commonly referred to as Card Builder but is also a powerful ad-hoc data exploration tool. Many card types also support interactive filtering while viewing from the dashboard.

We will focus on creating and navigating dashboards; working with Analyzer; and creating and editing cards.

Chapter 7, Working with Drill Pathways: Often when viewing cards users will want to drill down to see more details. Every card in Domo by default is enabled to drill down to the raw dataset. However, as a card designer, you may want to pre-define the drill path with custom visualizations that communicate more effectively than a raw table view. Creating these drill paths in Domo couldn't be simpler—actually, you just create another chart as a drill path under the main chart and Domo automatically figures out how to enable the drill down linkages! There is even the ability to drill down to a completely different dataset than the parent card's dataset provided the column names are the same. This ability to drill to a different dataset is particularly useful on large datasets where the parent dataset is aggregated and enables drilling to the larger dataset details.

We will cover creating card drill paths, drilling down to a different dataset, linking cards, and comparing parts to whole with segments.

Chapter 8, Interacting with Dashboards: One of the useful things a Domo dashboard can do is allow the user to start from a default dashboard view and then interact with the content, exploring the data in real-time. The Collections feature organizes the layout of cards on the page into logical sections like organizing slides in a PowerPoint. There is also Page Filters that enable filtering of all the cards on a page at once on any field including beast modes in the dataset. Filters once set can be named and saved as a group for re-use. Another advanced feature is Interactive Card Filtering that when enabled sets the page filters for all cards based on what is clicked on in a particular card. In other words, using one card to filter contents in other cards.

We will cover dashboard page interactions, working with collections, and using page filters.

Chapter 9, Interacting with Cards: Card interactions give us fine grained control of what we see in a card. These controls encompass changing the chart type, zooming in and out, filtering the data, exporting the data to Excel and even pivot table functionality. All these interactions are temporary in nature, as a browser refresh will reset the card to the default state set by the designer. These features are located on the card details page in Domo. There is also a lot of information about when the card data was last updated, who has access to, and who has viewed the card. This information provides context, and the interactivity is helpful for users when they need to do data discovery on the fly. These features enable the user to slice the information, experiment with different visuals or make permanent annotations of significant events that impact the data. There is even a feature to bring attention to any issues with the card to the card's owner.

We will cover using card interactions, using pivot tables, and exporting data to Excel.

Chapter 10, Telling Relevant Stories: Being relevant with data means bringing information to light that is important to current issues at hand. In a business context, a business model provides a framework for storytelling. We will learn about deciding what the relevant parts of the business are, and for organizing the story presentation using a pattern. A pattern for relevant storytelling seeks to explain the journey of where the story began, what happened, where things are now and where things could be in the future. A story is not meant to merely present facts and figures but to advocate for a position. Stories that don't decide what to advocate for or against may be interesting but are irrelevant and not actionable.

We will learn how to use a business framework to focus on a relevant story, then create a specific statement for the story to cover, and then apply a storytelling pattern in the Domo stories tool.

Chapter 11, Distributing Stories: This chapter will cover the many ways enabled by the Domo platform to disseminate the content. This includes native iOS and Android devices, scheduled emails, digital wall board presentations, secured mass distribution via URL, embedding in PowerPoint and embedding in bespoke web pages.

We are going to ensure the message gets out by practicing all these options for communications through channels enabled via Domo Platform.

Chapter 12, Alerting: So much time is spent with people working to make sense of data. Increasingly, technology is allowing us to turn that dynamic around and make the data work for us. A major way of putting the data to work is through alerting technology. Alerts allow continuous monitoring of data looking for conditions to be met, fail, crossed, removed, created and more. A well thought through alerting strategy can be more impactful on business performance than a great dashboard. In fact, and argument can be made that the main reason to have a monitoring dashboard is to automate alerts.

We will cover defining an alerting strategy, understanding the alert center, setting alerts on cards and datasets, subscribing to alerts created by others.

Chapter 13, Buzzing: Collaboration among users is a core part of making sense of information. The ability for users to dialog around content is critical. A big gap in the efficiency and effectiveness of most messaging solutions is that the users need to provide the content for the discussion from sources outside of the communication tool. For example, email attachments.

Buzz is Domo's topic-based messaging application that includes a content provided messaging solution! It goes one step further than tools like Slack (which provide self-defined topic threaded conversations) in that Buzz also provides contextually aware topics automatically. Say hello to automatic topics and goodbye to attachments and cut and paste insertions for data context.

We will cover the features to use Buzz effectively.

Chapter 14, Extending Domo with Domo Apps: There are many layers of architecture including data tiers, business logic tiers, security, user management, mobile apps, scalable infrastructure and more. What if we were to discover a platform where much of what a developer needs for building an enterprise grade application already existed and all that remained was to connect the data to the application functionality via HTML, CSS and client scripting. That would be fantastic.

We will learn what the Domo Dev Studio Framework delivers, how to leverage it to extend the Domo Platform via creating custom apps, and how to deploy those apps to the Domo App Store.

Chapter 15, Using Domo APIs in Python: Python as a language is very attractive for rapid application development, because it is simple, free to use, and easy to distribute. All these characteristics make it a great tool for democratizing extensions to the Domo platform via Domo APIs, directly or wrapped by Python libraries.

We will look at the two available options for using Python with Domo. The first is through the **Extract, Transform, and Load** (**ETL**) Scripting Tile for Python. The second is by using a local Python installation to access the Domo APIs.

Chapter 16, Using Domo Machine Learning: Democratizing ML, putting the power of ML into non-data scientist hands, is what Domo's AutoML feature does. Domo partnered with Amazon Web Services and their SageMaker Autopilot product to bring this capability to Domo platform users in a turnkey way. Domo also supports Jupyter Workspaces.

We will review the AutoML process, training Auto ML Models, Deploying AutoML models, and using Jupyter Workspaces.

Chapter 17, Securing Assets: Security is a primary concern when covering the governance of People, Groups, Roles, Content, Authentication, Network Security, Company Settings, Content Embedding, Feature Settings, and more.

We will cover a policy framework guiding tasks in securing platform assets, organizational culture considerations, authentication, network security, groups and roles, and handling **Personally Identifiable Information (PII)**.

Chapter 18, Organizing the Team: Four primary organizational roles exist to support the effective application of Domo in an organization: Executive Sponsor, MajorDomo, Domo Master, and Data Specialist. These roles should be formally defined and assigned. The roles can be held by a single person in a small organization or divided among many people in large organizations.

We will review organizational structure options including organization size, scope, and evolution; define the roles and responsibilities; and present job descriptions for hiring.

Chapter 19, Establishing Standard Procedures: The right amount of procedure is helpful; but having too much procedure is wasteful. When implementing Domo there are a few procedures to establish that are core to adoption and more effective use of the platform.

We will cover procedures for determining artifact ownership, accepting new requests; prioritizing and assigning request implementations; and developing, testing, and migrating changes to production. We will also look at formal content certification.

To get the most out of this book

This book requires use of a Domo instance. You can obtain a free trial of Domo using a sign-up page at `https://www.domo.com/start/free`. You will need access to a machine that you have administrative rights to and can download files and execute install programs. To complete some exercises, you will need to install or have access to Postgres and Python3. We walk through how to get these environments setup. As a few of the features covered are paid features, establishing a relationship with Domo Sales is advised to gain access.

Software/hardware covered in the book	Operating system requirements
Domo	Windows, macOS
Postgres	Windows, macOS
Python3	Windows, macOS

If you are using the digital version of this book, we advise you to type the code yourself or access the code from the book's GitHub repository (a link is available in the next section). Doing so will help you avoid any potential errors related to the copying and pasting of code.

Finally, to get the most out of this book, you should have a basic understanding of Domo, GitHub, SQL, JavaScript, HTML, and Python. Visual Studio Code is recommended for code editing.

Download the example code files

You can download the example code files for this book from GitHub at `https://github.com/PacktPublishing/Data-Democratization-with-Domo`. If there's an update to the code, it will be updated in the GitHub repository.

We also have other code bundles from our rich catalog of books and videos available at `https://github.com/PacktPublishing/`. Check them out!

Download the color images

We also provide a PDF file that has color images of the screenshots and diagrams used in this book. You can download it here: `https://static.packt-cdn.com/downloads/9781803232416_ColorImages.pdf`.

Conventions used

There are a number of text conventions used throughout this book.

`Code in text`: Indicates code words in text, database table names, folder names, filenames, file extensions, pathnames, dummy URLs, user input, and Twitter handles. Here is an example: "In the **Filename** field, select the `persons.csv` file you downloaded."

A block of code is set as follows:

```
-- Lead Source Standardization
case
when `LeadSource` = 'Patnes' then 'Partners'
when `LeadSource` = 'Diect' then 'Direct'
when `LeadSource` = 'Sales Ceated' then 'Sales Created'
when `LeadSource` = 'efeal' then 'Referral'
```

```
when `LeadSource` = 'Self-Souced' then 'Self-Sourced'
when `LeadSource` = 'Maketing Outbound' then 'Marketing
Outbound'
when `LeadSource` = 'Stategic Accounts Maketing' then
'Strategic Accounts Marketing'
when `LeadSource` = 'Patneing' then 'Partners'
else `LeadSource` -- If not mapped then keep as is
end
```

When we wish to draw your attention to a particular part of a code block, the relevant
lines or items are set in bold:

```
SELECT
DATE(`LastModifiedDate`) as 'Date',
SUM(case when `IsWon`= 'true' then 1 else 0 end) as 'Won',

-- Lost = Closed - Won
SUM(case when `IsClosed` = 'true' then 1 else 0 end) -
SUM(case when `IsWon`= 'true' then 1 else 0 end) as 'Lost',

SUM(case when `IsClosed` = 'true' then 1 else 0 end) as
'Closed'
FROM `opportunity_cleansed`
GROUP BY DATE(`LastModifiedDate`)
ORDER BY 1 DESC
```

Bold: Indicates a new term, an important word, or words that you see onscreen. For
instance, words in menus or dialog boxes appear in **bold**. Here is an example: "**Python
Script** is the design tile to drag and drop for use in a dataflow."

> **Tips or important notes**
> Appear like this.

Get in touch

Feedback from our readers is always welcome.

General feedback: If you have questions about any aspect of this book, email us at `customercare@packtpub.com` and mention the book title in the subject of your message.

Errata: Although we have taken every care to ensure the accuracy of our content, mistakes do happen. If you have found a mistake in this book, we would be grateful if you would report this to us. Please visit `www.packtpub.com/support/errata` and fill in the form.

Piracy: If you come across any illegal copies of our works in any form on the internet, we would be grateful if you would provide us with the location address or website name. Please contact us at `copyright@packt.com` with a link to the material.

If you are interested in becoming an author: If there is a topic that you have expertise in and you are interested in either writing or contributing to a book, please visit `authors.packtpub.com`.

Share Your Thoughts

Once you've read *Data Democratization with Domo*, we'd love to hear your thoughts! Scan the QR code below to go straight to the Amazon review page for this book and share your feedback.

https://packt.link/r/1-800-56842-8

Your review is important to us and the tech community and will help us make sure we're delivering excellent quality content.

Section 1: Data Pipelines

The hardest part made easy ... getting the data.

Getting to the data – no matter where it lives – is the hardest and most time-consuming task in analytics work. And the second most time-consuming is sculpting the data into a format that will provide insights. In this section, you will learn how data is stored, sculpted, and transformed within the Domo ecosystem. This includes intake paths that are going to get tactically easier. Your ability to sculpt the data will not require you to become a SQL guru. Imagine a wizard-driven solution that acquires data from sources and automatically creates the schema and provisions the data storage. Furthermore, imagine an **Extract, Transform, and Load (ETL)** tool that is schema-aware and provides a no-code, drag and drop experience that is so intuitive data transformation pipeline creation feels almost like magic.

After completing part one, you will be able to intake (acquire, store and sculpt) data in record time at cloud scale without needing highly technical resources.

This section comprises the following chapters:

- *Chapter 1, Domo Ecosystem Overview*
- *Chapter 2, Importing Data*
- *Chapter 3, Storing Data*
- *Chapter 4, Sculpting Data*
- *Chapter 5, Sculpting Data In-Memory*

1
Domo Ecosystem Overview

Before Domo, answering business questions was often a laborious, cross-disciplinary process requiring a diverse mix of technical and business resources to accomplish it and, far too often, was much too slow and expensive—falling short on the objective of informing decisions. Enter the **Domo ecosystem**, a **Platform-as-a-Service (PaaS)** solution consisting of five core integrated parts: **Acquire**, **Present**, **Communicate**, **Extend**, and **Govern**. This integrated architecture democratizes the process of answering business questions, which means that everything a user needs to use data to answer relevant business questions is available to them without having to wait for specialists.

You'll explore the Domo ecosystem by reading through the following topics:

- Introducing the Domo ecosystem
- Acquiring a data pipeline
- Presenting the story dashboard
- Communicating with Domo Buzz
- Extending with apps and APIs
- Governing security and operations

Introducing the Domo ecosystem

The **Domo ecosystem** was designed from its inception to be disruptive to how data is acquired, structured, stored, transformed, shared, used, extended, and governed. The overarching objective in its creation was to empower everyone to answer business questions at the speed of business. For example, no more waiting in the proverbial BI breadline for IT or technical resources. Understanding that many questions have a shelf life and are perishable in their utility is a foundational belief that translated into access, speed, and agility in the product. A large part of the product investment was made in data acquisition/pipeline capabilities to ensure that the most difficult and time-consuming activities in answering business questions, namely data intake, storage, and sculpting, are streamlined for the non-technical user while preserving appropriate security.

Therefore, to answer questions quickly, it follows that a typical user needs to be able to do the following:

- Intake data quickly, whatever the data volume, variety, and velocity.
- Provide data extraction to outtake data from Domo to other applications.
- Have the data automatically stored and indexed for fast queries at scale.
- Sculpt the data and automate data pipelines.
- Present relevant stories that drive favorable outcomes.
- Securely share information and communicate with context.
- Enable extensions with a crowd-sourced app store and custom apps.

This unique combination of pre-integrated and universally governed layers in the ecosystem enables non-technical users to quickly, accurately, securely, and relevantly deliver end-to-end solutions.

The following diagram visualizes the Domo ecosystem:

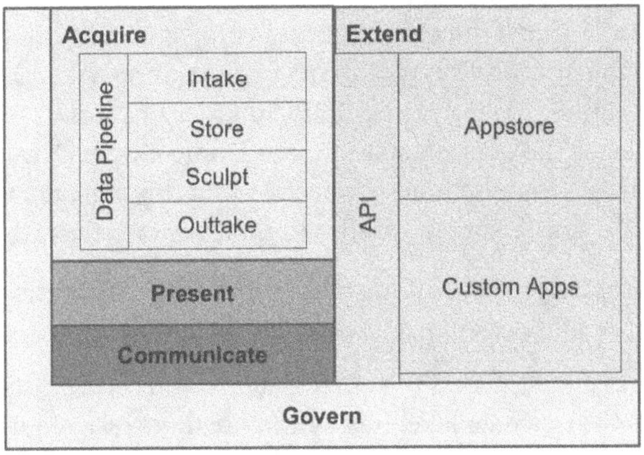

Figure 1.1 – The Domo ecosystem

Let's look at each of the layers shown in *Figure 1.1* in turn, starting with **Acquire**.

Acquiring a data pipeline

The **Acquire** layer contains the data pipeline processes for collecting data from a source system(s), storing the data, sculpting the data, and providing data back to other applications.

Intake tools – importing data from various data sources

This section will discuss the main options in the Domo ecosystem for data intake:

- A **connector** pulls data into the Domo cloud from files, cloud apps, and databases. It's a Domo application that, through a wizard-like interface, allows a user to select source systems to get data from. The user then simply enters a few configuration inputs and provides credentials in order to copy and store the data in the Domo cloud data warehouse as a dataset.

  There is a rapidly growing universe of cloud-based applications. As of the time of writing, the Domo ecosystem supports over 670 cloud app connectors. Domo's connector team constantly keeps the connectors updated to the vendors' current API versions. And they also created an API so that if you can't find the connector you need, you can create your own custom connector.

- **Workbench** pushes on-premises data from behind firewalls into the Domo cloud. It is a Windows-based application that you can download and install on a Windows machine, sometimes called the universal data connector as it can handle any flat file or database table upload. It typically sits behind a firewall and pushes the data up to the Domo cloud. It mainly uses ODBC queries to connect to databases, and essentially bulk loads the query result into the Domo cloud. Organizations that do not want to whitelist Domo's cloud connectors find this a great solution—avoiding the need for Domo connectors to reach into their systems from the cloud.

  Workbench has robust job scheduling and monitoring capabilities and can even listen to a directory and automatically upload new or changing Excel files.

- **Federated queries** enable access to source data from wherever the data currently resides, for example, in a data lake. It is Domo's data-at-rest acquisition method for organizations with large existing data lakes. It facilitates the need to have a way to use content where it lies rather than having to copy it. So, with federated queries, the data is not permanently persisted in the Domo cloud. Rather, with federated queries, the query data is cached in memory with a time-to-live setting. The time-to-live setting determines how long the query results will persist in memory before a new request causes a re-query or simply times out and is released. Federated queries effectively make Domo Analyzer a powerful data lake discovery tool and unlock data lake information without any data transport requirement.

- **Internet of Things (IoT)**: Machine-to-machine streaming communication data capture is the newest data acquisition type supported in the Domo ecosystem. This area addresses capturing the information flowing from internet devices and machine sensor data. Existing connectors include AWS IoT Core, AWS IoT Device Management, AWS IoT Device Defender, AWS IoT Analytics, AWS Kinesis, OPC, Matomo, Apache Kafka, MQTT, Particle.io, Beonic Traffic, and more. Domo has partnered with AWS and Verizon ThingSpace to bring innovative solutions to the market. Another example is how SharkNinja is using Domo IoT to improve the customer experience using sensor data: `https://bit.ly/32I44w7`.

- **Webforms** are your go-to tool when you just need a simple brute-force way to get data into the Domo cloud. You can enter or cut and paste tabular data into a webform. Think of it as a cloud spreadsheet lite. This is used for lookup table creation, test data, and other miscellaneous needs to create a dataset on the fly.

Store – automatic schema management and performance optimization

Figuring out where and how to store data can be a time-consuming and costly exercise. After you have decided on a structure and location to store the data, then the ongoing process of tuning the data structure so that it is performant requires specialized technical skills. All this means that the storage problem can become a significant obstacle. With Domo, that obstacle is removed via automatic data structure design, storage, and performance tuning, all enabled in a business user-empowering way.

Data architecture (Vault, Adrenaline, Tundra, Federated)

The Domo data architecture consists of four major parts:

- **Vault** is the persistent storage component that is just a bunch of disk storage. Everything is flattened and stored in tabular arrays called datasets. Vault is a collection of datasets whose data structures are created and cataloged automatically as the data is ingested. The acquired data is physically stored in leading cloud infrastructure providers such as Amazon S3 or Azure.

- **Adrenaline** is an in-memory data store indexing vault datasets for high query performance. The Domo visuals, beast modes, and filters run a query language called **DQL** (short for **Domo Query Language**) against Adrenaline to retrieve the data for the visual.

- **Tundra** is a specialized in-memory query optimizer companion to Adrenaline and handles queries that specifically fit its optimization algorithm.

- **Federated** storage is virtually mapped into the Domo ecosystem but is physically stored outside of Domo. If you already have a data lake, then federated queries, which work on federated storage, are a good option. All the preceding storage components work seamlessly together. Behind the scenes, the Domo operations team works to monitor and tune the performance for you.

Cloud data lakes

Importantly, UI components are built on top of the storage layer that showcases the cloud data lake user artifacts. The basic data lake user artifacts are data sources, dataflows, and datasets. Users can search, manage, and enhance the data lake user artifacts. Conveniently, the data lake catalog and data lineage information are auto-generated and can be tag enhanced by the users for the data dictionary. The UI even shows in real time how the data is flowing into the data lake. All the external data sources are represented in the outermost ring of the platter. The data source platter's inner ring shows the data sources created from data sculpting as shown in *Figure 1.2*:

Figure 1.2 – Data source platter

Additionally, to see datasets derived from the data source, you click on a particular data source and then another platter layer opens containing the dataset(s), as seen in *Figure 1.3*:

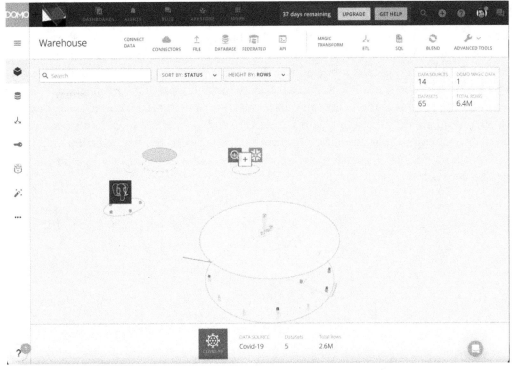

Figure 1.3 – Datasets platter

Data sculpting

Data sculpting tools are not new; for example, Informatica, Boomi, Alteryx, and so on have been around for years. Some have limitations regarding the lack of auto schema discovery, and the lack of auto-generated, non-siloed data cataloging. This leads to a cumbersome data sculpting experience. Domo's data sculpting tools are schema-aware and integrated from source to report. This enables the user to see potential reuse or do new transforms on data quickly. The fully integrated and schema-aware data sculpting toolset includes the ability to stitch data together and transform data, similar to well-known ETL tools. Primarily, it's a no-code, visual programming paradigm for greater adoption. Secondarily, for those SQL scripters out there, a SQL serialization tool is available as well. As you will see, the integration and ease of use truly democratize the data sculpting activity.

Magic ETL

Magic ETL is a drag-and-drop, no-code data sculpting tool that enables a typical user to sculpt the data they have acquired into Domo Vault. It leverages the Apache Spark framework and is a simple yet powerful visual data transformation tool. Transforms include text, date, and number operations. Custom formulas, parsing, and mapping operations are provided, as are filter and de-duplication operations. Of course, merge and join operations are included along with aggregation and pivot operations. Additionally, data science operations for classification, clustering, forecasting, outlier detection, multivariate outliers, and prediction are available and can be automated via Python and R scripting support.

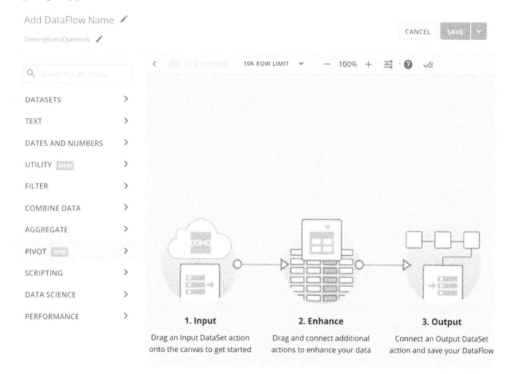

Figure 1.4 – Magic ETL

Magic SQL

The Magic sculpting tools have evolved over time from scripting to no code. Magic ETL is the newest no-code data sculpting tool while **MySQL Dataflows**, which is a **Structured Query Language** (**SQL**) tool supporting serialized SQL statement execution, was the original sculpting tool option. MySQL Dataflows runs on dynamic **MySQL** instances and is perfect for the SQL developer who doesn't want to transition to Magic ETL; although, it is not as scalable as Magic ETL.

Blend

Blend, formally **Fusion**, is the Domo equivalent of a VLOOKUP function in Excel creating a new virtual dataset by adding additional information to the primary dataset by combining data from the lookup dataset. It also has properties that allow functions to append datasets like a UNION operation in SQL.

Data Views

Data Views is a new feature that enables users to create aggregates and/or filtered views on an underlying dataset. It is very similar to views available on relational databases. Views are non-materialized and are only available to the Magic ETL dataflows.

> **Important Note**
> Because Domo is integrated from bottom to top, the data sculpting jobs can be triggered to run when any of the source data changes. No more worrying about batch window timing because the job will run when the source data has changed. Of course, the jobs can also run on a specific time schedule as needed.

Outtaking data

Don't leave your insights stranded. A crucial part of any data platform resides in the ability of other systems to **outtake** data. Several ways exist to export data from the Domo cloud.

Exporting

The first question many analysts ask when evaluating Domo as a platform is *What are the options to export data?* So, the product makes sure that getting data exported to Excel is simple, fast, and always possible.

In addition to Excel exports, there are PowerPoint exports and add-ins for Excel and PowerPoint live data embedding.

Finally, there is a RESTful data API for those who want to hit Vault directly to retrieve data.

Writeback connectors

A relatively new feature is writeback connectors. As the name implies, this allows Domo to write data in Domo directly into other systems. A quick search in Domo for writeback connectors reveals 22 connectors to many popular systems, as seen in *Figure 1.5*. Specific writeback connectors are part of Domo Integration Cloud and require your Domo account team to unlock the feature for your use.

Figure 1.5 – Writeback connectors

RESTful API

For those interested in taking a code-based approach to extracting data from Domo, a RESTful API exists to extract data from the Domo cloud as well: https://developer.domo.com/docs/dev-studio-references/data-api.

Now that we have examined the data acquisition pipeline (Intake, Store, Sculpt, and Outtake), it's time to discuss options for presenting the data.

Presenting the story dashboard

Presenting refers to the parts of Domo that are used to visualize and distribute content. The main presentation artifacts in Domo are **cards** (tables, charts, and graph visuals) and **pages** (dashboards/stories). The visuals are linked to the underlying datasets and update in real time as the datasets are updated. Beast Mode allows you to create powerful custom formulas on the fly.

Analyzer

Analyzer is the card creation and editing tool. This is where the visuals are created. It also happens to have solid real-time filtering and is a powerful ad hoc analysis tool—though it is not often thought of in that way. The drag-and-drop user interface will be very familiar to anyone who has created a chart or used a pivot table before.

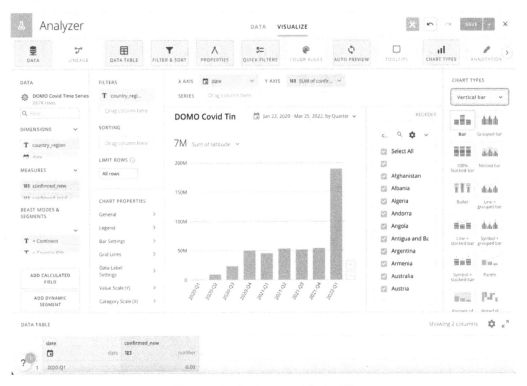

Figure 1.6 – Analyzer card design UI

Beast Mode

Beast Mode is a data sculpting tool that enables the user to create new fields from calculations on the fly. It's called Beast Mode because it enables you to overcome many data challenges with custom formulas. If you know about case statements and Excel formulas, this is where you go to do that kind of work. Beast Mode is accessed through the analyzer UI.

> **Important Note**
>
> Currently, Beast Mode is a one-pass calculation engine, which means it cannot perform post operations on aggregations such as `SUM()` or `MAX()`. For those of you familiar with SQL coding, this would be like using a `Having` clause and it is not supported in Beast Mode.

Dashboards and stories

Pages are the standard presentation artifact in Domo. The destination of any card is to a page. Pages typically contain several cards (sometimes more) and the pages can be layered up to three menu levels. Each page is commonly referred to as a **dashboard** and has built-in layout formatting and card placement capabilities. In order to tell a story as an analyst would typically do in a PowerPoint deck, a page can be converted into a **story** layout, which provides much more granular control of the presentation commentary, as well as the look and feel.

Pages and the card(s) they contain are updated in real time and typically address data from multiple datasets. Pages and cards on the page, either in standard or story layout, are instantly deployed to the Domo mobile app for iOS, Android, and mobile web.

> **Important Note**
>
> If you are considering developing a custom mobile app, take the time to evaluate the Domo mobile app as it may already have the ability to cover what you are looking to do out of the box. Search for Domo in the mobile app stores for Apple or Android.

Content distribution

Pages enable people to take the initiative and look at information of interest. But what are your options if you want to push content to people?

Scheduled reports

Fortunately, the Domo platform allows for the distribution of content as a scheduled email with links and attachments.

Public or private URLs

Domo has a publish feature that allows any page to become a URL presentation. The URL is visible to anyone with the link but can be password protected. A downside of this feature is the content is no longer interactive. This feature is often used to drive office **wallboards** that rotate through cards on a timer.

Domo Everywhere – Embed

Domo pages and cards are designed to be embedded or white-labeled in other portals and applications. Many Domo customers use Domo content in their external customer/ supplier portals as well as internal employee portals. Some customers enhance their product and increase their time to market and feature capabilities by embedding Domo best-of-breed visuals in their products. There are both private-embed and public-embed offerings for internal- and external-facing content respectively. Embedded content is interactive.

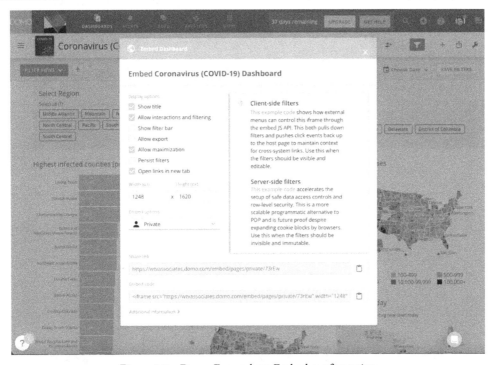

Figure 1.7 – Domo Everywhere Embed configuration

> **Important Note**
>
> If you can evaluate how content is presented to external stakeholders and your IT organization is considering a bespoke build, you will likely get to market faster, with more features, more securely, on mobile and desktop, with fewer costs, and less risk if the Domo Everywhere Embed feature is adopted rather than a custom build.

Next, let's peruse Domo's integrated communication tools.

Communicating with Domo Buzz

Presenting data is often just a step toward resolving an action. The dialog that ensues is important to the process and can sometimes be an ongoing process. To facilitate these kinds of communication in an integrated way, Domo has several communication tools.

Domo Buzz

Domo Buzz is a Slack-like tool that enables the dialog around topics and supports threads. In Domo, topics are available automatically around pages and cards. This means both the topic and the content are auto-linked and you don't have to bring your own content. No more digging through email threads to find the relevant data and wondering whether it is still accurate; Buzz is always available as a pop-out shelf in the UI, and conversations can be public or private.

Figure 1.8 – Domo Buzz card panel

Next, let's look at the alerting feature.

Alerts

Alerts in Domo are sourced from multiple places: cards, datasets, and Mr. Roboto, the Domo AI.

User alerts

User-originated alerts are created on cards and datasets, are supported through a publish and subscribe mechanism, and are visible in the alert center. When triggered, an alert can be received by push notification, text, or email.

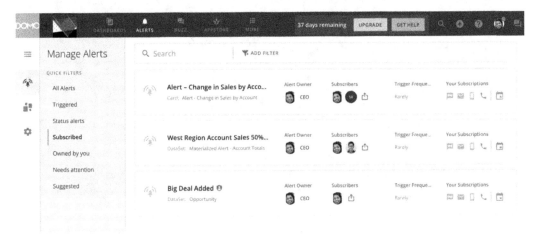

Figure 1.9 – Alert center

Next, we'll see some of Domo's AI capabilities around communications.

Mr. Roboto

Mr. Roboto insights are system-generated by the Domo AI engine and are found in two places: first, the **Discover console** under the **ALERTS** menu shows insights on activity in the Domo instance.

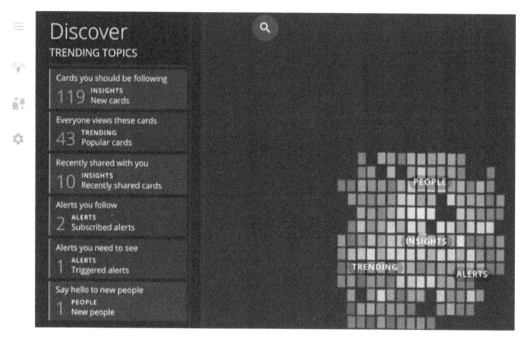

Figure 1.10 – Discover console

And second, the **Card Insights panel** on the card details page (which surfaces insights generated by the AI on the card).

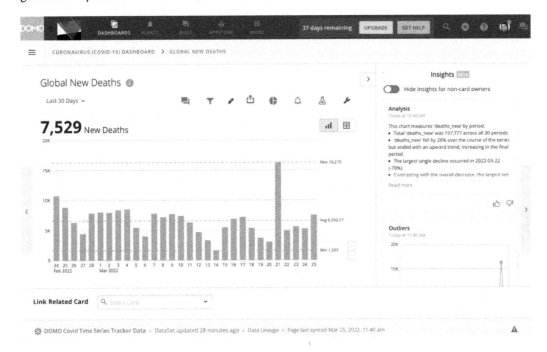

Figure 1.11 – Card Insights panel

Next, let's review the profile feature.

Profiles

Effective communicators realize that knowing what is important to the other person(s) and being able to build on common ground are crucial. The **profile** feature is a way to quickly learn about a person's interests and what is important to them in terms of the information they consume and the people they converse with. Profiles also function as an updatable user organization directory as seen in *Figure 1.12*:

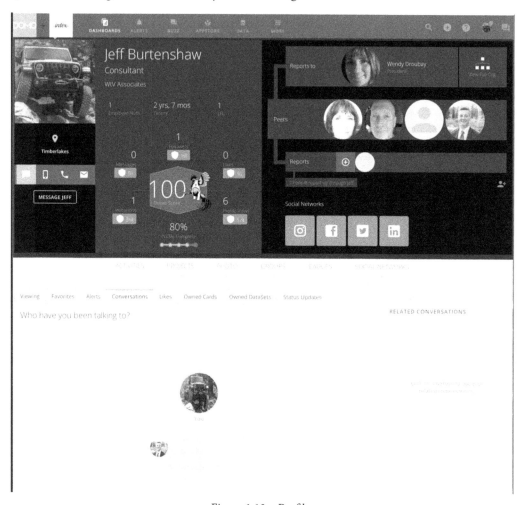

Figure 1.12 – Profile

Next, let's review the project management features.

Projects and Tasks

Projects and Tasks, as the name implies, is a lightweight Trello-like project task management tool. Any Domo Buzz message can be converted into a task to be assigned and tracked.

To enable analytics on Projects and Tasks, install the DomoStats Projects and Tasks App from the Appstore, which creates the datasets that contain the Projects and Tasks data and a dashboard. The dashboard content can be edited just like any other dashboard, as seen *Figure 1.13*:

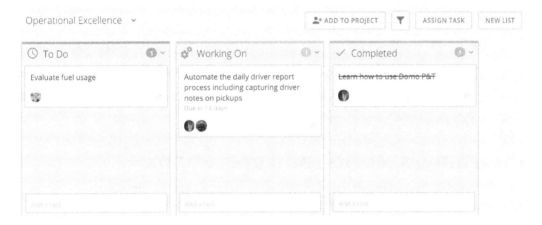

Figure 1.13 – Projects and Tasks project board

Important Note

DomoStats is a moniker used for internal data about Domo. Search *DomoStats* in the data connectors area to see a full list of DomoStats data sources and related apps.

Next, let's learn about how to extend Domo using apps and APIs.

Extending with apps and APIs

The Domo ecosystem is extensible to the crowd through the Appstore and through APIs.

Appstore

The Domo **Appstore** contains many pre-built, configurable applications that can be installed into your Domo instance from the Appstore. Many are free of charge while some are created by partners and have a charge to use them. One innovative feature is that any dashboard can be uploaded as an app into the Appstore. The uploaded app can be private to your instance or shared across all Domo instances.

APIs

APIs exist to use connectors, data, pages, accounts, users, groups, projects, and tasks in Domo. More information on the Domo APIs can be found here: `https://developer.domo.com/`.

The most common use for APIs is to create custom data pipelines and build custom applications.

Next, let's discuss governance and security.

Governing, security, and operations

To govern a system, you need to see what artifacts are in the system, who owns the artifacts, who can access the artifacts, and how the artifacts are being used. If you can't track it, then you can't govern it. Let's look at some of the tools enabling governance of the platform.

Tools

The Domo ecosystem has an extensive feature set supporting enterprise governance through all platform layers. A must-have for any **MajorDomo** Domo system administrator is the **Domo Governance Datasets Connector** available in the Appstore. This connector app allows you to create datasets with metadata and usage for Beast Mode, cards, pages, datasets, dataflows, users, groups, and more.

Data lineage is tracked from source to consumption, similar to many industries where it is important to understand the chain of delivery, for example, from field to fork.

User and group security roles and privileges are enterprise-grade, including single sign-on and row-level data access controls.

The **DomoStats People app** can be installed from the Appstore and shows user security and usage patterns.

Domo **CourseBuilder** is a downloadable app available to create training content. Search for CourseBuilder in the Appstore.

Visual indicators of artifact ownership and access sharing are pervasive, creating trust and facilitating communications.

For the audit-oriented, there is even a detailed activity log, as seen in *Figure 1.14*:

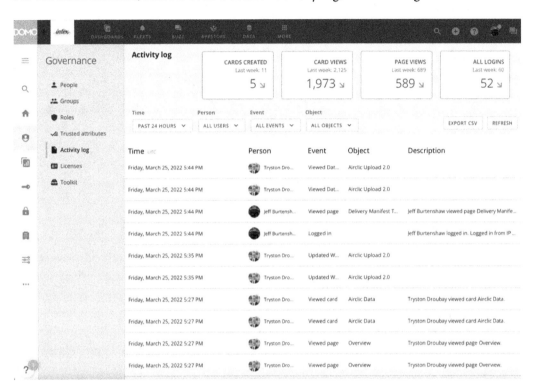

Figure 1.14 – Activity log

Impact tools for datasets, fields, and Beast Mode are at the ready to prune and standardize the experience, as seen in *Figure 1.15*:

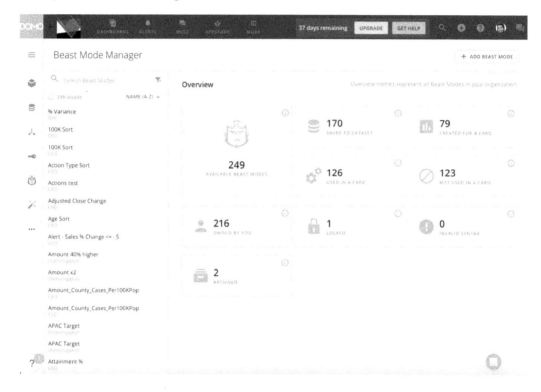

Figure 1.15 – Beast Mode Manager

Overall, the Domo ecosystem has the enterprise-grade tools to be a trusted and heavily utilized platform.

Next, let's discuss some people and organizational considerations.

People – organizational role design

Domo provides powerful technology, but the appropriate organization of people's responsibilities is also a critical factor in getting the most from the technology. Domo suggests the following roles be assigned in any Domo implementation:

- **Executive Sponsor**: A business leader who has the budgetary and result accountability for the Domo ecosystem in the organization
- **MajorDomo**: A person who has overall administrative ownership, driving clear artifact ownership and sharing policies, and is able to synthesize business needs with the Domo ecosystem's capabilities

- **Data Specialist**: Data architects who have overall responsibility for the data pipelines and data governance in specific subject areas

The following are additional roles that come with large/global organizations:

- **Domo Master**: A tactical expert on Domo platform features serving in consultative positions to execute business priorities enabled in the ecosystem
- **Team Champion**: Analytic talent that executes business strategy for a given business area primarily in the present and communicate layers of the ecosystem

That concludes our introduction to the architecture of Domo. Let's recap what we've learned.

Summary

In this chapter, we reviewed five major areas in the Domo ecosystem (acquiring, storing, presenting, extending, and governing). We learned all these areas are integrated, working together as seen in *Figure 1.1*. This integration simplifies the user experience, making it so that a typical business user can use the entire ecosystem without having to rely on other more technical resources. This democratizes the process while also tracking user activities so that proper governance can be enforced. We learned that it is important to not only create content but also have proper roles in place to get the most out of the technology and that as the organization's scope grows, the roles also expand.

In the following chapters, we will be diving deeper into all five areas, providing detailed steps for how to accomplish various tasks. We will start with getting data into Domo, then move on to sculpting the data, then address presenting and distributing content. Additionally, we will cover extending the system beyond Domo. And finally, we will walk through governing activities.

In the next chapter, we will take a hands-on approach to using the acquire pipeline intake features of the Domo ecosystem.

2
Importing Data

As a user, often, the first thing you may need to do to answer a question is to import data from wherever it resides. Acquiring data via the intake process is accomplished in Domo through data connectors, Domo Workbench, and other methods. So, let's dive right in and start acquiring data.

In this chapter, we will cover the following topics:

- Overview of intaking tools
- Importing data from Excel and CSV files
- Importing data from Google Sheet
- Importing data from email attachments
- Importing data from cloud apps
- Importing data from databases
- Importing data from on-premises systems
- Using federated data connections
- Using Webforms
- Outtaking data from Domo

Technical requirements

To follow along with this chapter, you will need the following:

- Internet access

- Your Domo instance and login

- Ability to download example files from GitHub

- Admin access to a computer or VM running Windows 10

- Ability to download and install the **Postgres** database

- Ability to download and install the **pgAdmin** utility

> **Important Note**
>
> If you don't have a Domo instance, get a free trial instance here: `https://www.domo.com/start/free`.

All of the code found in this chapter can be downloaded from GitHub: `https://github.com/PacktPublishing/Data-Democratization-with-Domo`.

Overview of intaking tools

The Domo platform greatly simplifies the process of getting structured data into the cloud; well, that is, after you have identified the right Domo tool for doing so. The data's source, volume, and security are key factors to selecting the right tool. The destination for all structured data coming into Domo is the dataset. The following figure illustrates some of the many options for data intake:

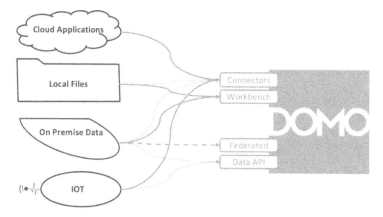

Figure 2.1 – Intake pathways into Domo

Let's look at the sources of data that can be used for import.

Data sources

Here are brief descriptions of the supported data sources:

- **Cloud applications** are applications such as Salesforce that run in the cloud, or pretty much any SaaS app with a data API. Connectors also can bring in files from email attachments, cloud databases such as Snowflake, and even public data sources such as census data or RSS feeds.

- **Local files** are Excel, CSV, JSON, and XML files stored on local directories/filesystems/laptops.

- **On-premises data** are owned database systems such as Oracle, SQL Server, Postgres, and many more.

- The **Internet of Things** provides machine sensor data and streaming data services such as Apache Kafka.

Now, let's discuss the tools for acquiring data from our data sources.

Tools

Here are brief descriptions of the data acquisition tools available in Domo:

- **Connectors** are Domo cloud applications that connect to specific data sources and intake the data by pulling the data into Domo using a wizard-based user experience.

- **Workbench** is an on-premise application that installs on a server behind your firewall and pushes data into Domo. It can connect to and load from a variety of data formats, including Excel, CSV, JSON, ODBC, JDBC, and MDX. I call Workbench the universal connector because if there is no connector, then Workbench is probably the tool to use.

- **Federated** is a data-at-rest tool for consuming data in Domo without moving the data physically into the Domo cloud. Typically, Federated is used to connect to existing data warehouses without needing to physically replicate all that data. While Federated does not move the data into Vault, it does still use the advanced in-memory caching ability of the Domo Adrenaline engine.

- The **Data API** is Domo's API for writing code to execute data actions. The Data API has massively parallel streaming methods for tackling even the largest data loads. It also enables the coding of users/groups/roles/privileges from other systems. For example, it can read Oracle EBS's security model and duplicate security configurations in Domo. Typical CRUD actions are available as well.

- The **Command Line Interface (CLI)** is a downloadable command-line utility that exposes the Data API as simplified commands.

The following table contains guidance for when to use which tool:

Source	Method	Tool	When to Use
Local files - Excel/CSV	Primary	Connector	One-time, quick load of data.
Local files - Excel/CSV	Primary	Workbench	Scheduled, recurring data loads.
Cloud Applications	Primary	Connectors	Whenever there is a connector available. Able to Whitelist Domo's Servers on your network. Volume <1 Billion Rows.
Data warehouses	Primary	Federated	Use in the data-at-rest approach where the source datawarehouse data is not moved or replicated into Domo Vault but stays at rest in the original data warehouse. When a fast, large scale, in-memory cache is needed to improve performance of a non-Domo source.
Source systems and Domo	Primary	API	When massive parallel data loads > 1 billion rows are needed. When a connector's baseline functionality is not enough. When enterprise batch administration tasks need to be automated. For extracting data from Domo datasets.
Source systems and Domo	Primary	CLI	When a true real-time view is needed against operational systems that essentially does a query passthrough to the source and returns the data. For running administrative housekeeping commands without having to code against the API.

Figure 2.2 – Data acquisition tool choices

Ok, now we have oriented you on how to choose the right tool, let's show you how simple the tools are to use.

Importing data from Excel and CSV files

One of the most common activities is to intake data from Excel or CSV files into Domo. This enables you to get data from a wide variety of sources. We'll see how to do that in the following sub-sections.

Loading your Excel data into the Domo cloud

We'll start with loading data from Excel files:

1. Download this example Excel worksheet of sample Shopify Order data: `https://github.com/PacktPublishing/Data-Democratization-with-Domo/blob/main/Shopify%20Orders.xlsx`.

2. Log in to your Domo instance: `Yourdomain.domo.com`

3. Click the **Data** option on the main menu.

4. Click on the **Cloud App** sub-menu item, which takes you to the **Data Connectors** page.

5. Enter `excel` in the **Search Connectors** box.

6. Click on **Connect** in the Microsoft Excel cloud app connector tile.

Figure 2.3 – Microsoft Excel connector tile

7. Then click on **GET THE DATA** button.

8. In the **Details** section of the **Upload Spreadsheet** panel, click on **browse your files** and choose the `.xlsx` file we downloaded in *step 1*. It will process for a few seconds and show a preview of the data.

> **Note**
>
> The **EDIT SELECTION** option in this section contains advanced options that allow you to set parameters for importing only certain sections, handling alternative data formats from column headers such as row headers and cross-tabbed data.

9. In the **Select tables** section, scroll down until you see and click the **NEXT** button at the bottom of the section. Be sure to scroll on the section area and not in the data table itself, or you will have to scroll to the end of the data before you see the end of the section.

10. In the **Update Mode** section, select either the **Replace** or **Append** method:

    • **Replace** will remove the previously uploaded data and do a full replacement with the new contents.

    • **Append** will preserve previous loads and append the data on each run.

> **Tip**
>
> Use **Append to Snapshot Spreadsheet Data** for trending. Append is a useful brute-force snapshot method if you have Excel data that you want to start to track historically or view trends over time, such as AR aging data on a report export from your financial system. When using append, I recommend checking the **Include batch upload metadata** in the **Details** -> **Edit Selection** area. This option adds batch ID and timestamp columns to your data. Append works on every execution and does not prevent duplication. So, be warned: if you import the same version of the spreadsheet multiple times you will have multiple duplicates. I will show you how to handle upserts and deletes in the data sculpting chapter.

11. Then, click **NEXT** to go to the **Name & Describe Your Dataset** section.

12. Name and describe your dataset, then click **SAVE**.

That's it! You will be taken to the dataset overview screen for your new dataset.

Updating the Excel-based dataset with new data from the spreadsheet

The following steps show how to update an existing dataset with fresh data from Excel:

1. Find the dataset on the **DataSets** page, as shown in *Figure 2.4*.

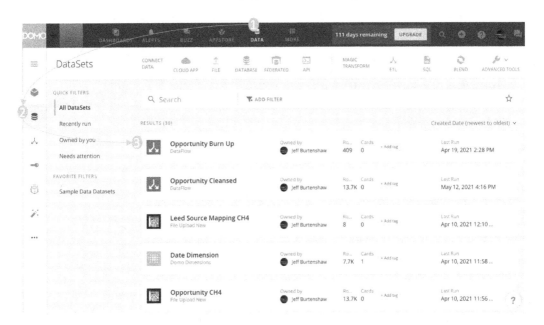

Figure 2.4 – DataSets page

2. Click on the name of the dataset you want to update; that will take you to the dataset details.

3. Click on the **SETTINGS** menu item.

4. In the **Details** section, click **browse your files** and choose the .xlsx file to update from. It will process for a few seconds and move to the **Select tables** section.

5. In the **Select tables** section, scroll down until you see and click the **SAVE** button at the bottom of the section.

6. You will see a temporary popup, **Sample Order History Saved**.

> **Important Note**
>
> Yes, there is an easy way to automate spreadsheet updates into Domo. Using the file connector to update the dataset is always a manual process. But there is a different way to automatically update the intake of spreadsheet data changes into Domo (see the Domo Workbench Directory Watch feature).

Now that we have a dataset, you may want to go ahead and create a card on the data.

Creating a card/dashboard from the dataset

If you want to create a dashboard on this dataset, refer to *Chapter 6, Creating Dashboards,* for directions. Or you can just click on the **CARDS** menu item and then click **Add Card** if you are feeling adventurous!

Next, let's see how to get data in from Google Sheets.

Importing data from Google Sheets

The process for importing a Google Sheet using a connector is very similar to Excel. However, a big advantage the Google Sheet connector has over the Excel connector is that it can run on a schedule as part of the connector.

Loading your Google Sheet data into the Domo cloud

The following steps demonstrate how to load Google Sheet data:

1. Click the **DATA** option on the main menu.
2. Click on the **CLOUD APP** menu item, which takes you to the **Data Connectors** page.
3. Enter sheets in the **Search Connectors** box.
4. Click on **Connect** in the Google Sheets connector tile.
5. Then click on the **GET THE DATA** button.
6. If it is the first time you are acquiring data from Google Sheets, you will need to set up account credentials by clicking **CONNECT** and following the process.
7. After allowing Domo to use your Google credentials, click **NEXT** in the **Credentials** section.
8. In the **Details** section, click on the **...** link in the **Google Sheet Choose a File** field, which will open a Google Sheets file list.

9. Enter the name of the Google Sheet in the **Select A Google Sheet** search box.

10. Click on the name of the desired Google Sheet.

11. Select the sheet/tab in the **WORKSHEET NAME** field.

12. Click **NEXT** to go to the **Scheduling** section.

13. Choose the **Update Interval** desired.

14. Click **NEXT** to go to the **Name & Describe Your DataSet** section.

15. Fill in the name and description for the dataset.

16. Click the **SAVE** button.

OK, now let's learn how to get data from an email attachment.

Importing data from email attachments

The **DataSet via Email** connector is used to acquire data from scheduled email attachments (Excel or CSV files) when source system APIs or other direct access paths are not available. Using the append option, each upload is added to the dataset, creating a historical record that enables trend analysis. The connector generates a unique email address that is associated with the dataset. After the email address is created, you can send an email with the attachment to that email address and the connector will process the email and update the dataset accordingly.

To use the email connector, follow these steps:

1. Click on the **DATA** main menu item.

2. Click on the **CLOUD APP** menu item.

3. Search for `email`.

4. Click on the **DataSet VIA Email** tile.

5. Click the **CONNECT** button in the tile.

6. Click the **GET THE DATA** button in the upper right to go to the **Details** section of the connector setup wizard.

7. Click the orange **COPY EMAIL** button.

8. Save the email address generated to a notepad (it will be a long alphanumeric string including your Domo domain) for use as the **TO:** address line in the source email.

9. In the **FILE TYPE** field, select the file format of the attached file.

10. In the **FROM EMAIL ADDRESS** field, enter either a full email address or just the domain of the source email, such as @wtvassociates.com, to set where the email can originate from. Enter NONE if you will accept emails from any address.

11. In the **EMAIL SUBJECT EXPRESSION** field, enter a regular expression to validate the email subject. Emails with a subject that does not match the expression will be rejected. Enter NONE to accept any email subject.

12. In the **EMAIL BODY EXPRESSION** field, enter a regular expression to validate the email body content. Emails with a body that does not match the expression will be rejected. Enter NONE to accept any email body.

13. In the **ATTACHMENT NAME EXPRESSION** field, enter a regular expression to validate the file name of the attachment. Enter NONE to accept any file name.

14. Click the checkbox next to **Show advanced options** to see more properties regarding the format of the attached file.

15. Click **NEXT** to go to the **Update Mode** section.

16. In the **Update Method** field, select **Replace**.

17. Click **NEXT** to go to the **Name & Describe Your DataSet** section.

18. Enter the **Name**, then click on the orange **SAVE** button.

19. Send an email to the saved email from *step 8* with the file to upload.

> **Important Note**
>
> Using the email attachment connector is not the most elegant way to intake data but it works when other alternatives are not available. Case in point, I was working with a financial analyst who had been struggling with another department to allow her to automate an extract of AR aging information. The other group's position was that she already had basic reporting capabilities from the ERP system to see that information and there were higher priorities. Conveniently, the ERP report could be scheduled as an email with a .csv attachment. She was able to automate her AR aging data acquisition process using the append option. For her, this meant no more waiting, no more senseless busywork stitching reports together to get trends.

Next, we will look at how to intake data from common cloud applications.

Importing data from cloud apps

Now that you understand how to get data in from the Excel and CSV files and Google Sheets connectors, there are hundreds of other cloud application connectors that follow the same wizard-based approach. Feel free to search in Domo for commercial and social applications you are using and go through the connector wizard to load in the data. To give you a better feel for these connectors, I'll suggest a few for you to try.

Follow these steps to find and connect to each of the connectors listed here:

1. Click on the **DATA** icon in the main menu area.

2. Enter the name into the **Search** box of the desired connector from the following connector list:

   - **US Census Connector**: Use the Domo US Census connector to receive data about the US Census, business dynamics, and economics of the United States of America.

   - **Domo Dimensions Connector**: Use the Domo Dimensions connector to create datasets as needed for the following data: **All Countries Geo; Area Codes; Calendar; AU Calendar; City Zip; Country Latitude Longitude; Countries; GeoLite2 City Blocks IPv4, IPv6; Standard Normal Distribution; US States; US Zips Cities States Counties Latitude Longitude; Zip2Fips2, Zip2Fips3; Zip2MSA; Zip Codes**; and more.

   - **NOAA Weather Connector**: This connector provides access to the NOAA public data repositories, where you can retrieve a weather forecast for a specified latitude and longitude.

   - **NOAA Historical Weather Connector:** This connector provides weather data as annual summaries, monthly summaries, daily summaries, daily averages, and so on.

   - **Strava:** Strava is a popular fitness app and website used to track athletic activity. Use the Strava connector to create dashboards about physical activities for you and your friends.

   - **Zoom:** Zoom is a leader in modern enterprise video communications, with an easy and reliable cloud platform for video and audio conferencing, chat, and webinars.

   - **CNN RSS**: Use this to connect to CNN.com feeds of story headlines in XML format for visitors who use RSS aggregators. The Domo CNN connector takes those headlines and creates a new data source that can be used to provide current headlines in your Domo instance.

3. Click on the desired connector tile in the search results area, which will take you to the connector's details.

4. Click on the **GET THE DATA** button.

5. Follow the wizard's steps from there.

Unfortunately, **deep web** sources such as **PubMed**, **Infomine**, and **Deep Web Monitor** are not yet covered by a Domo connector. However, if you want to get deep web data into your Domo instance, you can always write your own Domo custom connector. Instructions for creating custom connectors are found here: https://developer. domo.com/docs/custom-connectors/connector-dev-studio.

In the next section, we will explore how to load in data from common database systems.

Importing data from databases

Often, you need to connect to a database server. Domo has connectors for databases such as **Snowflake**, **Mongo**, **Vertica**, **Oracle**, **SQL Server**, **MySQL**, **Postgres**, **IBM DB2**, **Informix**, **Sybase**, **Redshift**, **Aurora**, **Google Cloud Storage**, **Google Big Query**, **Hive**, **Teradata**, and **Firebase**.

These connectors are found in the **DATA** menu and the **DATABASE** submenu. These connectors make it easy to connect to many popular databases. However, for these connectors to access the databases directly from the cloud, the database server firewalls may need to whitelist the Domo connector services' IP addresses. Here is a detailed list of the IP addresses to whitelist: https://bit.ly/3mp0vmc.

If you have direct access to one of the database systems listed, then follow these steps:

1. Click on the **DATA** main menu item.

2. Click on the **DATABASE** sub-menu item.

3. Search for the database.

4. Click on the database tile.

5. Click the **CONNECT** button in the tile.

6. Click the orange **GET THE DATA** button in the upper right to go to the **Credentials** section of the connector setup wizard.

7. Follow the wizard to connect.

For an example you can follow explicitly, let's install a local Postgres database and then connect to it using the database connector:

1. Download the Postgres relational database here: `https://www.postgresql.org/download/`.

2. Install **PostgreSQL** on your Windows machine or virtual machine by running the downloaded file.

3. Download the **pgAdmin** tool from here: `https://www.pgadmin.org/download/`.

4. Install and run the pgAdmin application.

5. In the **pgAdmin** window, navigate to the **Databases** -> **postgres** folder in the admin tool on the left panel.

6. Click on **Tools** in the main menu, then click on the **Query Tool** option.

7. Cut and paste the following code and run it in the pgAdmin Query Tool to create a persons table:

```
-- Table: public.persons
-- DROP TABLE public.persons;
CREATE TABLE public.persons
(
    first_name character varying(50) COLLATE pg_
catalog."default",
    last_name character varying(50) COLLATE pg_
catalog."default",
    dob date,
    email character varying(255) COLLATE pg_
catalog."default"
)
TABLESPACE pg_default;

ALTER TABLE public.persons
    OWNER to postgres;
```

You can also download the code from here: `https://github.com/PacktPublishing/Data-Democratization-with-Domo/blob/33b6d3f99983f98e68d11436e379e81209e70899/Postgres%20Create%20Person%20Table%20Query.txt`.

8. Download the `persons.csv` file from the GitHub repository: `https://github.com/PacktPublishing/Data-Democratization-with-Domo/blob/8bdc485ec260bc39751de7ddf964050df8caf822/persons.csv`.

9. In the **pgAdmin** window, navigate in the left panel to the **Databases** -> **postgres** -> **Schemas** -> **public** -> **Tables** and click on the **persons** table so it is highlighted.

10. In the **pgAdmin** main menu, click on **Tools** -> **Import/Export...**.

11. In the **Import/Export data – table 'persons'** popup window, toggle the first field to **Import**.

12. In the **Filename** field, select the `persons.csv` file you downloaded.

13. Set the **Delimiter** field to `,`.

14. Click **OK**.

15. Now, configure your Postgres database server so that Domo's connector servers can see it. Find the location of the Postgres server configuration files by running the following query in the **pgAdmin->Query Tool**:

```
SHOW hba_file;
```

16. Look down in the results grid and copy the full path to the file for the next step.

17. Using a text editor, open the `pg_hba.conf` file from the directory shown in the previous step, scroll to the `# Ipv4 local connections:` section, and add the following line:

```
host    all              all              0.0.0.0/0
                password
```

18. As Domo requires encrypted communications even when not using client certificates, setting up OpenSSL on the database server is required even for standalone localhost servers. To set up the SSL certificate on your standalone server, follow the directions here: `https://docs.microfocus.com/itom/Database_Middleware_Automation:10.61_Ultimate/SSLPostgreSQLClu`. Hint: the directory where your Postgres server was installed will be different and your hostname will likely be localhost. You will also need to make SSL property changes to the `postgresql.conf` file typically located in the data directory and stop and restart the Postgres server, as explained in the instructions.

19. Using a text editor, open the `postgresql.conf` file in the Postgres directory location, scroll to the `# - Connection Settings -` section, and set the `listen_addresses` property:

```
# - Connection Settings -
listen_addresses = '*'
```

Here is an article that also describes these changes to the Postgres configuration files: `https://bit.ly/2Ma4Ah0`.

20. Next, configure your network **router-port forwarding** so that the external WAN requests from Domo connectors are properly routed to your Postgres server's LAN IP address.

The Domo connector servers obviously don't reside on your LAN; consequently, they need to be able to talk to your Postgres database server through your router/firewall.

Additionally, since the Domo servers can only see your external WAN IP address, you will need to configure port forwarding on your router from the WAN IP address to your database server's LAN IP address. This link, `https://www.pcmag.com/how-to/how-to-find-your-ip-address`, shows you how to get both external and internal IP addresses. But for now, you just will need the internal IP address where you installed Postgres to configure port forwarding, as shown in the example in *Figure 2.5*. Scroll down in the article to the **Find Your Internal IP Address** section.

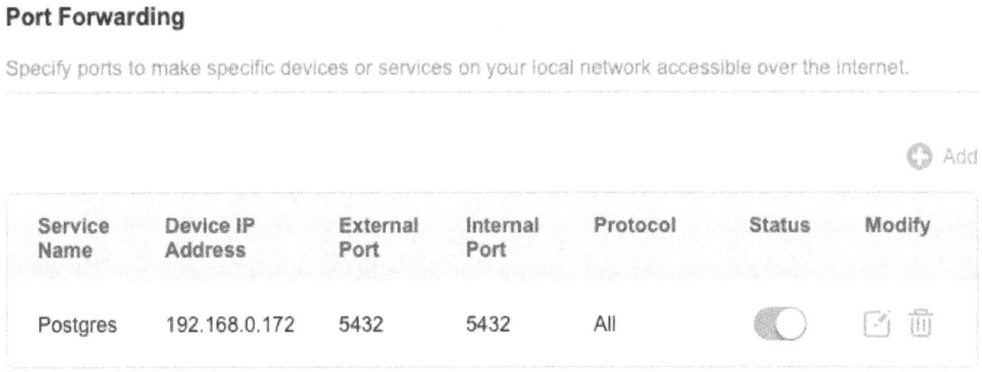

Port Forwarding

Specify ports to make specific devices or services on your local network accessible over the internet.

Figure 2.5 – Example of a router port forwarding entry

The default port for Postgres is 5432 and the router configuration entry will forward all incoming traffic on port 5432 to your Postgres server's LAN IP address. See the articles on connecting to your router, port forwarding, and how to see IP addresses in the *Further reading* section if you want additional information.

21. Finally, you may also need to allow the Domo servers to communicate through your firewall. This is called **whitelisting**. Here are the current Domo server IP addresses for whitelisting in your firewall: `https://bit.ly/3poodAc`.

Now, follow these steps to use the database connector to load the data into Domo:

1.   Click on the **DATA** main menu item.

2.   Click on the **DATABASE** sub-menu item.

3.   Search for the `Postgres` database.

4.   Click on the **PostgreSQL** database tile.

5.   Click on the **CONNECT** button in the tile.

6.   Click on the orange **GET THE DATA** button in the upper right to go to the **Credentials** section of the connector setup wizard.

7.   Fill in the **Credentials** section as shown in *Figure 2.6*.

8.   The connector's **Host** field must contain the WAN IP address of your router, which is the external IP address of your Postgres database. The following site will show the WAN IP address of your router: `https://whatismyipaddress.com/`. Enter this external IP address in the **Host** field.

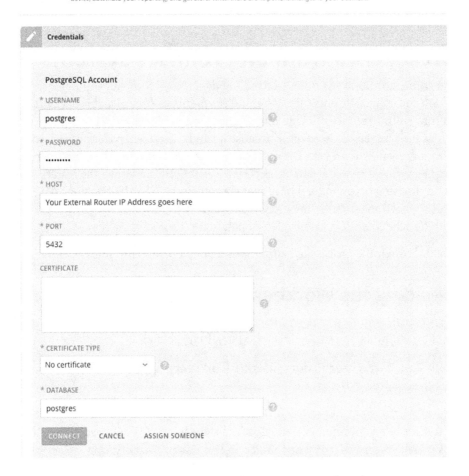

Figure 2.6 – PostgreSQL database connector credentials

9. Click on the orange **CONNECT** button to create the account's credentials.

10. Click **NEXT** to move to the **Details** section.

11. Click **persons** in **DATABASE TABLES** and check all the columns in **TABLE COLUMNS**.

12. Click on the **NEXT** button in the **Details** section.

13. Take the default schedule options and click the **NEXT** button in the **Scheduling** section.

14. Enter `Persons from Postgres` in the **DATASET NAME (REQUIRED)** box.

15. Click on the orange **SAVE** button.

The next section discusses how to use Workbench to get to ODBC/JDBC and other standard protocols for querying databases.

Importing data from on-premises systems

The Domo connector library is robust for pulling data into Domo, but sometimes, especially for non-cloud systems that sit behind on-premises firewalls, it is better to push data into Domo than to pull data into Domo. **Domo Workbench** is sometimes referred to as the **Universal Connector** because it can be used when a connector isn't available, or a firewall will prevent Domo connectors from accessing the source. Primarily, Workbench lets you query local databases directly via ODBC and push the results into the Domo cloud. It also lets you upload local files such as Excel, CSV, JSON, and XML, and access local applications such as Jira and Quickbooks. All Workbench jobs can run on a convenient job scheduler. It is extensible, via plugins, to protocols such as **MDX** as well.

So, let's walk through the process of getting Workbench up and running.

Downloading the Workbench agent

To use Workbench, you need to download the application onto a Windows machine. The following steps will walk you through how to install Workbench:

1. To get started, you need to download the Workbench agent onto a Windows machine.

   For Mac users, you will need to install a Windows emulator such as VM VirtualBox for free, or use a commercial emulator. The steps to get a Windows OS running on VirtualBox on a Mac can be found here: `https://www.youtube.com/watch?v=1EvM-No4eQo`. The Windows 10 OS image file can be found here: `https://www.microsoft.com/en-us/software-download/windows10ISO`.

2. Log into your Domo instance in your web browser on the Windows machine or VM.

3. Click on the **More** main menu item.

4. Click on the **ADMIN** item to go to the admin setting page.

5. On the left panel of the admin setting page, click **Tool Downloads**.

6. Click on the orange **DOWNLOAD DOMO WORKBENCH** button to install.

7. After installing it, you can start Workbench by using the Windows global search for Domo Workbench, as shown in *Figure 2.7*:

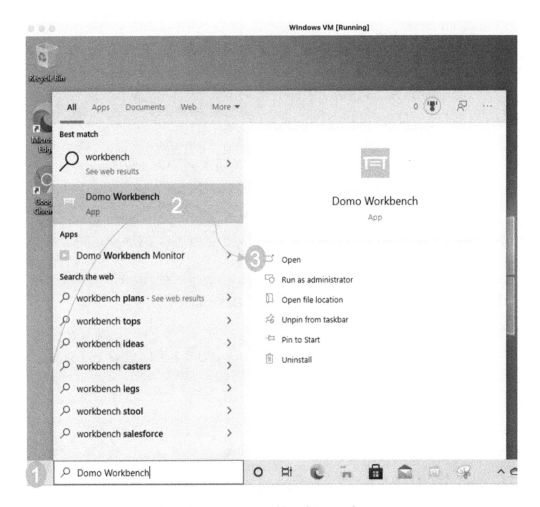

Figure 2.7 – Starting Workbench in Windows 10

Now that Workbench is installed and running, our next step is to set up the Domo credentials to where it will push the data.

Adding a Domo account to Workbench

You will need to add the Domo account to which the data will be loaded:

1. Click the icon of a key on the side menu ribbon.

2. Click on the + symbol to the right of **Accounts**.

3. Enter your Domo domain (just the part before `.domo.com`)

4. Click **Authorize**.

5. Click the clipboard icon to copy the authorization code.

6. Click on the authorization link provided in the box above the authorization code.

7. Log in to your Domo instance as prompted.

8. Paste the verification code into the box.

9. Click on the orange **SUBMIT** button.

10. Navigate out of the browser back to the Active Workbench window.

11. Click on the floppy disk icon in the top right to save the account.

12. The account will then appear on the left account panel.

OK! Your credentials are ready, so now it is time to push some data to Domo using a workbench job.

Adding a DataSet job to upload an Excel file

Let's upload an Excel file using a dataset job:

1. To create a new dataset job to push data into Domo, click the database icon on the left menu ribbon.

2. Click on the + to the right of the **DataSet Jobs** area.

3. Fill in the job details on the **Overview** tab:

    *   For **Transport Type**, select **Local File Provider**.

    *   For **Reader Type**, select **Excel** to upload a local Excel file.

4. Fill in the Domo details for **DataSet Type** and select **Workbench Excel**. Be aware that this field only becomes active when all the **Job Details** fields are entered.

5. Click **Create**, which will take you to the **Configure** tab.

6. In the **data** section, click on **Source Edit** to select the file.

7. Click on **Processing Edit** and select the desired **Sheet name** in the Excel file.

8. Click **Apply**.

9. Select **Replace** as the **Update Method**.

10. Click on the eye icon in the top menu to preview the job.

11. Close the preview window by clicking on the **X** to go back to the **Configure** tab.

12. Click on the **Schedule** tab to select scheduling options.

13. Click on the **Schema** tab to review/modify schema properties.

14. Rename fields in the **Destination** column.

15. Change data types in the **Data Type** column.

16. **Include** lets you exclude specific columns from the extract.

17. **Lock data type** will override the data type auto-detection.

18. Click on the **Notifications** tab to set notification options.

19. Click on the play icon in the top menu to run the job.

20. Close the run window by clicking on the **X** to go back to the job screen.

21. Click the floppy disk icon to save the job definition.

22. Log in to the Domo instance and find the created dataset.

Next, we will see how to push data from a database to Domo via Workbench.

Adding a DataSet job to upload an ODBC query result

If you can connect to a database locally via ODBC, then skip to the next section, *Creating an ODBC system DSN*; otherwise, install a local database, as explained here:

1. Download the Postgres relational database from here: `https://www.postgresql.org/download/`.

2. Install PostgreSQL on your Windows machine or VM by running the downloaded `.exe` file.

3. Run the pgAdmin application in the PostgreSQL folder.

4. Navigate to the `Databases -> postgres` folder in the admin tool's left panel.

5. Right-click on the **postgres** database and click on **Query Tool**.

6. Follow the instructions here to create a `persons` table: `https://www.postgresqltutorial.com/import-csv-file-into-posgresql-table/`.

Creating an ODBC system DSN

Workbench requires an ODBC connection, and setting up a system DSN is an essential step in the process. If you have already created a system DSN, then skip forward to the next subsection, titled *Creating a job from an ODBC connection*:

1. Download the ODBC connector for PostgreSQL from `https://www.postgresql.org/ftp/odbc`.

2. Go to the `versions/msi` folder and download and install the ODBC driver.

3. Open **Control Panel** and search for `ODBC`.

4. Click on **Set up ODBC data sources**.

5. In the **ODBC Data Source Administrator** window, click on the **System DSN** tab.

6. Click on the **Add...** button.

7. Select the driver and click on the **Finish** button.

8. Enter the **Driver Setup** properties, as shown in *Figure 2.8*:

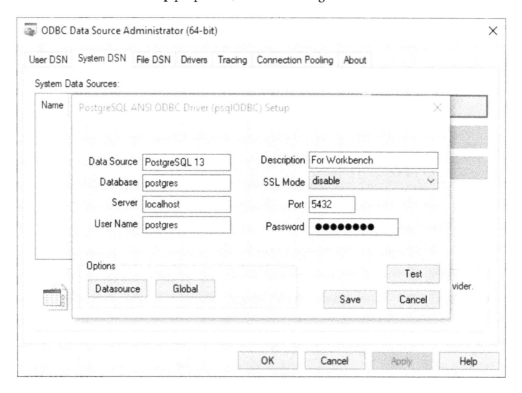

Figure 2.8 – Postgres ODBC driver setup

9. Click **Save**.

Next, we will see how to create a job that pushes the ODBC query results into Domo.

Creating a job from an ODBC connection

Follow these steps to create a new job that uses an ODBC connection:

1. Open Workbench by clicking on the Workbench icon on the Windows taskbar, as shown in *Figure 2.9*:

Figure 2.9 – Workbench icon in the Windows taskbar

2. To create a new dataset job to push data into Domo, click the database icon on the left menu ribbon.

3. Click on the + on the right side in the **DataSet Jobs** area.

4. Fill in the **Job Details** on the **Overview** section, as shown in *Figure 2.10*:

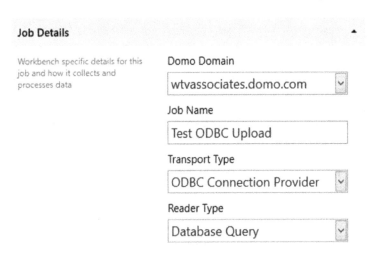

Figure 2.10 – Job Details section

5. Fill in the **Domo Details** section as shown in *Figure 2.11*:

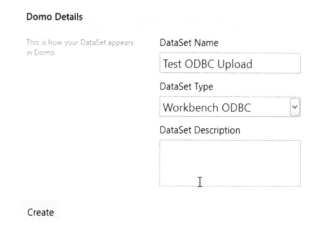

Figure 2.11 – Domo Details section

6. Click **Create**.

7. Click **Source Edit** and fill in the **Properties** as shown in *Figure 2.12*:

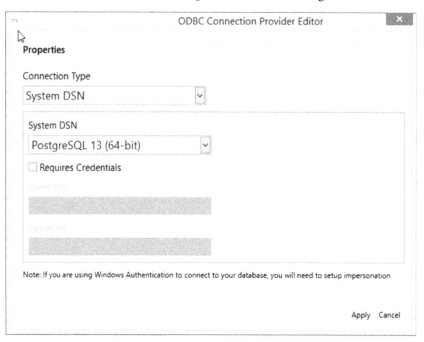

Figure 2.12 – ODBC Connection Provider Editor properties

8. Click **Apply**.

9. Click **Processing Edit** and fill in query properties as shown in *Figure 2.13*:

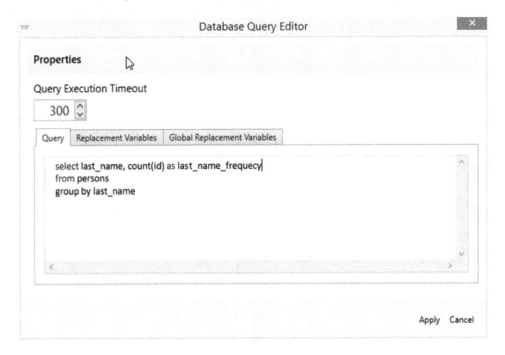

Figure 2.13 – Database Query Editor

10. Click **Apply**.

11. Click on the eye icon to see the **Data Preview** results, as shown in *Figure 2.14*:

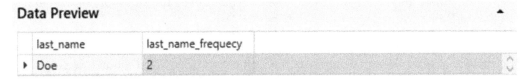

Figure 2.14 – Data Preview results

12. Close the preview window by clicking the **X**.

13. Click on the **Schedule** tab to adjust scheduling properties.

14. Click on the floppy disk icon to save the job.

15. Click on the play icon to run the job.

16. Click **X** to close the run results window.

17. Go to your Domo instance and search for the `Test ODBC Upload` dataset, as shown in *Figure 2.15*:

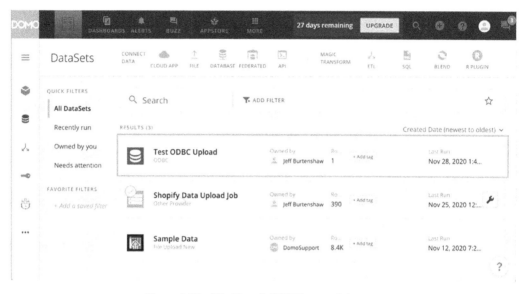

Figure 2.15 – Workbench ODBC created datasets

Next, we will learn how to apply the federated query feature to use Domo to virtually consume data at rest in data lakes external to Domo.

Using federated data connections

A good way to think of federated data is as an in-memory cache layer. Federated is a great fit when there are large existing data lake investments and infrastructure that can be accessed via Domo without replicating the data.

Follow these steps when you need to connect to a federated source that is accessing the data where it is currently stored outside of the Domo cloud. Federated sources supported at the time of writing are **Amazon Athena, Amazon Redshift, Google Big Query**, IBM **DB2, Microsoft Azure, SQL Server, MySQL, Oracle, PostgreSQL, Presto, SAP Hana**, and **Snowflake**:

1. Log in to your Domo instance.

2. Click on the **DATA** main menu item.

3. Click on the **FEDERATED** sub-menu item.

4. If you haven't created a federated account, do so now by clicking the **ADD NEW ACCOUNT** button in the bottom left corner.

5. Select the adaptor. In this case, let's choose **PostgreSQL**.

6. Click on the orange **NEXT** button.

7. Fill in the configuration information as shown in *Figure 2.16*:

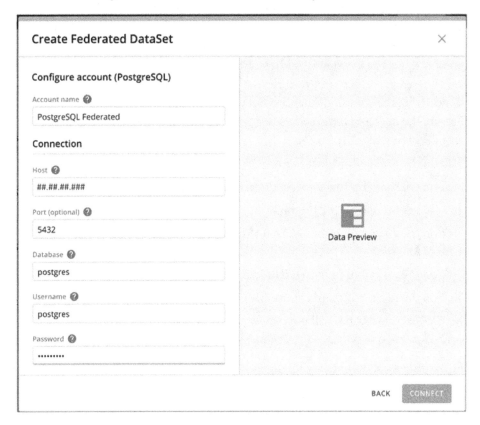

Figure 2.16 – PostgreSQL Federated DataSet configuration

8. Configure your database so Domo servers can see it.

   The **Host** field will contain the WAN IP address of your router. Consequently, you will need to expose your Postgres database outside your LAN so the Domo federated connector can see it. The Domo federated connector servers obviously don't reside on your LAN, but they need to be able to see your database server through your router/firewall.

   Here is an article that shows the changes that need to be made to the Postgres configuration files to do this: `https://bit.ly/2Ma4Ah0`.

9. Configure your router.

   Additionally, since the Domo servers can only see your external WAN IP address, you will need to do port forwarding on your router from the WAN IP address to your database server's LAN IP address. The default port for Postgres servers to listen on is 5432. The following site will show the WAN IP address of your router: https://whatismyipaddress.com/. Here is a good overview that also shows you how to get the internal IP address: https://www.pcmag.com/how-to/how-to-find-your-ip-address. Finally, you may also need to allow the Domo servers to communicate through your firewall. This is called **whitelisting**. Here are the current Domo server IP addresses for whitelisting in your firewall: https://bit.ly/3poodAc.

10. Click on the orange **Connect** button.

11. Click on the radio button to select the account you just created.

12. Click on the orange **NEXT** button.

13. Click on the > symbol in the **DATABASES** area. This will open the **SCHEMAS** area.

14. Click on the > symbol In the **SCHEMAS** area. This will open the **TABLES** list.

15. Click on the radio button to choose the **City** table.

16. The **TABLE** preview tab will populate showing the table data.

17. Click on the **SCHEMA** tab to see the table columns in the federated dataset.

18. Unchecking a column here will remove it from the federated dataset view.

19. Click on the orange **NEXT** button.

20. Enter the name of the dataset, something like Federated City, in the **DataSet Name** field.

21. Enter the **Time To Live (TTL)** in the **Cache TTL** field. This is the number of seconds the data will remain cached in memory until a new query refreshes the data. A TTL of 0 turns off the caching and forces a query to go back to the source each time. The right value for the TTL depends on the velocity of change in the source data.

22. Click on the orange **CREATE** button.

   You now have a Domo dataset based on a federated query against the Postgres database. This process is similar for other federated sources.

In the next section, you will see a manual way to enter tabular data straight into the Domo cloud.

Using Webforms

The **Webform** connector is a tabular data input mechanism, similar to a spreadsheet, that stores data directly into a dataset in Domo Vault. It is very useful for capturing user parameter inputs for calculations and lookup tables for Magic ETL and Blend processing.

Creating a Webform dataset

Follow these steps to create a dataset linked to a Webform:

1. Click on the **DATA** option on the main menu.
2. Click on the **CLOUD APP** sub-menu item, which takes you to the **Data Connectors** page.
3. Enter Webform in the **Search Connectors** box.
4. Click on **Connect** in the Domo Webform cloud app connector tile, as shown in *Figure 2.17*:

Figure 2.17 – Domo Webform connector tile

5. Enter the dataset title.
6. Click on **New Column** and enter the column name.
7. Set the data type of the column name by clicking on **abc** and selecting the type.

> **Important Note**
> The **Auto detect** type, when selected, will use the type of the column values in row 1 to cast the column types. So, if row 1, column 1 is a number but you want column 1 to be typecast as text, then explicitly select **Text** as the type.

8. Repeat the previous two steps for each column to change the typecast if needed.

9. Enter data into the data area. You can cut and paste:

    A. Right-click on the column heading to insert or remove columns.

    B. Click on the import icon, as shown in *Figure 2.18*, to import data into the Webform from an existing dataset:

Figure 2.18 – Import icon

10. Then click on the orange **Save & Continue** button to create the dataset.

After you have created a Webform dataset, you may need to edit it.

Editing a Webform dataset

Follow these steps to edit a Webform dataset:

1. Click the **DATA** option on the main menu.

2. In the search area, click on **ADD FILTER** and select **TYPE** -> **Domo Webform** to filter the datasets to show just those that are Webforms.

3. Find your dataset in the list and click on the dataset name to go to the dataset details page.

4. Click on **EDIT WEBFORM** in the sub-menu area.

5. Make your edits:

    A. To search for specific values, enter a value in the search field.

    B. To search and replace, click on the three vertical dots in the search field.

    C. Then click on the orange **APPLY** button.

    D. Right-click on the column heading to insert or remove columns.

6. Then click on the orange **SAVE & CONTINUE** button to save the changes.

Taking data from Domo

Yes! Thank heavens, taking data from Domo is possible via the Domo API, ODBC, or the Premium Writeback connectors. Any Windows ODBC query utility can read data from Domo datasets.

Using ODBC to query Domo datasets

To get a 30-day free trial of the ODBC drivers, follow these steps:

1. Click on the **Admin** main menu option.
2. Click on the **Tool Downloads** option in the admin settings area.
3. Scroll to the **Domo ODBC Driver** section.
4. Click on **DOWNLOAD ODBC DRIVER**.
5. Follow the additional instructions in the **GETTING STARTED** section.
6. Connect with any ODBC-compliant query tool.

Next, let's discuss using the CLI.

Using the CLI to query Domo datasets

To get the CLI app, follow these steps:

1. Click on the **Admin** main menu option.
2. Click on the **Tool Downloads** option in the admin settings area.
3. Scroll to the **Domo CLI** section.
4. Click on **TRY THE DOMO CLI** to download the domoUtil.jar file.
5. Follow the additional instructions in the **GETTING STARTED** section.
6. For more help on the CLI, see this chapter's *Further reading* section link on the CLI.

Next, let's take a look at the Google Sheets add-on for getting data into Google Sheets from Domo.

Using the Domo Google Sheets add-on to consume Domo datasets

A very useful tool for getting data from Domo datasets into a Google Sheet tab is the **Google Sheets add-on**. To install the Domo Google Sheets add-on, follow these steps:

1. Click on the **Admin** main menu option.

2. Click on the **Tool Downloads** option in the admin settings area.

3. Scroll to the **Domo Google Sheets add-on** section.

4. Click on **TRY OUR SHEETS ADD-ON** to install.

5. Follow the additional instructions in the **GETTING STARTED** section.

Perhaps a quick summary of what we have learned in this chapter would be helpful.

Summary

In this chapter, we walked through how to decide which tool to use for uploading data—the process of getting data into Domo. Detailed walkthroughs of the various methods were provided. We learned how to upload data from Excel, Google Sheets, cloud applications, databases, on-premises data via Workbench, federated data at rest from existing data lakes, and even Webforms. Clearly, the Domo platform can use your data assets no matter where they are. We also briefly touched on how to download data from Domo using ODBC, the CLI, and the Google Sheets add-on.

In the next chapter, we will go in-depth into the data storage mechanisms available in Domo.

Further reading

Here are related articles of interest:

- For information on using SSL with Postgres, check this out: `https://www.postgresql.org/docs/9.5/ssl-tcp.html`.

- To learn more about regular expressions, go here: `https://regexone.com/`.

- For more on Postgres table creation, take a look at this: `https://www.postgresqltutorial.com/import-csv-file-into-posgresql-table/`.

- How to log in to your router: `https://www.netspotapp.com/how-to-log-into-router.html`.

- How to get your router and LAN device IP addresses: `https://www.pcmag.com/how-to/how-to-find-your-ip-address`.

- Port forwarding basics: `https://portforward.com/`.

- CLI help: `https://domohelp.domo.com/hc/en-us/articles/360043437733-Command-Line-Interface-CLI-Tool`.

3
Storing Data

Now that you have created datasets, we will walk through how the data is stored and accessed in the Domo cloud. Domo, by design, enables an iterative and adaptive approach to data storage. The essential storage structure of a dataset is something that would look like a typical Excel spreadsheet. One advantage that this simplicity offers is that dataset schemas are derived from the data as it is ingested: when data is changed at the source, the schema is also automatically updated. That is kind of a big deal because in traditional data warehousing, it was all about rigid schema building and compliance – which, honestly, was slow and failed to deliver business value at cloud speed. So, although you can enforce highly normalized schemas in Domo, there is simply not a pressing business reason to do so. In fact, many Domo customers find that a year or so into the implementation, their Domo instance has organically accumulated more tabular data than their legacy data warehouse. This is primarily due to the low-hassle manner in which data is ingested and stored, which was enabled by the fundamental assumption (now proved over years) that the data itself contains all the information needed to create and maintain the raw schema.

Storing the data becomes useful when it is also easy to find, profile, and secure the data. We will walk through searching, browsing, data profiling, and securing datasets as well.

In this chapter, we will specifically cover the following:

- Understanding how data is stored

- Finding datasets

- Working with properties of a single dataset

- Working with properties of multiple datasets

Technical requirements

To follow along with this chapter, you will need the following:

- Internet access.

- Your Domo instance and login.

- The sample *Opportunity* dataset from GitHub: `https://github.com/ PacktPublishing/Data-Democratization-with-Domo/blob/main/ Opportunity.xlsx`.

- Install the Coronavirus (COVID-19) dashboard from the Domo Appstore.

- Access to a test email address other than the one used for your primary Domo login.

Understanding how data is stored

All tabular data brought into Domo via connectors, Workbench, or APIs is stored in **Vault**. Vault is a persistent storage layer that feeds both the in-memory caching engine Adrenaline and the cache optimizer Tundra. *Figure 3.1* illustrates these relationships:

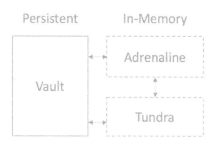

Figure 3.1 – Persistent versus in-memory storage

Vault resides on major cloud infrastructure file services from either AWS or Microsoft Azure. The default cloud service is AWS unless Azure is specifically requested. The physical data is stored and versioned in Vault, regardless of the intake mode (**Append** or **Replace**) on the dataset. In **Replace** mode, previous data versions are no longer queryable from the UI, but they are still stored as an older version of the data. In **Append** mode, all data versions are queryable because they are linked together as a single entity as illustrated in *Figure 3.2*:

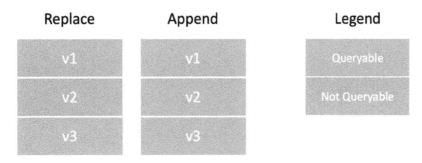

Figure 3.2 – Dataset version chaining in Domo

A great thing about the Domo cloud storage infrastructure is that it is maintained and optimized by Domo operations, freeing up time and resources for you to focus on other priorities. But there are also robust administration features on the datasets controlled by the user. In the following sections, we will discuss what features you can use to administer datasets.

Finding datasets

Once the datasets are created, the user will need to be able to find them by browsing or searching. To browse through available datasets, simply do the following:

1. Click the **Data** main menu item.
2. Click on the dataset icon on the left.

3.   Scroll through the list.

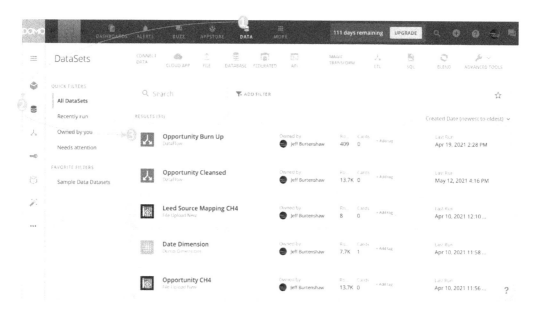

Figure 3.3 – Browsing datasets

To search the list of dataset names, do the following:

- Click in the **Search** field and enter op to see all datasets in the list with op anywhere in their name.

To filter the list of dataset names for all datasets of a certain connector type, take these steps:

1.   Click on the **ADD FILTER** area.

2.   Select **Type**.

3.   Select **Sample Data** to see all the datasets uploaded from Excel.

To save the filter options as a favorite for future use, do the following:

1.   Click on the favorite star icon.

2.   Enter the name of your filter – in this case, Sample Data Datasets.

3.   Apply the filter by clicking on it in the **FAVORITE FILTERS** area.

To apply one of the preset filters to see a list of datasets recently updated, take this step:

- Click on **Recently run** in the **QUICK FILTERS** area.

Next, we will examine the features available when viewing the properties page of a specific dataset.

Working with the properties of a single dataset

The dataset detail page has a robust set of features for interacting with and securing datasets. To view a dataset detail page view, click on the desired dataset name, in this case, **Opportunity**, in the dataset **Results** list as seen in *Figure 3.4*:

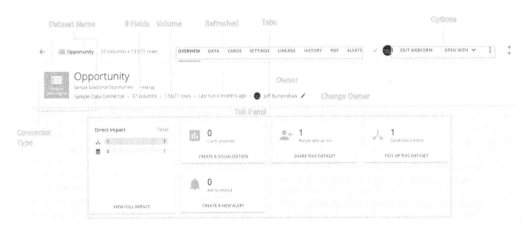

Figure 3.4 – Opportunity dataset detail page

The dataset page has all of the metadata you would expect: dataset name, owner, connector type, number of fields and rows, and data refresh date. Additionally, the **Tabs** and **Options** areas provide actions that can be taken on the dataset. Next, let's go through each of the tabs, starting with the **OVERVIEW** tab.

Using the OVERVIEW tab

The **OVERVIEW** tab has several tiles that communicate the use of the dataset and also enable actions to be taken:

- The **Direct Impact** tile shows the number of objects by object type (cards, datasets, dataflows, and alerts) that directly use the dataset. Clicking **VIEW FULL IMPACT** will show the data lineage of the dataset.

- The **CREATE A VISUALIZATION** tile shows the number of cards that have been created with this data source. And clicking on the **Cards powered** image takes you to a page that shows you all of those cards. From there, click on **Add Card** to go to the card analyzer for a new card; click on **Switch cards to a different DataSet** to change the source of all the cards to a different dataset; or click on an existing card tile to open that card in Analyzer. Don't worry, Analyzer is covered in depth in *Chapter 8, Interacting with Dashboards*, on creating visuals.

- The **SHARE THIS DATASET** tile shows the number of people with access to the dataset. Clicking on the **People with access** image takes you to a dialog to share the card with other users or groups. In this case, we will share the card with the **DomoSupport** group by entering DomoSupport in the entry field and selecting **Can Edit** instead of **No Access** in the dropdown. Click on the **ADD PEOPLE** button to go to a dialog where you can enter an email address and invite that person into your Domo instance with the specified access to the card.

- The **TIDY UP THIS DATASET** tile shows the number of dataflows that use the dataset as an input. Clicking on the **DataFlows created** image takes you to the **DataFlows** page with the **Has Input** filter set to the dataset name, showing only dataflows that have the dataset as an input. Clicking on **TIDY UP THIS DATASET** takes you to the new dataflow page with the dataset preset as an input in the dataflow. We will walk through how to create a dataflow in *Chapter 4, Sculpting Data*.

- The **CREATE A NEW ALERT** tile shows the number of alerts on the dataset. Clicking on the **Alerts created** image takes you to the **Alerts** page where you create and manage alerts. Alerts are covered in depth in *Chapter 12, Alerting*. Clicking on **CREATE A NEW ALERT** takes you directly to the new alert creation dialog.

Now that you understand the dataset overview tab, let's explore the **DATA** tab.

Using the DATA tab

The **DATA** tab is where you can discover details about the actual data and schema of the dataset. Search, filter, and statistics features about each column's data are located here. By default, the page comes up in the **Table View** as seen in *Figure 3.5*:

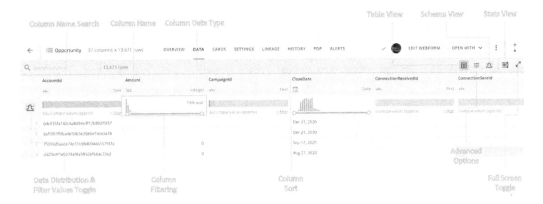

Figure 3.5 – Opportunity dataset DATA tab

The following are descriptions of each of the features on the **DATA** tab:

- The **Search Columns** area enables you to easily find columns in datasets. Enter date in the **Search Columns** field to see just the columns with date anywhere in their name.

- To sort the dataset by the values of a particular column, click on the sort arrows to the left of the column name. Click on the sort arrows again to change the direction of the sort.

- Click on the **Data Distribution & Filter Values** toggle to hide or show the data distribution charts and the column filtering areas.

- To see the distribution of column data values that contain text values, hover the pointer over the bar charts; for numeric and date values, filter by sliding the range viewer endpoints.

- To filter values in a column, click in the filtering area (but not on the chart) and a filter dialog will open. Clicking on the chart will apply a column filter value based on what was clicked. For example, in the **StageName** column, clicking on the **Closed Won** bar in the horizontal bar chart will filter the dataset to just the **Closed Won** opportunities as seen in *Figure 3.6*. Remember, the column search above is a completely different feature from the column filter values.

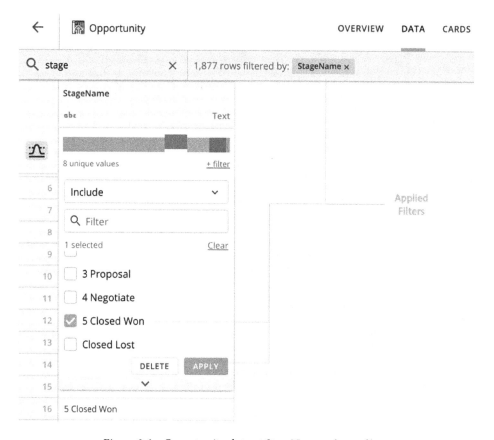

Figure 3.6 – Opportunity dataset StageName column filter

To switch to **Schema View**, click on the **Schema** icon. This displays a table of the dataset schema values: **Column Name**, **Column Type**, **Description**, and **Tags**. This is also where the **Data Dictionary** functionality of entering detailed column descriptions and tags resides.

> **Important Note**
>
> **Column Type** is not editable here. Changing **Column Type** requires using the **Magic ETL EDIT COLUMNS Set Column Type** tile in a dataflow. This can be accomplished from here by using the **OPEN WITH Magic ETL** option. An alternative approach to changing column types is to use **Workbench**, which has the ability to change column types. Or the easiest method is to use the DataSet Views feature.

- Click on the **Stats** icon to bring up **Stats View** containing the data profile or descriptive statistics on each column in the dataset. The total number of values – unique values and null values, as well as shortest/longest or minimum, median, average, maximum, lower quartile, upper quartile, and standard deviations – are presented. This is a great way to get an idea of the data value profile for each column in a dataset that you may not be familiar with.

- Click on the **Advanced Options** icon to bring up the **Options** dialog for setting the **DISPLAY TEXT**, **NULL HANDLING**, and **DECIMAL PRECISION** settings. The **NULL HANDLING** setting of **Advanced** allows comparison of null values to evaluate as true or false, in contrast to **Basic** mode where null comparisons are not evaluated.

> **Important Note**
>
> Another way to get null values to evaluate in a comparison operator even when in **Basic** mode is to use **Beast Mode Manager** to create a calculated field on the dataset to explicitly convert the null value. For example, on the *Opportunity* dataset, create a *Stage* Beast Mode using `IFNULL(`StageName`, 'Blank')`. The Beast Mode can be added as a new column via the Domo DataSet Views feature too.

Next, we will learn how to use the **CARDS** tab.

Using the CARDS tab

The **CARDS** tab shows thumbnails of all cards that are based on the chosen dataset. You can go to the card detail view from here by clicking the thumbnail image of the card. Also, the option to **add a card** to the dataset can be used here as well. Once the dataset is in use by at least one card, the option to **switch cards to a different dataset** appears. The **Switch cards to a different DataSet** feature will attempt to migrate all the cards for the dataset from the existing dataset to another dataset. If the column names and data types of the fields used in the existing charts are in the new dataset, it will work without modification. If there are name and data type differences, we can still complete the migration but cleanup on the cards and Beast Modes may be required.

Next, we will see how to adjust dataset settings.

Using the SETTINGS tab

The **SETTINGS** tab is where the connector properties are viewed and updated. For example, to update the dataset values from an Excel file connector, just drag and drop the Excel file into the details area. Be sure to review the **Update Mode** settings to **replace** or **append** as desired. Remember to hit the **SAVE** button to save any changes. Of course, each connector type will have its own full set of properties displayed here. Connectors are covered in depth in *Chapter 6, Creating Dashboards*.

Next, we'll learn about the data lineage of datasets.

Using the LINEAGE tab

The **LINEAGE** tab is a powerful feature that, as the name implies, shows both parent and child relationships for how the dataset is related to other objects where it is used, such as other datasets, dataflows, cards, and alerts. This is a beginning-to-end, source-to-consumption view of how the dataset is used. And because Domo is used to acquire, sculpt, and present the data, a 360° view is possible.

For a full example of dataset lineage, let's use a dataset from the Domo Coronavirus (COVID-19) Tracker app:

1. To install it, click on the **APPSTORE** icon on the Domo main menu bar.

2. Then, search for `Covid` and click on the **Coronavirus (COVID-19) Dashboard** tile.

3. Next, click the **GET** button to install the app and wait until it is installed.

4. After it is installed, click the **CONNECT** button to import live data and wait until the data is connected.

5. Click on the **DATA** main menu, click the **DataSets** icon, then enter `Covid` in the **Search** field.

6. Click the dataset named **DOMO Covid Current Snapshot Tracker Data**, then click on the **LINEAGE** tab.

7. In the **LINEAGE** tab, click on the **Cards** icon to see the cards using the dataset, as seen in *Figure 3.7*:

Figure 3.7 – DOMO Covid Current Snapshot Tracker data lineage

Next, we will examine how to use the dataset history.

Using the HISTORY tab

The **HISTORY** tab gives a complete log of each time an attempt is made to update the dataset and whether it was successful or not. A user can see how long ago the last successful update occurred, the average time it takes for each successful update, and the percentage rate of successful updates.

In addition to seeing the update history, there are several very useful actions for working with the dataset history and managing dataset versions. Continuing with the example **DOMO Covid Current Snapshot Tracker Data** dataset, click on the **HISTORY** tab and then the **Action** wrench icon showing a selection list as seen in *Figure 3.8*:

Figure 3.8 – Dataset HISTORY tab actions

The following are the action option descriptions:

- **Preview Updated Rows** will display the last version updated.
- **Download** will download the queryable dataset version(s).
- **Delete this update** will remove the selected version from the dataset version chain.
- **Revert to this point** will reset the dataset chain back to that point in time.

Next, we will explore how to secure datasets by setting security privilege policies down to the row level.

Using the PDP tab

Personal Data Privacy (**PDP**) is the security framework for creating policies around who can access data in a dataset, which can be controlled down to row-level granularity.

Perhaps the best way to understand the basics of PDP is to walk through a scenario. Heads-up: there are quite a few steps here. In this case, let's configure the **Opportunity** dataset to allow a test user to only see **Opportunity LeadSources** of type **Email** by creating a new PDP policy to restrict the rows the test user assigned to the policy can access.

First, you need to set up a test user account for which you can set security policies and see the impact:

1. Invite yourself to Domo using a test user email address.
2. Click on the **MORE** main menu icon and then click **ADMIN**.
3. Click **People** on the left side.
4. Click the **ADD NEW PERSON** button.
5. Enter the name and email of the test user.
6. Select the role of **PRIVILEGED**.
7. Click **Invite**.
8. Check the email inbox of the test user and follow the instructions in the *You Got Domo'd* email invite.
9. Click the **GET STARTED** button in the email and set up your password.
10. Sign out the test user from your Domo instance.

    Now, we'll set up the PDP policy for the test user on the **Opportunity** dataset.

11. Sign in to your Domo instance as your primary user.
12. Click on the **PDP** tab of the **Opportunity** dataset detail page.
13. Slide the **ENABLE PDP** switch to the right and then click the **ENABLE** button in the pop-up dialog. **ENABLE PDP** should now be checked.
14. Click the **+ ADD POLICY** button.
15. Enter `Email Lead Viewer` in the **Policy Name** field.
16. Click the **+ ADD DATA** button.
17. Select **LeadSource** in the **Column Name** dropdown.
18. Select **Simple Filter** for **Access Type**.
19. Click in the **Search/Add row values** field and select **Email** to filter the rows.
20. Click **Apply**.

21. To add the test user to the policy, click on the person icon to the right in the row.

22. In the **Add groups & people** dialog, click in the **Search/Add groups and people** field.

23. Start typing the test username and watch for the full test username in the pop-up list. Select it when it appears.

24. Click **Apply**.

25. Click **SAVE**. *Figure 3.9* shows the results:

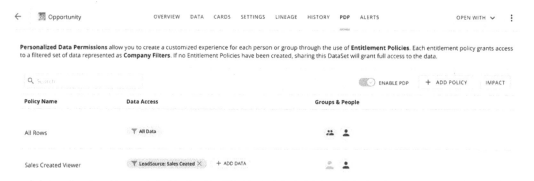

Figure 3.9 – PDP policy page with Email Lead Viewer policy created

Now we will create a card to share the content with the test user:

> **Important Note**
> Domo requires that non-admin users have shared access to at least one card on a dataset before they can view/access a dataset they did not create.

1. Click on the + icon on the main menu and click **Card**.

2. Click the **SELECT** button under **Visualization**.

3. Click the **SELECT** button under **Existing data**.

4. Select **Opportunity** in the dataset list and click the **CHOOSE DATASET** button.

5. This will take you into the analyzer (or card editor); click the **SAVE** button.

6. Click **Create New Dashboard**, enter Opportunity Dash as the dashboard name, and leave **Take me to the Dashboard when I click save** checked.

7. Click **SAVE**.

8. Click on the **Share Dashboard** icon as seen in *Figure 3.10* and select **Share** in the dropdown:

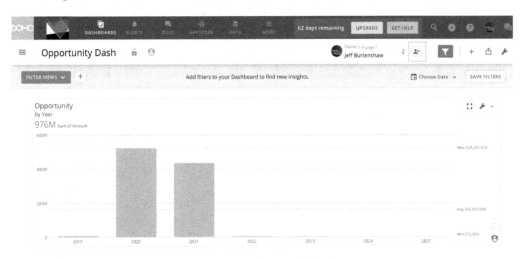

Figure 3.10 – Share Dashboard

9. In the **Share "Opportunity Dash"** dialog, enter the test username and select the test user from the dropdown.

10. Click the **SHARE** button. The test user should now be in the shared list.

11. Click the **X** in the upper-right corner to close the dialog.

12. Make a note of the total amount on the card to compare to the test user view of the same card.

    Next, we need to test whether the PDP policy is working.

13. Sign out of your Domo instance as your primary user by clicking the profile picture in the main menu and clicking **SIGN OUT**.

14. Sign in to your Domo instance using the test user email and password.

15. Click on the **DASHBOARDS** icon in the main menu.

16. Hover over the three horizontal lines to the left of **Overview**.

17. In the dashboards side drawer that pops out, select **Opportunity Dash**.

18. Notice that when we view the **Opportunity** card as the test user, the total amount is much lower than the total when it was being viewed as the primary user, as seen in *Figure 3.11*. This is due to the PDP policy automatically applying the filter to limit the dataset rows the test user has access to:

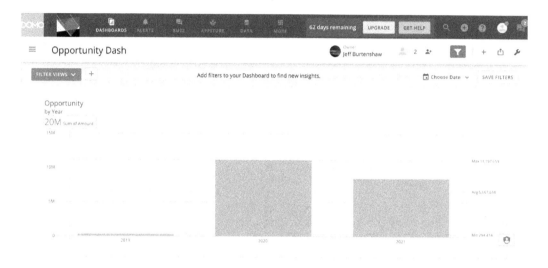

Figure 3.11 – PDP filtered opportunities

19. Congratulations, you made it through all the steps to automatically filter the rows in a dataset via a PDP policy! Now that you have mastered the basics of PDP, let's move on to how to set up alerts on a dataset.

Using the ALERTS tab

Domo has the capability to set alerts on dataset data that are evaluated anytime the dataset is updated. For example, using the **Opportunity** dataset, let's say each time an opportunity greater than $100,000 came into the pipeline that we wanted to have an alert that people could subscribe to:

1. Click on the **ALERTS** tab on the **Opportunity** dataset detail page.

2. Click on the **+ NEW Alert** button.

3. Enter `Big Deal Added` in the **Add an alert name** field.

4. Under **Personalized Data Permissions**, leave the setting as **Contextual**, which means apply the PDP permissions based on the currently logged-in user's policies. There are options for the alert to be on **All Dataset Rows** or to be for a specific PDP policy only.

5. Under **Metric**, select **Any row is added**.

6. Click **+ ADD FILTER**.

7. Under **Metric**, in the **Select a column** field, choose the **Amount** column.

8. Under **Meets this condition** in the **Select a condition** field, choose **Is greater than or equal to (≥)**.

9. Under **For this value**, enter 100000 in the field.

10. Under **What column or combination of columns identifies a row as unique?**, select **Id**.

11. For **You'll be notified when this alert is in a triggered state**, select **every time your data updates**. There are fixed options for other periodic updates but in most circumstances, triggering the alert on dataset update makes sense. However, there may be times when you want the alert not to be real-time but daily or monthly instead.

12. If you want the alert to resend the notification every time the dataset updates and the alert condition still evaluates to **True**, then check **Keep sending notifications while in a trigger state**. This would then be an alert on every deal >= 100,000 and would alert on every deal in that state on every update. In this case, we want to be alerted only the first time a big deal is added so *uncheck* this option.

13. Click the **NEXT** button.

14. In the **Compose a notification** message dialog, you can customize the alert message. For this exercise, take the defaults here, but this is where you can customize the message. It works on a template and variable substitution basis. The field names enclosed in the square brackets will be replaced with the values from the dataset where the alert condition was met.

15. Click the **NEXT** button.

16. In the **Share this alert** dialog, under the **People** tab, you can enter users and groups or emails you want to subscribe to the alert. As the alert creator, you are automatically subscribed to the alert, but this is where you subscribe others. Enter the test user in the **Enter users, groups or emails** field. There is also the option in the **Buzz** tab to send the alert to a Buzz channel or person for the alert to show in the chat interface.

17. Click the **SAVE** button.

18. You may want to create a similar alert to fire when any data changes as the above alert only evaluates new opportunities added to the dataset.

19. Dataset alerts also have the capability to trigger actions in **Domo tasks** or in third-party systems via **webhooks** as a result of an alert. For example, with the `Big Deal Added` alert created previously, we could have the alert also create a Domo task for the head of sales to follow up on the big deal.

20. Click on the **+ ATTACH ACTION** button in the **Message Preview** area.

21. Click **Task**.

22. Choose a project (you will have to create at least one project first).

23. Select the task owner.

24. Click the **SAVE** button.

Next, we will cover the Dataset Options Actions menu.

Using the dataset OPTIONS actions menu

Toward the upper-right corner of the dataset detail page is the **OPTIONS** actions menu icon – the three stacked dots as seen in *Figure 3.12*.

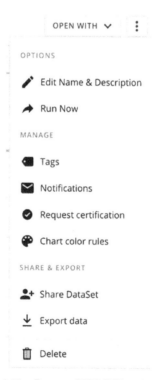

Figure 3.12 – Dataset OPTIONS action menu

This menu provides access to important actions/options on the dataset:

- To update the **data dictionary** information on the dataset level, click on the **Edit Name & Description** item.

- To force the dataset to refresh data from its source, click the **Run Now** item.

- Click **Tags** to manage the tags on the dataset.

- Click **Notifications** to set whether or not the owner is notified when a dataset update fails.

- Click **Request certification** to start the process of certifying the dataset. **Certification** is a process to indicate that a dataset has been through a review and certification process.

- Click **Chart color rules** to set up color rules that will be applied to any card using the dataset. For example, if you have a brand category that you always want to be a specific color no matter where it is used, this is where you can set the rules to dictate that.

- Click **Share DataSet** to share the dataset with others.

- Click **Export data** to export the dataset contents to Excel or CSV.

- Click **Delete** to remove the dataset.

That covers the major functionality of working with datasets. Now we will examine how to apply batch actions to more than one dataset at a time.

Working with properties of multiple datasets

Sometimes it is helpful and saves time to perform the same action on multiple datasets together. Such as sharing, tagging, changing ownership, deleting, or running a data refresh. To access the multi-dataset action menu, do the following:

1. Click on **All Datasets** in the **Quick Filters** area on the **DataSets** page.

2. In the datasets list, hover over the dataset type icon left of the dataset name on any dataset in the list and a check mark icon option will appear on the upper left of the data type icon.

3. Click the check mark icon to select the dataset for a batch operation.

4. When the first dataset is selected, the batch operation ribbon will appear as seen in *Figure 3.13*:

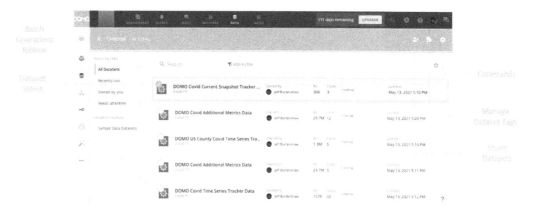

Figure 3.13 – Batch dataset operations

5. Enter Covid in the **Search** field to see a list of the COVID datasets.

6. Click on the first select dataset icon in the dataset list.

7. Click on **Select All** on the batch operations ribbon.

8. Click on the **Commands** icon on the batch operations ribbon.

9. Click **Run Now** to update all the COVID datasets.

10. Click the **RUN** button.

Summary

In this chapter, we picked up from where the data is acquired and introduced how the data is stored and accessed. The storage architecture and the physical storage approach were discussed. We also introduced the Domo DataSet object and walked through numerous data catalog and dictionary features, and actions to create, manage, secure, profile, monitor, and alert on the data. You are now empowered to populate data storage at a massive scale, which is essentially a query-optimized data warehouse. Auto-populated metadata and a data dictionary are included, all without ever having to create a schema or optimize a physical database structure. Wow, that's a game changer!

In the next chapter, we will go through the data sculpting tools available to transform and blend data in Domo.

Further reading

Here are some articles of interest:

- Domo documentation on creating alerts: `https://knowledge.domo.com/Optimize/Notifications_and_Alerts/Creating_an_Alert_for_a_DataSet`

- Wikipedia article on webhooks: `https://en.wikipedia.org/wiki/Webhook`

4
Sculpting Data

Artistic sculpting requires the ability to see the potential for a work of art in raw material. The sculptor has a range of tools at their disposal to shape raw stone into something more valuable. Data analysts, similar to sculptors, can see potential in raw data to answer questions. Analysts also require tools to shape and refine the raw data. The Domo platform has many tools to enable data sculpting.

If you are not a technical data manipulation guru, then don't worry. Domo has a very approachable graphical drag-and-drop tool that requires no coding and any business analyst can use. However, if you are a SQL expert, Domo also has tools to enhance your scripting skills for sculpting data.

In this chapter, we will cover the following topics:

- Introducing the sculpting tools for persistent datasets
- Using ETL dataflows
- Using the DataFlows page
- Using SQL dataflows

Technical requirements

To follow along with this chapter, you will need the following:

- Internet access

- Your Domo instance and login details

- The ability to download the example files from GitHub

> **Important Note**
>
> If you don't have a Domo instance, you can get a free trial instance here:
> `https://www.domo.com/start/free`.

All of the code for this chapter can be downloaded from GitHub at `https://github.com/PacktPublishing/Data-Democratization-with-Domo`.

Introducing the sculpting tools for persistent datasets

In Domo, the sculpting tools are called **Magic** because they make it simple to transform data. The tools allow you to create dataflows that output new persistent datasets – ones that are stored on physical media. Let's look at the data transformation tools that are available for creating persistent datasets:

- **Extract, Transform, and Load** (ETL) **dataflows** establish a sequence of operations through a visual programming interface that takes input datasets and performs transformations when executed that create and/or update output datasets.

- **SQL dataflows** use SQL commands and sequence them together into a dataflow that, when executed, reads the input datasets, transforms the data according to the SQL commands, and outputs datasets.

The following table contains guidance for when to use which tool:

Tool	When to Use	Expertise Level	Availability
ETL DataFlows	The primary tool for creating persistent transformed datasets in Vault. It is the first choice for creating dataflows until you reach a limitation in each volume, performance, or transformation type.	End user	Standard with the platform
SQL DataFlows	A tool for advanced persistent transforms that results in datasets in Vault. Use this when a transform is not available in ETL DataFlows or if you just prefer SQL scripting over visual programming. Note that these are slower in terms of performance and handle smaller volumes than ETL DataFlows.	Knowledge of SQL	Standard with the platform

Figure 4.1 – Magic transform persistent tools guidance

Now that we know how to choose the right tool, let's see how simple the tools are to use.

Using ETL dataflows

ETL dataflows are the primary tools for joining or merging multiple datasets and creating data transformation pipelines. Domo datasets are the only supported inputs; the output of the dataflow is one or more datasets. So that you don't get confused, ETL is a misnomer in the sense that the sequence is extract, load, and transform, or ELT – it starts with the datasets being created by connectors. In this section, we'll walk through how to create an ETL dataflow.

Creating an ETL dataflow

The scenario we are going to walk through will use the Opportunity dataset. We want to add date dimension information and cleanse the lead source type to produce a new enhanced/cleansed dataset.

Let's start by making sure we have all the input data sources we need in the necessary datasets:

1. Download the `Opportunity.xlsx` file from `https://github.com/` `PacktPublishing/Data-Democratization-with-Domo/blob/` `e92849ebfd14321be6410f88f473dac7838d18b1/Opportunity.xlsx`.

2. Search for **Excel connector** in **APPSTORE** and use it to load `Opportunity.xlsx`.

3. Name the dataset `Opportunity CH4`.

4. Search for **Domo Dimension Connector** in **APPSTORE** and use it to load the `calendar.csv` file by choosing it from the **FILES** option in the connector.

5. Name the dataset `Date Dimension`.

6. Download the `Mapping - Lead Source.xlsx` file from `https://` `github.com/PacktPublishing/Data-Democratization-with-Domo/` `blob/09c7d2c8d5708e6d2c702db7b8ebadd3c050715a/Mapping%20` `-%20Lead%20Source.xlsx`.

7. Use **Excel connector** from **APPSTORE** to load `Mapping - Lead Source.xlsx`.

8. Name the dataset `Lead Source Mapping CH4`.

    Now that the input datasets are in place, let's create the ETL dataflow.

9. Click on the **DATA** item in the Domo main menu bar.

10. Click **ETL** in the **MAGIC TRANSFORM** area.

11. Click the pencil icon next to the **ADD DataFlow Name** label and enter `Opportunity Cleanse` as the dataflow name.

    Now, add the `Opportunity CH4` dataset to the dataflow.

12. From the left panel, under the **DATASETS** area, click and drag the **Input DataSet** tile onto the canvas to the right. Notice that the tile properties area is directly below the canvas area.

13. Click **SELECT DATASET** in the tile properties area.

14. Find and click the `Opportunity CH4` dataset in the **Select a Dataset** popup.

15. Find and click **CHOOSE DATASET** in the lower-right corner of the **Select a Dataset** popup.

16. Click on the **DETAILS** tab in the tile properties menu to see metadata information about the dataset.

17. Click on the **DATA** tab in the tile properties menu to see a data preview.

18. Click **DONE** via the tile properties menu to close the tile properties area.

19. Add the `Date Dimension` and `Lead Source Mapping CH4` datasets to the canvas by applying *Steps 12* to *18* for each.

At this point, your canvas will look something like this:

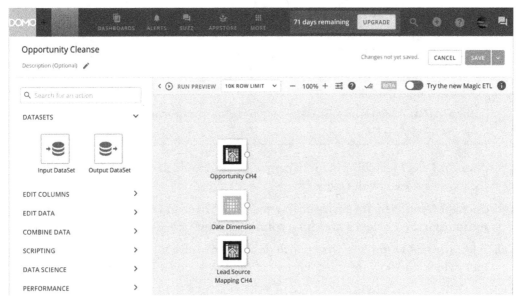

Figure 4.2 – Dataflow canvas with the input sources defined

Now, we'll add the tiles to join the datasets to the dataflow. First, let's join the date dimension attributes.

20. From the left panel, click **COMBINE DATA** to see the tiles underneath.

21. Click and drag the **Join Data** tile onto the canvas to the right.

22. Connect the `Opportunity CH4` tile and the **Join Data** tile by clicking and dragging from the circle on the middle right-hand side of the `Opportunity CH4` tile to the triangle on the middle left-hand side of the **Join Data** tile. You will see a line connecting the tiles.

23. Connect the `Date Dimension` tile and the **Join Data** tile by clicking and dragging from the circle on the middle right-hand side of the `Date Dimension` tile to the triangle on the middle left-hand side of the **Join Data** tile. A line will connect the tiles.

24. Click the **Join Data** tile to ensure it is highlighted.

25. In the tile properties menu, click the pencil icon, enter `Oppty Left Join DateDim` as the join name, and press *Enter*. Changing the default name is a good practice to make the dataflow more readable.

26. Click the **Select Input** combo box in the **1 Select an identifying column** section and select `Opportunity CH4`.

27. Click the **Select identifying column** combo box and select the **CloseDate** field.

> Note – Compound Joins
>
> The **MATCH ANOTHER COLUMN** button is used to add compound join fields when necessary.

28. Click the **Select matching column** combo box in the **2 Select a matching column** section and select the **dt** field.

29. Click the Venn diagram circles in between the **1 Select an identifying column** section and the **2 Select a matching column** section.

30. Select the **Left Outer** join type to include all the rows from the `Opportunity CH4` dataset and only the matching rows from the `Date Dimension` dataset.

At this point, the canvas will look like this:

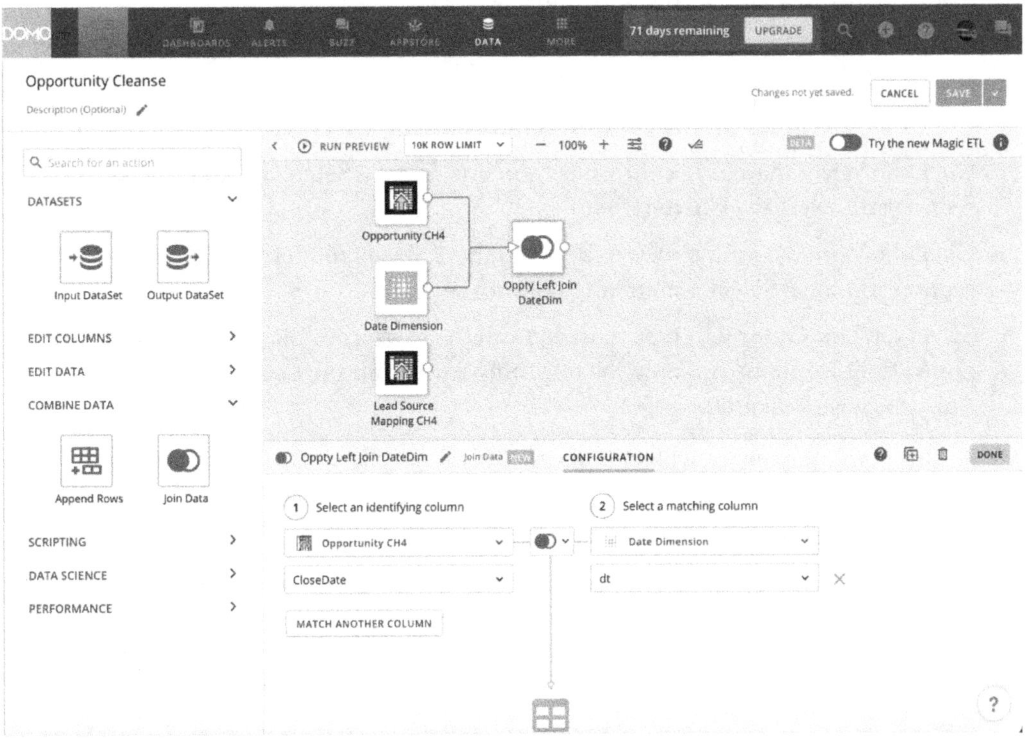

Figure 4.3 – Dataflow canvas with one join

Next, we will add another join to handle standardizing the lead source type values in the opportunity table from the mapping table.

31. Click and drag the **Join Data** tile from the left panel, under **COMBINE DATA**, onto the canvas to the right.

32. Connect the **Oppty Left Join DateDim** tile and the new **Join Data** tile by clicking and dragging from the circle on the middle right-hand side of the **Oppty Left Join DateDim** tile to the triangle on the middle left-hand side of the **Join Data** tile. You will see a line connecting the tiles.

33. Connect the `Lead Source Mapping CH4` tile and the **Join Data** tile by clicking and dragging from the circle on the middle right-hand side of the `Lead Source Mapping CH4` tile to the triangle on the middle left-hand side of the **Join Data** tile. A line will connect the tiles.

34. Click the **Join Data** tile to ensure it is highlighted.

35. In the tile properties menu, click the pencil icon, enter `OpptyDateDim Left Join Lead Source` as the join name, and press *Enter*.

36. Click the **Select Input** combo box from the **1 Select an identifying column** section and select **Oppty Left Join Date Dim**.

37. Click the **Select identifying column** combo box and select the **LeadSource** field.

38. Click the **Select matching column** combo box in the **2 Select a matching column** section and select the **Current** field.

39. Click the Venn diagram circles in between the **1 Select an identifying column** section and the **2 Select a matching column** section.

40. Select the **Left Outer** join type to include all the rows from the **Oppty Left Join Date Dim** tile output and only the matching rows from the `Lead Source Mapping CH4` dataset.

At this point, the canvas will look like this:

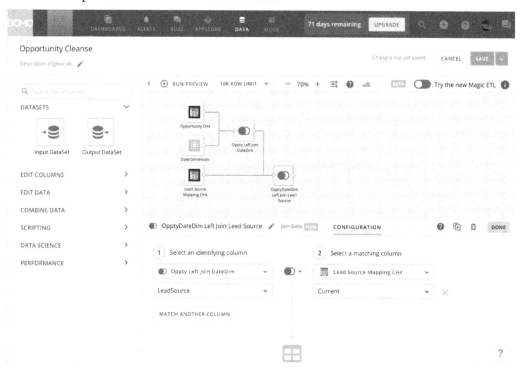

Figure 4.4 – Dataflow canvas with two joins

Next, we'll add the output dataset for the dataflow.

41. Click and drag the **Output DataSet** tile from the left panel, under **DATASETS,** onto the canvas to the right.

42. Connect the `OpptyDateDim Left Join Lead Source` tile and the new **Output Dataset** tile by clicking and dragging from the circle on the middle right-hand side of the `OpptyDateDim Left Join Lead Source` tile to the triangle on the middle left-hand side of the **Output Dataset** tile. You will see a line connecting the tiles.

43. Click the **Output Dataset** tile. In the tile properties area, click on the pencil icon, name the dataset `Opportunity Cleansed`, and press *Enter*.

44. Click the **DONE** button to the right of the tile properties menu bar to close the tile properties area.

At this point, your canvas will look like this:

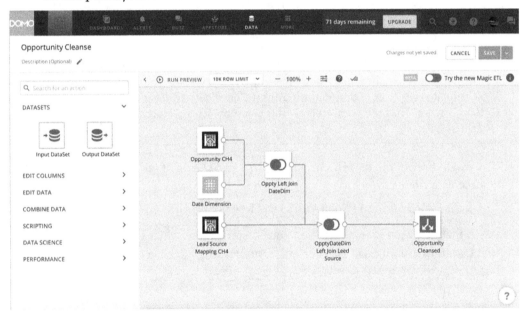

Figure 4.5 – Dataflow canvas

Now, let's run a preview to see the transforms each step of the way.

45. Click on the default **10K ROW LIMIT** combo box and select **100K ROW LIMIT** to bring in all the data from the input datasets.

46. Click on the **RUN PREVIEW** option via the menu bar. This will start running the process. Each step that is running will show a little blue circle icon in the top-right corner of each tile. This will turn into a green circle with a white checkmark inside it when completed.

47. To see the data that's been produced for each tile, simply click on the tile and click the **PREVIEW** or **DATA** tab via the tile properties.

48. A good check of successfully cleansing the dataset is to compare the row count from the `Opportunity CH4` input dataset with the `Opportunity Cleansed` output dataset: they should match. You can do this by clicking on each tile and then clicking to the **DETAILS** tab via the tile properties to see the row count.

Next, save the dataflow.

49. Click the down-pointing chevron to the immediate right of the **SAVE** button.

50. Select **Save and Run** to save and run the dataflow.

51. Enter `First Draft` into the **Version Description (Optional)** box in the **SAVE DataFlow** popup.

52. Select **Only when DataSets are updated** in the **When should this Dataflow run?** combo box.

> **Note – Only run dataflow when datasets are updated**
>
> This is a powerful scheduling feature that eliminates the timing headaches of scheduling runs. This option simply reruns the dataflow whenever the source data changes. Trust me – you will love this common-sense feature for keeping data in sync.

53. Check all three of the dataset checkboxes to rerun the dataflow when any of the source data changes.

54. Click the **SAVE** button in the **SAVE DataFlow** popup.

With that, you have created and run your first dataflow, which will have left you on the **Dataflow Detail** page for your new **Opportunity Cleanse** dataflow:

Figure 4.6 – Opportunity Cleanse dataflow page

Next, we will walk through the **DataFlows** page so that you know how to find and manage dataflows.

Using the DataFlows page

The **DataFlows** page is where data transformations are managed in Domo.

Once a dataflow has been saved, it is administered through the **DataFlows** page, as shown in the following screenshot:

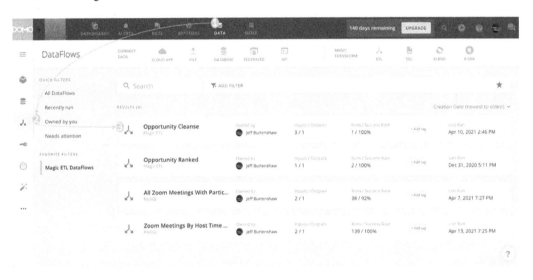

Figure 4.7 – Browsing DataFlows

To search the list of dataflow names, follow these steps:

1. Click inside the **Search** field and enter op to see all datasets in the list with op anywhere in their name.

2. To filter the list of dataflow names by type, click on the **ADD FILTER** area.

3. Select **DataFlow Type** and click **Magic ETL** to see all the dataflows that were created using the Magic ETL tool.

To save the filter options as a favorite for future use, follow these steps:

1. Click on the favorite star icon to the far right in the search bar.

2. Enter the name of your filter – in this case, Magic ETL DataFlows – and click **SAVE**.

3. Apply the filter by clicking on it in the **FAVORITE FILTERS** area.

To apply one of the preset filters to see a list of recently updated datasets, follow these steps:

1. Click on **Recently run** in the **QUICK FILTERS** area.

2. To run a dataflow from the **DataFlows** page, hover your mouse over the right-hand side of the desired opportunity, click the gear icon that appears, and click **Run**, as shown in the following screenshot:

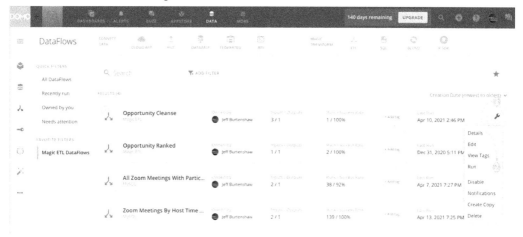

Figure 4.8 – Running a dataflow from the DataFlows page

Next, we will examine the features that are available when viewing the details page of a specific dataflow.

Using the DataFlow Detail page

The **DataFlow Detail** page has a robust set of features for administering dataflows. To view a **DataFlow Detail** page, click on the desired dataflow name – in this case, Opportunity Cleanse – in the dataflow **Results** area. You should see something like the following:

Figure 4.9 – Opportunity Cleanse DataFlow Detail page

The dataflow page contains various metadata: the dataflow's name, owner, dataflow type, number of inputs and output datasets, and dataflow refresh date. Additionally, the tabs and options areas provide actions that can be taken on the dataflow. Next, we'll look at each of the tabs, starting with the **HISTORY** tab.

Using the HISTORY tab

The **HISTORY** tab provides a complete log of each time an attempt is made to update the dataflow and whether it was successful or not. A user can see how long ago the last successful update occurred, the average time it takes for each successful update, and the successful percentage rate of updates.

In addition to seeing the update history, several actions are provided for working with the dataflow's internal execution steps and reverting to prior dataflow versions.

Continuing with the example `Opportunity Cleanse` dataflow, click on the **HISTORY** tab and then the ... icon to see the row actions:

Figure 4.10 – Dataflow HISTORY tab actions

The following row action options are available:

- **View Details** displays the step-by-step execution details of the dataflow's execution.
- **Make Version Current** will revert the current dataflow definition to the version on that execution.

Next, we will explore how to view the version history of the dataflow definition.

Using the VERSIONS tab

The **VERSIONS** tab is where you can see each saved version of the dataflow definition.

Click on the **VERSIONS** tab and then the ... icon to see the row action. In this case, there is only one action for changing the row to the current version, so in the first row, no action is available.

Now, let's review the **SETTINGS** tab.

Using the SETTINGS tab

The **SETTINGS** tab simply allows you to control the schedule options for running the dataset. Let's take a look:

1. Click on the **SETTINGS** tab.
2. Click on the **When should this Dataflow Run?** combo box and select **On a schedule**.

3. In the **Frequency** box, select **Weekly**.

4. Choose the day of the week to run the dataflow on by clicking on one of the day boxes and setting the time to run here as well.

5. Click the **APPLY** button to set the schedule.

Next, let's look at the **DATASETS** tab.

Using the DATASETS tab

The **DATASETS** tab shows the input and output datasets in the dataflow:

1. Click on the **DATASETS** tab.

2. To see the dataset's details page, simply click on the desired dataset tile. Use the browser's back button to return to the dataflow's detail page.

The last tab to review is the **LINEAGE** tab.

Using the LINEAGE tab

The **LINEAGE** tab displays the datasets that have been used in the dataflow and is useful when you're performing impact analysis to see how changes in the artifacts that have been used in the dataflow might impact other artifacts. This tab has three sections, as shown in the following screenshot:

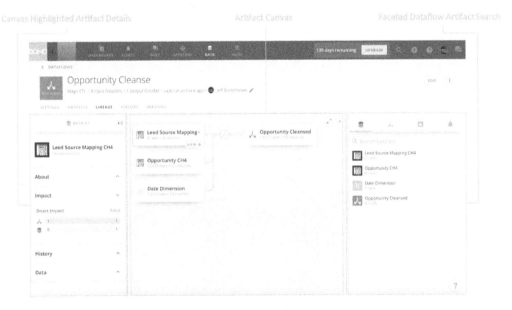

Figure 4.11 – The DataFlow Detail page's LINEAGE tab

Let's look at these in more detail:

- **Artifact Canvas** is the center display layout area where the dataset and dataflow artifacts are viewed.

- **Canvas Highlighted Artifact Details** is the left panel and is where the selected canvas artifact properties are displayed. This also includes the **Impact** section, where you can specify the card, dataset, and dataflow artifacts the selected artifact is used in.

- **Faceted Dataflow Artifact Search** is the right panel and is where the artifact types are represented as icons for datasets, dataflows, cards, and alerts. These are facets of a dataflow-wide search, no matter what is highlighted in the canvas. This is very useful for finding specific artifacts in dataflows with many artifacts.

Next, we will look at the options menu's features.

Using the options menu

The options menu is in the top-right corner of the page and is used to access administrative functions on the dataflow. Let's take a look:

1. Click the stacked dots icon on the far right to show the list of options, as shown in the following screenshot:

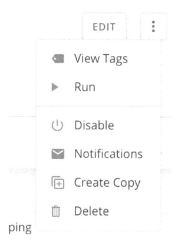

Figure 4.12 – Dataflow options

The following are brief descriptions of each item:

- **View Tags**: This enables you to see and manage tags on the dataflow. Tags are used in searches.

- **Run**: This manually runs the dataflow.

- **Disable**: This is a toggle feature that manually disables the dataflow from running until it is toggled back to enabled.

- **Notifications**: Here, you can choose whether you want to receive email notifications when the dataflow run fails.

- **Create Copy**: Here, you can create a duplicate copy of the dataflow.

- **Delete**: Here, you can permanently remove the dataflow.

- Now, let's edit the dataflow. In this exercise, we will remove the irrelevant Domo Fiscal Calendar-related fields from the dataset that were joined in from the Domo **Date Dimension** dataset.

2. Click the **EDIT** button in the options menu.

3. Hover over the line connecting the OpptyDateDim Left Join Lead Source tile and the Opportunity Cleansed tile and click the **X** icon to remove the line.

4. In the left panel, click **EDIT COLUMNS** to see the tiles.

5. Click and drag the **SELECT COLUMNS** tile onto the canvas between the OpptyDateDim Left Join Lead Source tile and the Opportunity Cleansed tile.

6. Click and drag from the circle on the right middle side of the OpptyDateDim Left Join Lead Source tile to the triangle on the left-middle side of the **Select Columns** tile to connect these tiles.

7. Click on the **Select Columns** tile. Then, in the tile properties pane, click the **ADD ALL COLUMNS** button.

8. Scroll down the list of columns and click the **X** icon to the far right of the **domofiscalyear** row; also, click the **X** icon to the far right of the **domofiscalquarter** row.

9. Click **Done** in the top-right corner to close the tile properties.

10. Click and drag from the circle on the middle right-hand side of the **Select Columns** tile to the triangle on the middle left-hand side of the Opportunity Cleansed tile to connect these tiles.

11. Click the down-pointing chevron to the immediate right of the **SAVE** button.

12. Select **Save and Run** to save and run the dataflow.

13. Enter a version comment stating `Removed Domo fiscal year fields` in the **Version Description (Optional)** box in the **Save And Run DataFlow** popup.

14. Click **Save and Run** to save and run the dataflow.

Now, you are capable of creating, editing, running, monitoring, and deleting ETL dataflows to sculpt datasets into what you need.

Now, we will investigate another powerful data sculpting tool: SQL dataflows.

Using SQL dataflows

SQL dataflows is a tool for those of you who know the SQL scripting language. It enables you to chain multiple sequential SQL commands into a transformation pipeline. Domo datasets are the only supported inputs, and the output is one or more datasets. Let's go through the steps of how to create a SQL dataflow.

Creating a SQL dataflow

Let's walk through a scenario where we must create a SQL dataflow to produce a new aggregated dataset that supports **burnup charting**:

1. Click on the **DATA** item in the Domo main menu bar.

2. Click **SQL** in the **MAGIC TRANSFORM** area.

3. Enter `Opportunity Aggregate` in the **DataFlow Name & Description** box.

4. In the **Input DataSets** area, click **SELECT DATASET**.

5. Click **Opportunity Cleansed** in the **Select a Dataset** popup.

6. Click the **CHOOSE DATASET** button.

At this point, the page should look as follows:

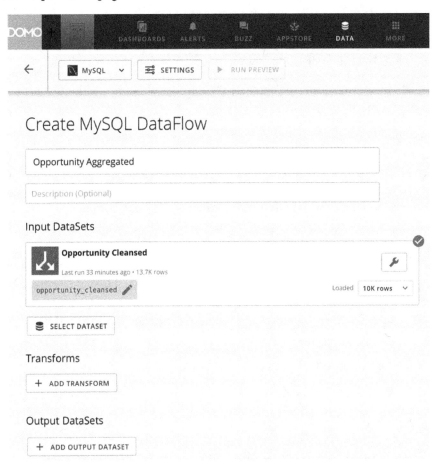

Figure 4.13 – SQL dataflow with an input dataset

Let's add the **Date Dimension** dataset as input as well.

7. In the **Input DataSets** area, click **SELECT DATASET**.

8. Click **Date Dimension** from the **Select a Dataset** popup.

9. Click the **CHOOSE DATASET** button.

Now, let's add a transform to aggregate the number of opportunities that are created each day.

10. Click on the + **ADD TRANSFORM** button.

11. Click on the **TABLE** option. We are doing this since we are going to use a `select` statement to create a table that can be indexed for performance in the subsequent steps. Note that the **SQL** option is used for SQL language commands other than `select`, also known as **data definition language** or **DDL** commands.

12. In the transform window, which can be found in the top left, rename the transform `groupbycreateddate`. It will convert whatever you enter into lowercase and will substitute "_" for spaces. This enforced name standardization occurs because the name of the transform is also the name of the temporary table that was created from the SQL's execution.

13. Click on `opportunity_cleansed` to see the schema of the table that was created from the input dataset step.

14. Enter the following code in the transform canvas:

```
SELECT date(`CreatedDate`) as 'Date', count(`Id`) as
'New'
FROM `opportunity_cleansed`
GROUP BY date(`CreatedDate`)
ORDER BY 1 DESC
```

This SQL command will return a row for each date with a count of the number of rows created on that date in descending date order. If you need a refresher on SQL, there is a link to a good tutorial in the *Further reading* section at the end of this chapter.

15. Click the **RUN SQL** button.

The page will look as follows when the query completes:

Figure 4.14 – SQL dataflow transform step with query results

16. Click on the **INDEXING** tab.

17. Click on the **ADD INDEX** button.

18. Click the **Select Type** combo box and select the **Primary** index option.

19. Click **Select columns to define the index** and select **Date**.

20. Click the **DONE** button.

21. Click the **APPLY** button.

Now, add a transform to aggregate the number of won, lost, and closed opportunities each day.

22. Click on the **+ ADD TRANSFORM** button.

23. Click on the **TABLE** option.

24. In the transform window, which can be found in the top left, rename the transform `GroupbyClosedWonDate`.

25. Enter the following code in the transform canvas:

```
SELECT
DATE(`LastModifiedDate`) as 'Date',
SUM(case when `IsWon`= 'true' then 1 else 0 end) as
'Won',
```

```
-- Lost = Closed - Won
SUM(case when `IsClosed` = 'true' then 1 else 0 end) -
SUM(case when `IsWon`= 'true' then 1 else 0 end) as
'Lost',

SUM(case when `IsClosed` = 'true' then 1 else 0 end) as
'Closed'
FROM `opportunity_cleansed`
GROUP BY DATE(`LastModifiedDate`)
ORDER BY 1 DESC
```

This SQL command aggregates the number of opportunities that have been won, lost, and closed for each date, sorted in descending date order.

26. Click the **RUN SQL** button.

27. Click on the **INDEXING** tab.

28. Click on the **ADD INDEX** button.

29. Click the **Select Type** combo box and select the **Primary** index option.

30. Click **Select columns to define the index** and select **Date**.

31. Click the **DONE** button.

32. Click the **APPLY** button.

Here are a few things to keep in mind when using SQL dataflows:

- The output of a table transform is a temporary table with the same name as the transform. This is only available within the scope of a particular execution for the dataflow.

- A transform can reference all the prior transform temporary tables, but cannot reference later transform outputs in the dataflow.

- The output from the RUNSQL command is limited to a preview of 100 rows, and as you may recall, the input datasets are limited as well. So, don't get confused if the output you are expecting is limited.

- If you want to persist the full results of a given transform, add an output dataset to the dataflow, along with the transform's temporary table, using the Select * from [transform name] method.

Now, let's add a transform to combine the results from the previous transforms to produce a new output with created, closed, won, and lost totals by day in a single table. We don't want any gaps in our dates, so we will also add the **Date Dimension** dataset.

33. Click on the + **ADD TRANSFORM** button.

34. Click on the **TABLE** option.

35. In the transform window, which can be found in the top left, rename the transform `opportunityburnrate`.

36. Enter the following code in the transform canvas:

```
SELECT d.`dt` as 'Date', cd.`New`,
cwd.`Closed`, cwd.`Won`, cwd.`Lost`
FROM `date_dimension` d
LEFT JOIN `groupbycreateddate` cd
ON d.`dt` = cd.`Date`
LEFT JOIN `groupbyclosedwondate` cwd
ON d.`dt` = cwd.`Date`
WHERE d.`dt` >= (SELECT MIN(`Date`) FROM
`groupbycreateddate`)
AND d.dt <= (SELECT MAX(`Date`) FROM
`groupbyclosedwondate`)
ORDER BY 1
```

This SQL command places the date dimension with the other tables to make sure every opportunity date has the necessary dimension information. The dates are constrained from the earliest opportunity creation date through the latest opportunity closed date for performance and size optimization.

37. Click the **RUN SQL** button.

38. Click the **Apply** button.

Now, let's add a transform to add a column for the running total of open opportunities.

39. Click on the + **ADD TRANSFORM** button.

40. Click on the **TABLE** option.

41. In the transform window, which can be found in the top left, rename the transform opportunity_burnup.

42. Enter the following code in the transform canvas:

```
SELECT t.`Day`, t.`New`, t.`Closed`, t.DayNet, t.`Won`,
t.`Lost`,
        @running_total:=@running_total + t.`DayNet` AS
Open
FROM
(SELECT
   `Date` AS 'Day', `New`, `Closed`, `Won`, `Lost`,
   (CASE WHEN `New` IS NOT null THEN `New` ELSE 0 END)-
   (CASE WHEN `Closed` IS NOT null THEN `Closed` ELSE 0
END) AS 'DayNet'
   FROM `opportunityburnrate`
   GROUP BY `Day`
) t
JOIN (SELECT @running_total:=0) r
ORDER BY t.`Day` ASC
```

This SQL command is advanced, and if you don't understand exactly how it works, that's OK. In brief, it creates a temporary structure to loop through each day and increments a variable in the temporary structure to accumulate the running total. The best way to see what this is doing is to run the statement and study the output.

43. Click the **RUN SQL** button.

44. Click the **Apply** button.

At this point, the **Transforms** section will look as follows:

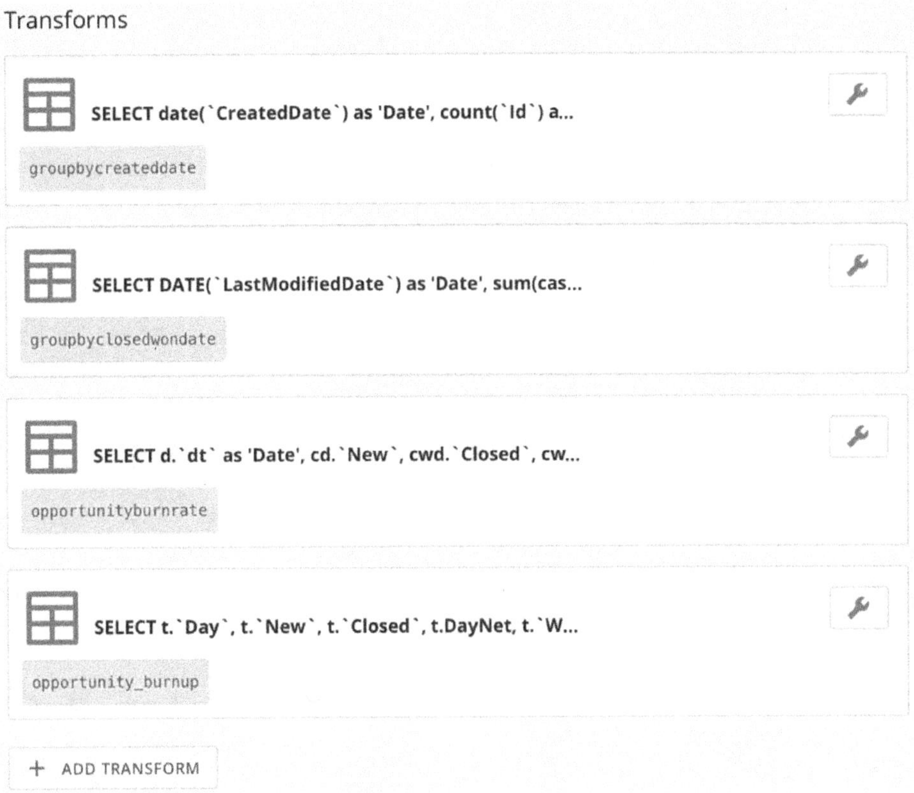

Transforms

Figure 4.15 – SQL dataflow with multiple transforms

Now, let's create the output dataset.

45. Click on + **ADD OUTPUT DATASET**.

46. Enter Opportunity Burn Up in the **Add Output DataSet Name** box.

47. Enter the following code in the transform canvas:

```
SELECT * FROM `opportunity_burnup`
```

48. Click the **RUN SQL** button.

49. Click the **Apply** button.

Next, let's save and run the dataflow.

50. Click the down-pointing chevron to the immediate right of the **SAVE** button.

51. Select **Save and Run** to save and run the dataflow.

52. Enter Burnup in the **Version Description (Optional)** box in the **SAVE DataFlow** popup.

53. Click **SAVE AND RUN** to run the dataflow and create the **Opportunity Burn Up** dataset.

The dataflow's lineage will look as follows:

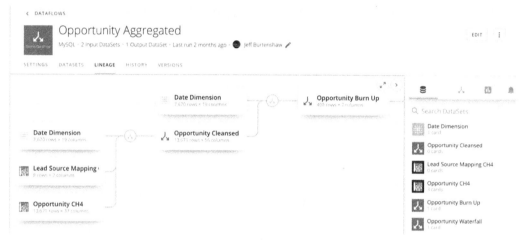

Figure 4.16 – Dataflow data lineage

Fantastic – you have sculpted a new dataset using a **SQL dataflow**, which will allow you to create amazing burnup reports.

Summary

In this chapter, we reviewed the data sculpting tools available in Domo for creating persistent datasets and how to decide which data sculpting tool to use for data transformation. First, we toured the dataflows page's features before creating new datasets using both the ETL dataflow and SQL dataflow tools. The output of both tools creates persistent datasets that can be stored in Vault. Between the Magic ETL and SQL dataflow tools, there is no persistent data sculpting shape that cannot be created.

In the next chapter, we will look at the in-memory sculpting tools for creating virtual datasets.

Further reading

For more information regarding what was covered in this chapter, take a look at the following resources:

- Burnup charting explained: `https://www.modernanalyst.com/Careers/InterviewQuestions/tabid/128/ID/3433/What-is-a-Burn-Up-Chart-and-how-does-it-differ-from-a-Burn-Down-Chart.aspx`.

- Here is a good tutorial on the SQL language: `https://www.w3schools.com/sql/`.

- To learn more about all the Magic ETL DataFlow tiles, go to `https://knowledge.domo.com/Prepare/Magic_Transforms/ETL_DataFlows`.

- For a general overview of magic transforms, go to `https://knowledge.domo.com/Prepare/Magic_Transforms`.

- Additional material on SQL DataFlows can be found here: `https://domohelp.domo.com/hc/en-us/sections/360007295474-SQL-DataFlows`.

5
Sculpting Data In-Memory

While working with data, sometimes, for performance or convenience purposes, we must perform operations on the data that are not persistently stored to disk. For performance, an example is combining billions of rows of transactions with millions of rows of customer data. For convenience, an example might be a simple code mapping lookup or applying business logic. These fine-tuning adjustments are enabled by tools that work on in-memory caches and query technology. In Domo's case, this is **Adrenaline**, which is a queryable **in-memory cache**. These in-memory transforms are orders of magnitude faster than persistent transforms via ETL and when fast is what you are after, these tools are the right choice. Occasionally, the use of Adrenaline dataflows becomes critical in cases when the rows of data being processed reaches billions of rows, joins are required, and materialization is not practical. If you are familiar with creating non-materialized views on a relational database or using formulas in Excel, then this is a similar technology.

In this chapter, we will cover tools that are part of the Domo **MAGIC** transforms family (magic because they are so easy to use) that operate on in-memory caches. The following topics will be covered in this chapter:

- Reviewing the in-memory sculpting tools
- Using **Views Explorer**
- Using Data Blends

- Leveraging Beast Mode
- Finding, editing, and archiving Dataset Beast Modes
- Adrenaline DataFlows

Technical requirements

To follow along with this chapter, you will need the following:

- Internet access
- Your Domo instance and login details
- The ability to download this chapter's example files from GitHub

> **Important Note**
> If you don't have a Domo instance, you can get a free trial instance here:
> `https://www.domo.com/start/free`.

All of the code in this chapter can be downloaded from this book's GitHub repository: `https://github.com/PacktPublishing/Data-Democratization-with-Domo`.

Reviewing the in-memory sculpting tools

In Domo, there are several tools for executing data transforms in the Adrenaline cache. These transforms are very performant because they are executed in the cache. They are as follows:

- **Data Views** can be used to filter or aggregate dataset data in Adrenaline, just like a typical relational database view. They are also capable of performing some data transformations that may otherwise need to run a dataflow, such as changing data types, removing columns, changing column names, and adding calculations. The Data View's virtual dataset appears in the dataset catalog.

- **Data Blends** are basic dataset joins that are equivalent to an Excel VLOOKUP function that's executed in the Adrenaline in-memory cache.

- **Beast Modes** are stored calculations on datasets that execute in Adrenaline when a visualization requests the data.

- **Adrenaline DataFlows** is an advanced performance tool that's used to run scripted sequential transformation pipelines on very large datasets in the Adrenaline in-memory cache. Due to the resource-intensive nature of transforming billions of rows, this tool is available as an enterprise extension for an additional subscription fee.

The following table contains guidance for when to use which tool:

Tool	When to Use	Expertise Level	Availability
Data Views	Used to create virtual data for real-time transformation either within a single dataset or across multiple datasets; for example, renaming/limiting columns or creating aggregations that aren't persisted in Vault. Datasets can be joined in Data Views.	End user	Standard with platform
Data Blends	Used to look up the values from another dataset to convert the values in the main dataset. This is essentially the equivalent of the VLOOKUP function in Excel. It has the benefit of caching the resulting dataset in Adrenaline and is much faster than performing extract, transform, and load (ETL) on large datasets.	End user	Standard with platform
Beast Modes	Used to extend an existing dataset by creating new calculated columns within that dataset.	End user	Standard with platform
Adrenaline DataFlows	Used when large (billions of rows) transformation pipelines are needed to execute on one or multiple large datasets in the in-memory cache instead of moving and storing the data persistently via a data-at-rest strategy. The user interface is like SQL DataFlows but runs in Adrenaline and does not write to Vault. It is a fast aggregation method for very large datasets.	End user	Additional subscription required

Figure 5.1 – Magic transform in-memory tools guidance

Now that you have learned how to choose the right tool, let's learn how to use them.

Using Views Explorer

Views Explorer is a data sculpting tool that creates non-materialized, in-cache views of your datasets. Views are useful when you want to reduce the columns users might see or create non-materialized data aggregates or mashups. Views are dataset definitions and once saved, they can be viewed in the dataset catalog. Views can be created by choosing the **Views Explorer** option on the dataset detail page under the **Open With** feature. Views live in the Adrenaline cache, so they are in-memory fast.

Next, let's take a tour of **Views Explorer**.

Touring the Views Explorer page

Before we proceed, let's get oriented with the features of the **Views Explorer** page, as seen in the following screenshot:

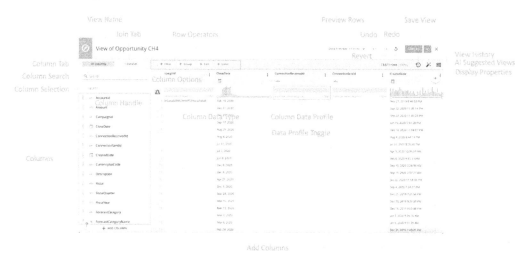

Figure 5.2 – The Views Explorer page's features

Let's look at the features that are available on the **Views Explorer** page:

- **View Name**: Enables you to name and change the name of the view. This name is what will appear in the dataset catalog.

- **Join Tab**: Enables you to add datasets and join datasets to create advanced views.

- **Row Operators**: Enables you to filter rows, group by rows, sort rows, and limit the number of rows that are returned in the view.

- **Preview Rows**: This allows you to select full or sample row counts to view while editing the view.

- **Undo/Redo**: Provides you with a way to undo and redo changes that you have made to the view definition.

- **Revert**: This allows you to undo all changes and reset the view back to the original settings.

- **Save View**: Saves the view, which also creates the non-materialized dataset.

- **View History**: This shows the revision history of the view.

- **AI Suggested Views**: This allows you to select view configurations that the AI engine has calculated that might be helpful. For example, you can create a view on the dataset that has removed the outliers.

- **Display Properties**: Enables you to change the display settings.

- **Column Tab**: Provides access to the columns on the dataset where the view is being created.

- **Column Search**: This lets you search the columns in the dataset for selection.

- **Column Selection**: This allows you to select one or more columns and move them to the top or bottom of the view or remove the columns that have been selected from the view.

- **Columns**: Displays a list of available columns.

- **Column Handle**: Enables you to drag and drop the columns into the preferred order in the view.

- **Column Options**: Opens a pop-up menu of actions you can perform on the column.

- **Column Data Type**: This is an iconic representation of the column's type of data, namely **Date, Text, or Numeric**.

- **Column Data Profile**: This shows the distribution of data values in the column.

- **Data Profile Toggle**: Hides or shows the Column Data Profiles.

- **Add Columns**: This allows you to manage the views columns from the datasets or calculated columns.

Now that the tour is over, it's time to learn how to create a Data View.

Creating a Data View

Let's walk through a basic scenario of creating a **Data View** to reduce the number of columns that are displayed and rename some of the columns to make the data more consumable. Then, we will learn about more advanced steps, including how to add additional datasets and create joins in the view, as well as how to standardize data values through a mapping table. Finally, we will walk through how to use a Data View to see **leader and laggard boards**.

Follow these steps to create a Data View on a dataset:

1. Click on the **DATA** item in the Domo main menu bar.
2. Click on the datasets icon on the left-hand side toolbar to go to the **DataSets** page.
3. Click **Opportunity CH4 dataset** to go to the dataset detail page.
4. Click the **OPEN WITH** combo box and select **Views Explorer** to enter the **Views Explorer** page.

Now, let's do some tidying up and remove, rename, and reorder some columns:

1. In the **Columns** area, enter Conn in the search box.
2. Click **SELECT ALL** to select the two columns.
3. Click the trashcan icon in the **Column** tab toolbar to remove the columns from the view.
4. Click **X** in the search box to clear the search.
5. Scroll to the right in the preview pane and look for the **IsDeleted** column.
6. See how the data profile for **IsDeleted** shows that there is only 1 unique value of **false** by hovering over the blue bar in the data profile area. You might have to click the data distribution icon in the table header to toggle the data distribution row.
7. Click on the three stacked dots icon in the **Column Options** menu in the column header.
8. Click **Remove** at the bottom of the pop-up menu.
9. Scroll to find the **Name** column in the preview pane.
10. Click on the **Name** label, enter OpportunityName, and press *Enter* to rename the field and make this less ambiguous.
11. Move the **OpportunityName** column position by clicking the three stacked dots icon in the **Column Options** menu in the column header.
12. Select **Move...** and choose **First** in the pop-up menu to move it to the first position in the table.

13. Click the **SAVE AS** button.

14. Click the **Save** button in the **Save as New DataSet View** popup. This will take you to the **Dataset Detail** page for the view you just created, as shown in the following screenshot:

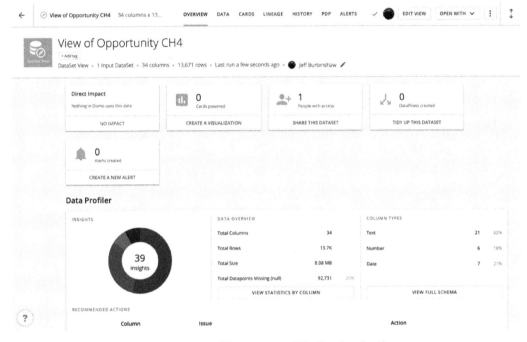

Figure 5.3 – View of Opportunity CH4 DataSet detail page

Next, let's filter out the unwanted rows in the view:

1. Click the **EDIT VIEW** button.

2. Click the **+ Filter** button in the **Row Actions** area.

3. Enter am in the search box in the popup.

4. Select **Amount**.

5. Click the filter condition combo box and select **Is Greater Than**.

6. Enter 0 in the **Value** box.

7. Click the **APPLY** button.

8. Let's change the view name by clicking on the dataset name and renaming it View of Opportunitys with Amounts >0 CH4.

9. Click the **SAVE** button.

10. Click the **X** icon to the right of the **SAVE** button to return to the **Dataset Details** page.

11. Notice that the DataSet View has been filtered so that it now shows only **5,761** rows.

Superb! Now, let's go crazy and use Data Views to standardize the **Lead Source** field values by joining it to a mapping table:

1. Click the **EDIT VIEW** button.

2. Click the **DataSet** tab to access the **+ ADD JOIN** and **+ ADD UNION** features. Click the **+ ADD JOIN** button.

3. In the **Join new DataSet** popup, under the **DATASETS** section, click **Select a DataSet** and choose the **Lead Source Mapping CH4** dataset.

4. Click the **CHOOSE DATASET** button.

5. Under the **JOIN KEYS** section, click **SELECT COLUMN** on the left and choose **LeadSource**.

6. Under the **JOIN KEYS** section, click **SELECT COLUMN** on the right and choose **Current**.

At this point, your screen will look as follows:

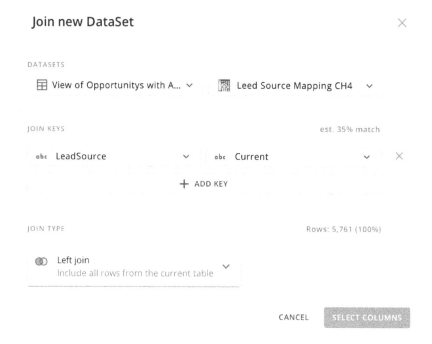

Figure 5.4 – Join new DataSet

7. Click the **SELECT COLUMNS** button.

8. In the **Edit Join Configuration** popup, under **Lead Source Mapping CH4** and on the right-hand side, select **Lead Source**.

9. Click the **SAVE** button.

10. Now, we need to rename the original lead source to avoid confusion by clicking the **Columns** tab.

11. Enter Lead in the column search box.

12. On the **LeadSource** field, without the space between **Lead** and **Source**, click the three stacked dots to the right and select **Rename** from the pop-up menu.

13. Enter LeadSource Original for the field name.

14. On the **Lead Source** field, click the three stacked dots to the right and select **Move** and then **First**.

15. Click **SAVE**.

16. Click **X** to close and return to the **View of Opportunities with Amounts >0 CH4** dataset detail page.

Well done! Now, we have consistently mapped values for **Lead Source** at the front of the View.

Next, let's see how we can use views on a view to identify the top five leaders and bottom five laggards in terms of Owners and their Closed Won revenue totals. Yes, that's right – you can create views on views. Let's start with the **Leaders** view:

1. Click on the **OPEN WITH** combo box and select **Views Explorer** on the **View of Opportunities with Amounts >0 CH4** dataset detail page.

2. Rename the view from **View of Opportunities with Amounts >0 CH4** to Opportunity Leaders.

3. Click the **+ ADD COLUMN** button to open the **Add Column** panel on the right of the page.

4. Click the **ADD CALCULATED COLUMN** button in the right panel.

5. Enter Board Type in the **Column** name box.

6. Enter 'Leader' in the **FORMULA** section.

7. Click the **SAVE AND CLOSE** button.

8. Click the **+ Group** button.

9. Enter bo in the search box and drag and drop the **Board Type** field into the **CATEGORIES** section.

10. Enter ow in the search box and drag and drop the **Owner Id** field into the **CATEGORIES** section.

11. Enter am in the search box and drag and drop the **Amount** field into the **AGGREGATIONS** section. Then, click the **AVG** button and select **Sum**.

12. Click the **FINISH** button.

13. Click the **+ Sort** button.

14. Click the **+ ADD SORT** button.

15. Select **Sum Amount**.

16. Select **High-Low**.

17. Click outside the popup to exit the dialog.

18. Click the **+ Limit** button.

19. Enter 5 in the **LIMIT** box.

20. Click the **APPLY** button.

21. Click outside the popup to exit the dialog.

22. Click the three stacked dots icon in the **Sum Amount** field.

23. Click **Rename**, enter Total Sales, and press *Enter*.

At this point, your page should look as follows:

Figure 5.5 – Leaderboard view configuration

24. Click the **SAVE AS** button.

25. Click the **SAVE** button.

Next, let's create the **Laggards** view:

1. Click on the **DATA** item in the Domo main menu bar.

2. Click on the datasets icon on the left-hand side of the toolbar to go to the **DataSets** page.

3. Enter `view` in the search bar.

4. Click the **View of Opportunities with Amounts >0 CH4** dataset to go to the dataset detail page.

5. Click on the **OPEN WITH** combo box and select **Views Explorer** on the **View of Opportunities with Amounts >0 CH4** dataset details page.

6. Rename the view from **View of Opportunities with Amounts >0 CH4** to `Opportunity Laggards`.

7. Click the + button and then + **ADD COLUMN** to open the **Add Column** panel on the right of the page.

8. Click the **ADD CALCULATED COLUMN** button in the right panel.

9. Enter `Board Type` in the **Column** name box.

10. Enter `'Laggard'` in the **FORMULA** section.

11. Click the **SAVE AND CLOSE** button.

12. Click the + **Group** button.

13. Enter `bo` in the search box and drag and drop the **Board Type** field into the **CATEGORIES** section.

14. Enter `ow` in the search box and drag and drop the **Owner Id** field into the **CATEGORIES** section.

15. Enter `am` in the search box and drag and drop the **Amount** field into the **AGGREGATIONS** section.

16. Click the **AVG** button and select **Sum**.

17. Click the **FINISH** button.

18. Click the + **Sort** button.

19. Click the + **ADD SORT** button.

20. Select **Sum Amount**.

21. Select **High-Low**.

22. Click outside the pop-up to exit the dialog.

23. Click the + **Limit** button.

24. Enter 5 in the **LIMIT** box.

25. Click the **APPLY** button.

26. Click outside the popup to exit the dialog.

27. Click the three stacked dots icon in the **Sum Amount** field.

28. Click **Rename**, enter `Total Sales`, and press *Enter*.

29. Click the **SAVE AS** button.

30. Click the **SAVE** button.

Next, we'll learn how to use the Data Blend tool.

Using Data Blend

Data Blend is a data sculpting tool that creates non-materialized views of your datasets. One way to think of a blend is a highly scalable equivalent of Excel's VLOOKUP function. Blends are useful when you want to do lookup translations on large tables in memory via a mapping table. Blends are also useful for appending (merging or performing a union) datasets into a single dataset. Blends are dataset definitions and, once saved, can be viewed in the dataset catalog. Blends can be created via the **BLEND** option on the **Datasets** page. Blends live in the Adrenaline cache, so they are in-memory fast.

Let's learn how to create a Data Blend.

Creating a Data Blend

Picking up from where we left off in the *Creating a Data View* section, let's use a blend to merge the **Opportunity Leaders** and **Opportunity Laggards** datasets into a single non-materialized dataset:

1. Click on the **DATA** item in the Domo main menu bar.

2. Click on **Blend** in the **MAGIC TRANSFORM** section.

3. Enter `Leader Board` in the title box.

4. Click the **SELECT DATASET** button in the **DataSets** section.

5. Enter `opp` in the **Search DataSets...** box.

6. Select the **Opportunity Leaders** dataset.

7. Click the **CHOOSE DATASET** button.

8. Click the **SELECT DATASET** button in the **DataSets** section.

9. Select the **Opportunity Laggards** dataset.

10. Click the **CHOOSE DATASET** button.

11. In the **FUSION TYPE** section, click the **ADD ROWS** button.

12. Click on the **Click to load preview** button to see the combined result, as shown in the following screenshot:

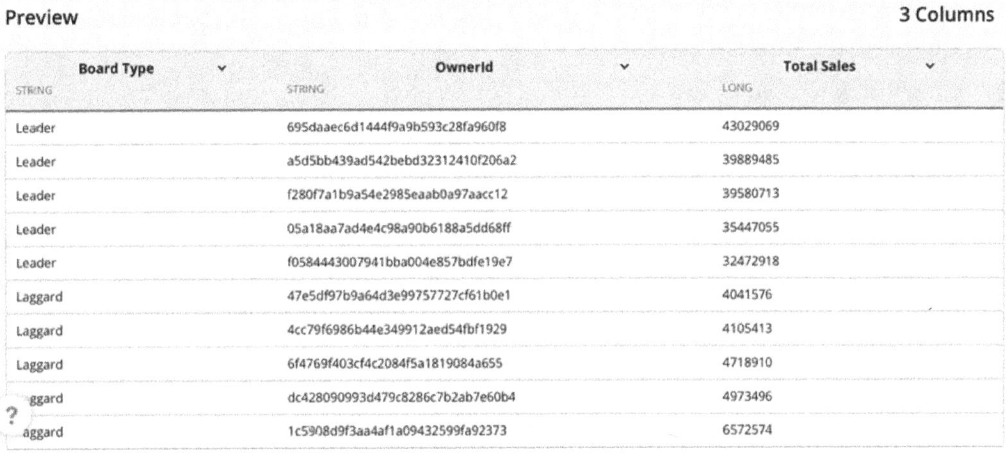

Preview   3 Columns

Board Type	⌄	OwnerId	⌄	Total Sales	⌄
STRING		STRING		LONG	
Leader		695daaec6d1444f9a9b593c28fa960f8		43029069	
Leader		a5d5bb439ad542bebd32312410f206a2		39889485	
Leader		f280f7a1b9a54e2985eaab0a97aacc12		39580713	
Leader		05a18aa7ad4e4c98a90b6188a5dd68ff		35447055	
Leader		f0584443007941bba004e857bdfe19e7		32472918	
Laggard		47e5df97b9a64d3e99757727cf61b0e1		4041576	
Laggard		4cc79f6986b44e349912aed54fbf1929		4105413	
Laggard		6f4769f403cf4c2084f5a1819084a655		4718910	
ggard		dc428090993d479c8286c7b2ab7e60b4		4973496	
aggard		1c5908d9f3aa4af1a09432599fa92373		6572574	

Figure 5.6 – Blend leaderboard preview

13. Click the **SAVE** button.

Well done! Next, let's learn how a blend can use a mapping dataset to standardize lead source values in the **Opportunity** dataset:

1. Click on the **DATA** item in the Domo main menu bar.

2. Click on **Blend** in the **MAGIC TRANSFORM** section.

3. Enter Blend Opportunity Mapping in the title box.

4. Click the **SELECT DATASET** button in the **DataSets** section.

5. Enter opp in the **Search DataSets...** box.

6. Select the **Opportunity CH4** dataset.

7. Click the **CHOOSE DATASET** button.

8. Click the **SELECT DATASET** button in the **DataSets** section.

9. Enter map in the **Search DataSets...** box.

10. Select the **Lead Source Mapping CH4** dataset.

11. Click the **CHOOSE DATASET** button.

12. In the **Fusion Type** section, click the **ADD COLUMNS** button.

13. In the **Link Columns** section, click the **Select a Field** combo box on the left side of the page and select the **Opportunity CH4** dataset. Then, scroll down and select the **LeadSource** column.

14. In the **Link Columns** section, click the **Select a Field** combo box on the right side of the page and select the **Lead Source Mapping CH4** dataset. Then, scroll down and select the **Current** column.

15. Click **Click to load preview**.

16. In the preview pane, scroll to the right until you see the **LeadSource** column header.

17. Click the selector to the right of the **LeadSource** name and select **Rename Column**.

18. Enter `LeadSource Original` in the **New Column Name** box.

19. Click the **Save** button.

20. Scroll to the right so that you reach the end of the preview table, click on the selector to the right of the **Current** column, and select **Exclude Column**.

21. Click the **SAVE** button at the top right of the blend page.

Great! Now that we have worked with blends, let's try out another powerful sculpting tool – Beast Mode.

Leveraging Beast Mode

Beast Mode is a data sculpting tool that works within the scope of a single dataset and enables a mixture of SQL/Excel-like functions to be applied to datasets to create new calculated columns. Beast Mode values only live in the Adrenaline cache. They are a good place to put business logic and transformations that are closely associated with a single dataset. Examples of common Beast Mode function types are aggregate, date and time, mathematical, string, and logical.

Next, let's take a tour of Beast Mode Manager.

Touring Beast Mode Manager

Beast Mode Manager is where you can create and find Beast Modes across all datasets. Let's quickly walk through the Beast Mode Manager page, as shown in the following screenshot:

Figure 5.7 – The Beast Mode Manager page

Let's look at the main features of the Beast Mode Manager page:

- **Beast Mode Manager Access**: This is the icon you click on to get to Beast Mode Manager. It is on the left-hand side of the menu bar, under the **DATA** main menu.

- **Beast Mode Search**: This is the search function for all Beast Modes on all datasets.

- **Faceted Search Tiles**: Each tile shows the number of Beast Modes that fit its respective criteria. When each tile is clicked, it executes a pre-defined search and shows the results in the **Search Results** area.

- **Search Results**: This is the area where the search results appear.

- **+ Add Beast Mode**: This is where you can create new Beast Mode functions on a dataset.

In the next section, we will learn how to add columns to a dataset using Beast Mode.

Adding Beast Mode columns to a dataset

The following scenarios will give you a better feel for some of the more common columns that can be created using Beast Modes.

Renaming a dataset column

In this scenario, you will create a new column called `Stage` to use instead of the existing `StageName` column. Yes – with Beast Mode, you don't rename the existing column; instead, you create a new column with the desired name:

1. Click the + **ADD BEAST MODE** button.

2. Select the **Opportunity CH4** dataset and click **CHOOSE DATASET**.

3. In the **DETAILS** box, enter `Stage` as the Beast Mode's column name.

4. In the **FORMULA** box, enter `` `StageName` ``.

5. Click **Validate Formula** to perform a syntax check.

6. Click **SAVE & CLOSE**.

    This will bring you to the Beast Mode details page, as shown in the following screenshot:

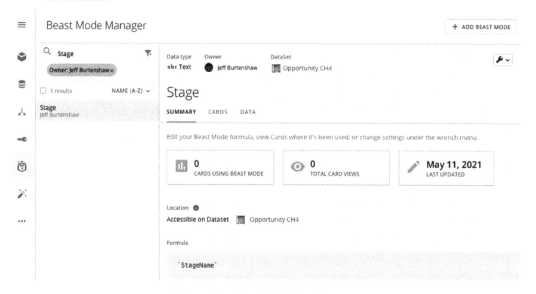

Figure 5.8 – Beast Mode details page

7. Click on the **DATA** tab to see a preview of the results.

Next, let's use Beast Mode to create a column that sums a value.

Summing a dataset column

In this scenario, you will create a new column that is the sum of the existing **Amount** column:

1. Click the + **ADD BEAST MODE** button.
2. Select the **Opportunity CH4** dataset and click **CHOOSE DATASET**.
3. In the **DETAILS** box, enter `Total Sales` as the Beast Modes' column name.
4. In the **FORMULA** box, enter `sum(`Amount`)`.
5. Click **Validate Formula** to perform a syntax check.
6. Click **SAVE & CLOSE**.

Now, let's create an average on a column.

Averaging a dataset column

In this scenario, you will create a new column for calculating the average sales:

1. Click the + **ADD BEAST MODE** button.
2. Select the **Opportunity CH4** dataset and click **CHOOSE DATASET**.
3. In the **DETAILS** box, enter `Average Sale` as the Beast Modes' column name.
4. In the **FORMULA** box, enter `avg(`Amount`)`.
5. Click **Validate Formula** to perform a syntax check.
6. Click **SAVE & CLOSE**.

Now, let's cover some basic data hygiene.

Remapping/cleansing column data values

In this scenario, you will create a new column for **Lead Sources** that remaps misspelled lead sources to be consistent:

1. Click the + **ADD BEAST MODE** button.
2. Select the **Opportunity CH4** dataset and click **CHOOSE DATASET**.
3. In the **DETAILS** box, enter `Lead Source Standardized` as the Beast Modes' column name.
4. In the **DATASET FIELDS** area, enter *Lead* in the search box.

5. Double-click on **LeadSource** in the search results directly underneath the search box. This will add `LeadSource` to the **FORMULA** box.

6. Replace `LeadSource` in the **FORMULA** box with the following code:

```
-- Lead Source Standardization
case
when `LeadSource` = 'Patnes' then 'Partners'
when `LeadSource` = 'Diect' then 'Direct'
when `LeadSource` = 'Sales Ceated' then 'Sales Created'
when `LeadSource` = 'efeal' then 'Referral'
when `LeadSource` = 'Self-Souced' then 'Self-Sourced'
when `LeadSource` = 'Maketing Outbound' then 'Marketing
Outbound'
when `LeadSource` = 'Stategic Accounts Maketing' then
'Strategic Accounts Marketing'
when `LeadSource` = 'Patneing' then 'Partners'
else `LeadSource` -- If not mapped then keep as is
end
```

7. Click **Validate Formula** to perform a syntax check.

8. Click **SAVE & CLOSE**.

9. Click on the **DATA** tab to see a preview of the results.

Now, let's learn how to add Aging Buckets as a column.

Creating Aging buckets

In this scenario, you will create a new column with Aging Buckets for how long an opportunity has been open:

1. Click the + **ADD BEAST MODE** button.

2. Select the **Opportunity CH4** dataset and click **CHOOSE DATASET**.

3. In the **DETAILS** box, enter Open Opportunity Aging as the Beast Modes' column name.

4. Enter the following code in the **FORMULA** box:

```
case
when `IsClosed` = 'false' and `IsWon` = 'false' and
`IsDeleted` = 'false' -- Open Opportunity
then
    case
    when DATEDIFF(CURRENT_DATE(),`CreatedDate`) >= 0 and
DATEDIFF(CURRENT_DATE(),`CreatedDate`)<=30 then '30 Days
or Less'
    when DATEDIFF(CURRENT_DATE(),`CreatedDate`) > 30 and
DATEDIFF(CURRENT_DATE(),`CreatedDate`)<=90 then '30 - 90
Days'
    when DATEDIFF(CURRENT_DATE(),`CreatedDate`) > 90 and
DATEDIFF(CURRENT_DATE(),`CreatedDate`)<=180 then '3-6
Months'
    when DATEDIFF(CURRENT_DATE(),`CreatedDate`) > 180 and
DATEDIFF(CURRENT_DATE(),`CreatedDate`)<=360 then '6-12
Months'
    else '> 1 Year'
    end
end
```

5. Click **Validate Formula** to perform a syntax check.
6. Click **SAVE & CLOSE**.
7. Click on the **DATA** tab to see a preview of the results.

Next, let's use Beast Mode to concatenate several values together.

Concatenating several values together

In this scenario, you will create a new column that combines the **Type** and **Opportunity** columns into a single column:

1. Click the **+ ADD BEAST MODE** button.
2. Select the **Opportunity CH4** dataset and click **CHOOSE DATASET**.
3. In the **DETAILS** box, enter Opportunity Type | Name as the Beast Modes' column name.

4. Enter the following code in the **FORMULA** box:

```
concat(`Type`,' | ',`Name`)
```

5. Click **Validate Formula** to perform a syntax check.

6. Click **SAVE & CLOSE**.

7. Click on the **DATA** tab to see a preview of the results.

We can also use Beast Mode to calculate percentages. We'll look at this next.

Calculating a percentage

In this scenario, you will create a new column that calculates the win percentage:

1. Click the + **ADD BEAST MODE** button.

2. Select the **Opportunity CH4** dataset and click **CHOOSE DATASET**.

3. In the **DETAILS** box, enter `Win %` as the Beast Modes' column name.

4. Enter the following code in the **FORMULA** box:

```
concat(round(
sum(case when `IsWon` = 'true' then 1 else 0 end)
/
sum(case when `IsClosed`='true' then 1 else 0 end) *100
, 1),'%')
```

5. Click **Validate Formula** to perform a syntax check.

6. Click **SAVE & CLOSE**.

7. Click on the **DATA** tab to see a preview of the results.

Next, let's learn how to create custom sorts using Beast Mode.

Creating a custom sort

In this scenario, you will create a new column that holds a custom sort order:

1. Click the + **ADD BEAST MODE** button.

2. Select the **Opportunity CH4** dataset and click **CHOOSE DATASET**.

3. In the **DETAILS** box, enter `Forecast Category Sort Order` as the Beast Modes' column name.

4. Enter the following code in the **FORMULA** box:

```
case
when `ForecastCategory` = 'Pipeline' then 1
when `ForecastCategory` = 'Forecast' then 2
when `ForecastCategory` = 'BestCase' then 3
when `ForecastCategory` = 'Closed' then 4
when `ForecastCategory` = 'Omitted' then 5
else  6
end
```

5. Click **Validate Formula** to perform a syntax check.

6. Click **SAVE & CLOSE**.

7. Click on the **DATA** tab to see a preview of the results.

With that, you have created a variety of Beast Modes You may have also surmised that they are very powerful and flexible. Check out the *Further reading* section at the end of this chapter if you want to dive even deeper into this. Now, let's move on to finding, editing, and archiving existing Beast Modes, which we will cover in more detail in *Chapter 6, Creating Dashboards*.

Finding, editing, and archiving dataset Beast Modes

To access administrative functions on an existing Beast Mode you must find the Beast Mode select it, and then use the wrench menu, as shown in the following screenshot:

Figure 5.9 – The Beast Mode details page showing the wrench menu and its options

Let's look at the Beast Mode details page's wrench menu options in more detail:

- **Create Beast Mode**: This is where you can make new Beast Modes.
- **Edit formula**: This is where you can edit an existing Beast Modes' formula.
- **Lock**: This is where you can lock the Beast Mode so that only the owner or admins can change it.
- **Change Owner**: This is where you can change the owner of the Beast Mode.
- **Buzz Owner**: This will be grayed out/inactive if you are the owner. If you are not the Beast Modes owner, you can message the owner via Buzz.
- **Archive**: This is where you can archive the Beast Mode.

There is also a multi-Beast-Mode action menu that you can access by clicking the box next to the desired Beast Modes in the search results area, as shown in the following screenshot:

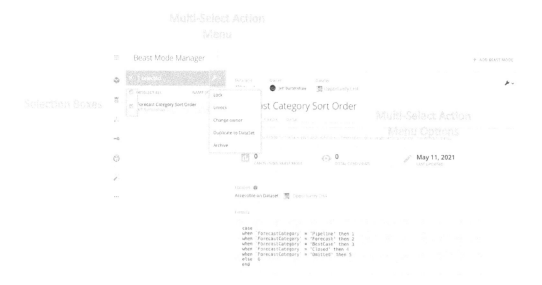

Figure 5.10 – Beast Mode – multi-select action menu options

Duplicate to Dataset is where you can copy the selected Beast Modes to another dataset.

The next section will show you how to use the **Wrench** and **Action** menu options.

Finding, editing, and administering an existing Beast Mode

In this exercise, you will find the **Forecast Category Sort Order** Beast Mode and learn how and where to edit the Beast Modes formula:

1. Click the **Beast Mode Manager Access** icon under the **DATA** main menu.
2. Enter forecast in the **Search Beast Modes** box.
3. Click the **Forecast Category Sort Order** Beast Modes.
4. Click the **Wrench** menu and select the **Edit formula** option.
5. Make some changes to the formula and click **Save & Close**.
6. Click the **Wrench** menu, select the **Lock** option, and click **LOCK**.
7. Click the **Wrench** menu and select the **Change owner** option.
8. Enter the name of a new owner in the **Search users** box and click **Assign**.

Now, let's learn how to perform actions on multiple Beast Modes:

1. Let's use **Multi-Select Action** to copy this Beast Modes to a different dataset. Click on the **Forecast Category Sort Order** Beast Modes in the search results panel.
2. Click the check box to the left of **Forecast Category Sort Order**.
3. Click the **Multi-Select Action** menu and select the **Duplicate to Dataset** option to copy the selected Beast Modes to another dataset.

This has been a great start – you are now reasonably well versed in **Beast Mode**! Now, let's turn our attention to **Adrenaline DataFlows**.

Adrenaline DataFlows

Adrenaline DataFlows is an advanced tool that you can use to perform data transformations on very large datasets (with billions of rows). The UX is very similar to SQL DataFlows but the process is executed in the Adrenaline cache. Since it is a resource-hungry tool that uses lots of memory, it is not an option that's included in the basic platform. However, if you need this scale of performance, such as if your organization is dealing with large volumes of data, you can contact your Domo account representative to switch it on in your instance.

Summary

In this chapter, we reviewed the in-memory data sculpting tools that are available in Domo and how to decide which tool to use. Data Views are the most flexible and powerful but also the more complex to use. Then, there are Data Blends, which allow you to perform VLOOKUPs at scale. Additionally, Beast Modes are a powerful way to extend existing datasets with calculated fields. All of these tools work in the Adrenaline cache, so they are extremely fast. Finally, for those working with datasets with billions of rows, we made you aware of the Adrenaline DataFlows tool for handling data transformation at a mind-boggling scale. Having this in-memory data sculpting power is a key feature that democratizes the process so that an end user doesn't have to wait for specialized resources to assist.

In the next chapter, we will move up from the data transformation layer to the data visualization layer features of the platform, including dashboards.

Further reading

To learn more about the topics that were covered in this chapter, take a look at the following resources:

- For reference material on example Beast Mode calculations, go to `https://knowledge.domo.com/Visualize/Adding_Cards_to_Domo/KPI_Cards/Transforming_Data_Using_Beast_Mode`.

- For the Beast Mode functions reference guide, go to `https://knowledge.domo.com/Visualize/Adding_Cards_to_Domo/KPI_Cards/Transforming_Data_Using_Beast_Mode/02Beast_Mode_Functions_Reference_Guide`.

- For additional material on Adrenaline DataFlows, go to `https://knowledge.domo.com/Prepare/Magic_Transforms/SQL_DataFlows/Creating_an_Adrenaline_DataFlow`.

Section 2: Presenting the Message

Get the fundamentals right.

In this section, we learn about the basic features for creating content in Domo. This includes cards, drill paths on cards and interactions with cards and dashboards. You will go through an iterative process to create a sales monitoring dashboard. We will go through examples of how to interact with cards and dashboards.

After completing this section, you will be proficient in using datasets to create visual content in Domo.

This section comprises the following chapters:

- *Chapter 6, Creating Dashboards*
- *Chapter 7, Working with Drill Pathways*
- *Chapter 8, Interacting with Dashboards*
- *Chapter 9, Interacting with Cards*

6
Creating Dashboards

In the previous chapters, you learned how to acquire and sculpt your data. Now, let's turn our attention toward creating cards and dashboards with the data. **Analyzer** is a tool that is used to create individual cards that are contained in dashboard pages. For navigation purposes, the dashboard pages are arranged into a logical tree structure of up to three layers deep, organizing how the content is presented. A **Card** is simply a chart contained inside of a tile on a dashboard page. Analyzer is commonly referred to as a **Card Builder**, but it is also a powerful ad hoc **Data Exploration Tool**. Additionally, many card types support interactive filtering while being viewed from the dashboard.

In this chapter, we will focus on the card-building functionality of Analyzer along with its interactive card filter features. Analyzer controls the settings of specific visual characteristics related to the charts, such as the chart type, filters, sorting, the summary number, the date grain, and other chart properties. Additionally, the card builder provides easy access to create and edit **Beast Mode** calculations. **Collections** are subsections on a dashboard page that enable the further grouping of related cards on a page—kind of like slides in a presentation deck in terms of organization.

In this chapter, we will cover the following topics:

- Navigating dashboards
- Working with Analyzer
- Creating and editing cards

Technical requirements

To follow along with this chapter, you will need the following:

- Internet access
- Your **Domo** instance and login details
- The ability to download example files from the GitHub repository
- Access to the datasets that were created in *Chapter 4*, *Sculpting Data*

> **Important Note**
>
> If you don't have a Domo instance, you can get a free trial instance at
> `https://www.domo.com/start/free`.

All of the code found in this chapter can be downloaded from GitHub at `https://github.com/PacktPublishing/Data-Democratization-with-Domo`.

Navigating dashboards

A typical dashboard requires many cards to answer the variety of questions being addressed. As such, it is necessary to consider a navigational tree structure for all the dashboards to organize the cards and, thus, avoid **card sprawl**—dashboards with 20+ cards and no logical organization. Often, the first level of dashboard organization in an enterprise uses the departments in the organization such as sales, marketing, HR, finance, and operations. The second level tends to be role and process areas of focus such as the executives, managers, and front-line workers. However, some organizations take a process-based approach for the first level, such as order to cash, procure to pay, hire to retire, accounting to reporting, concept to launch, while others take a geo-based approach. That said, others approach the organization of the dashboard via job descriptions and leverage page filtering and PDP policies to constrain different levels of access within the page role. There is no one best way to organize the dashboard navigation tree, and the good news is that the tree is very pliable should any changes be required. In the next section, let's walk through how the dashboard navigation is set up.

Working with the dashboard panel

Dashboards are accessed by clicking on the **DASHBOARDS** main menu item. This makes a pinnable panel available on the left-hand side, containing the dashboard navigation.

The following steps will take us through the process of how to use the dashboards panel:

1. Click on **DASHBOARDS** in the main menu.

2. Hover and click on the three parallel bars in the upper-left corner to see the **Dashboards** panel.

3. Click on the pushpin in the upper-right corner of the dashboard panel to pin the panel into place.

    You should see something similar to *Figure 6.1*:

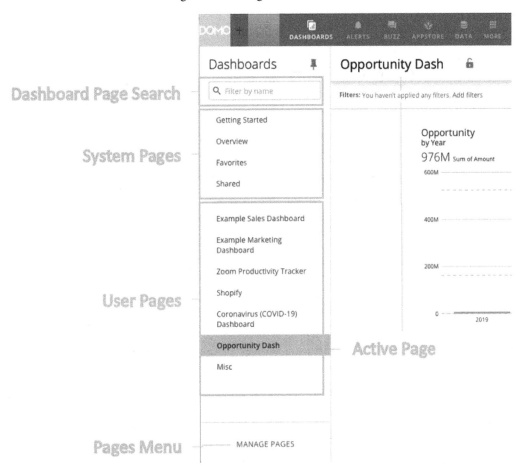

Figure 6.1 – The dashboard panel

Here is a list of the dashboard panel features, along with descriptions, from *Figure 6.1*:

- **Dashboard Page Search**: This feature filters down the list of dashboards to match the criteria.

- **System Pages**: This feature refers to the list of permanent system dashboard pages required by Domo.

- **User Pages**: This feature refers to the list of dashboard pages created by users that the logged-in user has access to.

- **Active Page**: This is the current dashboard page that is being displayed.

- **Pages Menu**: This gives you access to the page administration for managing the dashboard pages.

Next, let's examine the items in the **Pages Menu** feature, which contains features in which to manage pages.

Using the Pages Menu feature

Use the pages menu to add pages and sub-pages, rename pages, hide/show pages, and reorder page positions.

To create a new page, perform the following steps:

1. Click on **MANAGE DASHBOARDS** at the bottom of the **Dashboards** panel.
2. Click on + **Add a Dashboard**. Enter `Sales` in the **Enter a name...** box that appears at the bottom of the list in the **MANAGE DASHBOARDS** dialog.
3. Click on the **SAVE** button.

   Now, let's move the new **Sales** dashboard's position to the top since it is so important!

4. Click on the **Sales** row, and then click and drag the row handle up to the top and drop it.

   While we are here, let's add a sub-page that will hold the cards intended specifically for sales reps.

5. While on the **Sales** row, click on the > and then click on + **Add a sub-Dashboard**.
6. Enter `Rep` in the **Enter a name...** box that appears just below the **Sales** row.
7. Click on **SAVE**. Now you have a sub-dashboard page.
8. Click on **DONE** to exit the **MANAGE DASHBOARDS** dialog.
9. Next, let's make the new sales dashboard the active dashboard.

10. Click on the **Sales** dashboard in the **Dashboards** panel. Notice that there is also a **Rep** sub-page.

Now you can create dashboard pages and sub-dashboard pages! In the next section, we will review the dashboard page features.

Using the dashboard page

The dashboard page is where you interact with the dashboard content, and it has many options. Let's go through a quick overview:

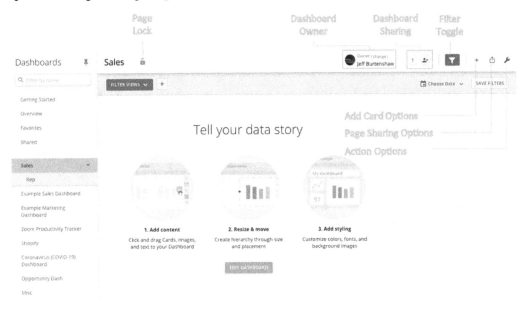

Figure 6.2 – Dashboard options

The following list includes descriptions of the dashboard options and features, as shown in *Figure 6.2*:

- **Page Lock**: This locks a page so that only the owner or a user with admin rights can change it.

- **Dashboard Owner**: This shows the current owner.

- **Dashboard Sharing**: This shows the number of people that have access to the dashboard and enables you to change who has access.

- **Filter Toggle**: This turns on and off page-level filtering.

- **Add Card Options**: This menu allows you to choose between the different ways to create a new card. You can add an existing card to the dashboard, create a new card from scratch, or assign someone else to create a card.

- **Page Sharing Options**: This menu enables you to manage who can view the page, view the page as a slideshow, publish the page to a URL as a slideshow, export to PowerPoint or PDF, schedule the page as an emailed report, and send the page as an email immediately.

- **Action Options**: This menu provides the different ways to edit the dashboard, convert the dashboard from a story into the standard format, add a dashboard to the favorites area, edit report schedules, open a buzz conversation on the dashboard, save a copy of the dashboard to another page, view the dashboard in fullscreen mode, and manage the page filter options.

> **Important Note**
>
> The standard and story dashboard formats are the effects of evolution, and currently, they are in a transitionary period. Eventually, I speculate that the older standard format will be deprecated for an ever-improving story experience. However, for now, I find it more efficient to initially create the cards in the standard view and more cumbersome to create the first draft of the card views in the story format. This might be a personal preference. Feel free to begin and stay in the story format as that is the future. The story format is covered, in greater depth, in *Chapter 10, Telling Relevant Stories*.

Now you are familiar with how to use the dashboard options.

In the next section, we will start using the Analyzer tool to create cards for the dashboard.

Working with Analyzer

Analyzer is the main tool for creating and managing the contents of cards on dashboards. Let's do a quick walk through of the available features:

1. To access Analyzer, click on the **Sales** dashboard in the **Dashboards** panel.

2. Click on the + icon in the **Add Card Options** menu, in the dashboard header, and select **Create new Card**.

3. You will see three choices. Click on **VIEW MORE** to see all six card types. They are as follows:

    - **Visualization** is the card type for all the various chart types that can be applied as visuals.

    - **Doc Card** is the card type that allows you to upload a file as a card to the dashboard.

    - **Notebook** is a card type that shows a rich text editor in card view. It is helpful for adding commentary.

    - **Sumo Table** is a pivot table in a card.

    - **Poll** is a card type that lets you take simple polls and present the results in a card.

    - **Enterprise App** is a card type that lets you use Domo's dev tools to create a custom card.

4. Click on the **SELECT** button under the **Visualization** card type, and you will see four choices. They are as follows:

    - **Existing Data** allows the new card to use data from an existing dataset.

    - **Spreadsheet upload** takes you through the file connector to upload a spreadsheet to a dataset and then use that dataset for the new card.

    - **Connect live data** lets you select a connector to get data and then use that data for a new card.

    - **Online data editor** lets you enter data into a simple tabular format stored in a dataset and then use that data for the new card.

5. Click on **SELECT** under **Existing data**.

6. In the **Select a DataSet** dialog, click on **Opportunity CH4** and then click on **CHOOSE DATASET**.

Great job! You can refer to the following screenshot:

Figure 6.3 – The Analyzer toolbar

Now, let's review the toolbar options, as shown in *Figure 6.3*.

Toolbar controls the hiding/showing panels (such as **DATA, DATA TABLE, FILTER & SORT, PROPERTIES, QUICK FILTERS, TOOLTIPS,** and **CHART TYPES**) in Analyzer and provides access to additional functions (such as **LINEAGE, COLOR RULES, AUTO PREVIEW, ANNOTATION, BEAST MODE, SEGMENTS,** and **YOY**).

The following list offers a brief description of each toolbar option:

- The **DATA** toolbar button: This hides/shows the **DATA** panel and the dataset the card is using. Clicking on the dataset name allows you to choose a different dataset for the card.

- The **LINEAGE** toolbar button: This opens the **Data Lineage** dialog for the dataset.

- The **DATA TABLE** toolbar button: This hides/shows the **DATA TABLE** panel.

- The **FILTER & SORT** toolbar button: This hides/shows the **FILTERS** and **SORTING** panels where you can drag and drop fields into the filter or sort the card.

- The **PROPERTIES** toolbar button: This hides/shows the **CHART PROPERTIES** panel where you set the various chart properties for the chart type.

- The **QUICK FILTERS** toolbar button: This hides/shows the quick filters panel where you can set up filters that are viewable and editable to the user while viewing the card. **REORDER** allows you to control the order in which the filters are presented. Setting the quick filter values here will be the default filter setting when viewing.

- The **COLOR RULES** toolbar button: This opens the **Color Rules** dialog, which allows you to configure color rules for just this card or apply the color rules to all of the cards using the dataset. For instance, you can set up a color rule that will set the applicable chart series color to green anytime the lead type is **Sales Created**.

- The **AUTO PREVIEW** toolbar button: This controls whether the chart updates in real time as you make changes to the builder. When working with very large datasets, turn this off.

- The **TOOLTIPS** toolbar button: This hides/shows the **TOOLTIP FIELD** panel. Tooltip fields provide a way to show, in the chart's data labels/hovers, values for columns on a chart that are not in the chart axis or series. Up to three fields from the dataset can be added.

- The **CHART TYPES** toolbar button: This hides/shows the quick filters panel where you can set up filters that are viewable and editable to the user while viewing the card. **REORDER** allows you to control the order in which the filters are presented. Setting the quick filter values here will be the default filter setting when viewing.

- The **BEAST MODE** toolbar button: This opens the beast mode dialog to create and manage beast modes.

- The **SEGMENTS** toolbar button: This opens the **Create a segment** dialog to create a segment.

- The **YOY** toolbar button: This opens the year-over-year dialog to set up the year-over-year chart settings.

Let's continue our review of additional Analyzer features, as shown in *Figure 6.4*:

Figure 6.4 – Analyzer features

The following list provides the names of the preceding features and their descriptions:

- **Icon**: This is a glass beaker symbolizing Analyzer.

- **Tabs**: This area provides access to the **DATA** and **VISUALIZE** tabs; the **DATA** tab is used for creating views, and the **VISUALIZE** tab is used for configuring the visual aspects of the card.

- **Toolbar Toggle**: This turns the toolbar on or off.

- **Undo** and **Redo**: These features enable the user to back out of or reapply any changes.

- **Save**: This saves the changes to the card.

- **X**: This closes the Analyzer toolbar.

- **Title**: This is the name of the card.

- **Summary**: This is the summary number/text on the card.

- **Date Range Selector**: This allows you to set the range and grain of a selected date field for the chart.

- **Annotation**: This is accessed via the toolbar and enables you to choose what period to annotate and comment on. This is useful for creating a permanent reminder of a key event associated with the chart, such as a change in product or a big sales event.

- **Dataset**: This shows the current dataset and allows you to change to a different dataset.

- **Chart Fields**: This area is where you drag and drop the fields you want on the chart.

- **Field Finder**: This enables you to search and drag and drop fields. Fields are organized into groups of **DIMENSIONS**, **MEASURES**, and **BEAST MODES**.

- **Filters**: This is the panel containing the fields that the card is filtered on.

- **Sorting**: This is the panel that contains the fields for sorting the card.

- **Row Limits**: This restricts the number of rows returned in a query. Typically, this feature is utilized in top/bottom visualizations.

- **Chart Properties**: This contains all the chart-type properties organized into groups that are related to the property settings.

- **Beast Modes**: The **ADD CALCULATED FIELD** button is another access path to create/edit beast modes.

- **Segments**: The **ADD DYNAMIC SEGMENT** button is an access path in which to create/edit dynamic segments.

- **Quick Filters**: This feature shows the fields that have been added as quick filters and allows you to set up the default filter choices.
- **Chart Type Selector**: This allows you to browse and select the best chart type.
- **Data Table**: This is the preview area for the underlying dataset for the card.
- **Table Column Chooser**: This allows you to select which columns to show in the data table.
- **Table Settings**: This allows you to show/hide totals on the table and see either the aggregated data for the table or the raw data.
- **Table Fullscreen Toggle**: This turns the fullscreen view of the data table on or off.

Wow, that is a lot of features to be aware of! Let's practice using these features by creating some cards.

Creating/editing cards

At first glance, the **Analyzer** toolbar looks overwhelming, but it is quite simple to create cards. Let's start by creating some cards to fill in our **Sales** dashboard.

Creating a monthly sales trend card

Let's create the proverbial *Up and To the Right* trend view card that sales teams want to see.

To create the card, follow these steps:

1. Click on the **DASHBOARDS** item in the main menu.
2. Click on **Sales** in the **Dashboards** panel.
3. Click on + in the main menu, and select **Card** in the **ADD TO DOMO** section.
4. Click on **SELECT** under **Visualization**.
5. Click on **SELECT** under **Existing Data**, click on **Opportunity CH4**, and click on **CHOOSE DATASET**.
6. In **Analyzer**, Domo tries to guess at a relevant chart, but we are going to change the default properties starting with the name. Double-click on **Opportunity CH4** in the chart preview area and enter `Sales Trend`.
7. Click on the **SERIES** field of **CurrencyIsoCode**, and then click on **REMOVE** to remove the currency type as a series.
8. Click on the **CHART TYPES** selector. Choose **LINE** in the drop-down menu and then select **Symbol Line**.

9. Click on the date selector and in the **Graph by** selector. Choose **Month**.

10. While you are still in the date selector, click on the **Date Range** selector and choose **Previous**. Then, choose **Last...** in the **Date Range** selector, enter 24, and select **Month**.

11. Now, let's filter the amounts down to just closed-won deals by dragging and dropping the **Stage** beast mode field into the **FILTERS** area. Then, click on the **Display as Quick Filter** switch and click on **Apply**.

12. In the quick filters panel, click on **Select All** to deselect all of the current selections. Click on **5 Closed Won** to set the default view to be filtered on closed-won only.

13. Now, let's clean up the formatting a bit. Click on the **Y AXIS** field of **SUM of Amount**, and in the **Data Label** box, enter Sales. Then, in the **Goal** box, enter 10000000 and click on **Format**. Choose **Currency** and then choose **0 Decimals**.

14. Let's add a left-hand *y* axis title by clicking on **Value Scale (Left)** in the **CHART PROPERTIES** panel and entering Sales in the **Title** box.

15. Having value labels on the chart view would be helpful, so let's add them by clicking on **Data Label Settings** in the **CHART PROPERTIES** panel and then clicking on + in the **Text** property. Finally, select **Value**. This will place the %_VALUE substitution token into the **Text** property. Additionally, let's shorten up the numbers by checking the **Use Scale Abbreviation** box. In the **Decimal Places** property, select **.0**.

16. Finally, click on **Summary Number** and enter Total Sales in the **Label** box. Then, click on **Show Formatting Options** and select **Currency** in the **Display as** selector. Select **1** in the **Decimals** selector. Click anywhere away from the dialog box in the chart preview area to close the dialog.

17. Click on **SAVE** and select **Sales** in the **Choose a Dashboard** field. Then, click on **SAVE**.

You should see the new **Sales Trend** card on your **Sales** dashboard.

18. Let's convert this into a standard dashboard layout by clicking on the wrench icon for page actions and choosing **Convert to Standard Page**.

Your **Sales** dashboard should look similar to *Figure 6.5*:

Important Note

This card is using a relative date filter for the last 24 months. This date window shifts over time, so we might need to go back even further, depending on the current date, to see the actual data since the date data in the dataset is static.

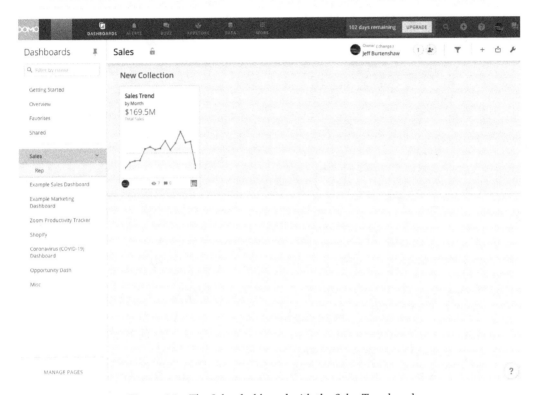

Figure 6.5 – The Sales dashboard with the Sales Trend card

Great work! You have created a card for your dashboard. The process of creating and editing cards in Analyzer is a skill that you will use often in Domo.

Next, let's edit the *Sales Trend* card and add some additional information to the card.

Using tooltip fields to create multi-value hovers

In Domo, tooltips are the popups that appear when you hover over a data point. Typically, a tooltip shows a single value. However, sometimes, it is useful to show multiple values on a hover label, for example, both the sales and the average sale values. Refer to *Figure 6.6*:

Figure 6.6 – Chart with tooltip fields used in labels

To customize the tooltip, let's walk through the following steps:

1. In the **Sales** dashboard, click on the **Sales Trend** card to go to the card detail view. Then, click on the Analyzer icon.

2. In Analyzer, activate the tooltip area by clicking on the **TOOLTIPS** toolbar item so that you can see the **TOOLTIP FIELD 1** area.

3. In the **DATA** panel, click and drag the **Amount** field from **MEASURES** into the **TOOLTIP FIELD 1** area. Select **Average** as the **Aggregation** amount and choose **Currency** and **0 Decimals** as the format.

4. Click on **Hover Text Settings** in the **CHART PROPERTIES** panel.

5. Enter the following code into the **Hover Text** box:

    ```
    Sales %_VALUE \\N Avg Sale %_TOOLTIP1
    ```

    The %_ prefix indicates a substitution of the particular value, while \\N adds a newline to the output.

6. Check both the **Use Scale Format** and **Use Scale Abbreviation** properties.

    Amazing! Now in the tooltip area, you can see both the sales amount and the average sale.

Next, let's add a gauge chart to track the progress of this month's sales target.

Creating a sales target gauge card

Gauges are a useful visual to show where you are in relation to a goal. Let's go through the steps to create a gauge:

1. In the **Sales** dashboard header, click on + and select + **Create new Card**.

2. Click on **SELECT** under **Visualization**, and click on **SELECT** under **Existing data**. Select the **Opportunity CH4** dataset and click on **CHOOSE DATASET**.

3. In Analyzer, rename the chart by double-clicking on **Opportunity CH4** in the chart preview area and replacing it with `Sales Attainment`.

4. In the **DATA** panel, enter `stage` in the filter search box, and drag and drop the **Stage** beast mode into the **FILTERS** panel. Check **5 Closed Won** and then click on **Apply**.

5. In the **CHART TYPES** panel, select **Gauges** and click on **Filled**.

6. Click on the date range selector, and in the **Date Range** selector, pick **Previous** and select **Last....** Then, enter `24` and **Month**. We might need to extend the month selection to more than 24 months depending on the current date being viewed as the dataset data is static.

7. Now, let's reset the default fields by removing them from the chart. Click on the field in **GAUGES VALUES** and click on **REMOVE**. Click on the field in **TARGET VALUE** and click on **REMOVE**. Click on the field in **OPTIONAL GROUP BY** and click on **REMOVE**.

8. In the **DATA** panel, enter `amount` in the search. Then, drag and drop the **Amount** field, underneath **MEASURES**, into **GAUGES VALUES**. Enter `Sales` in the **Data Label** box, and click on **Format**. Select **Currency**, and then select **1** for **Decimals**.

9. In the **CHART PROPERTIES** panel, click on **General** and enter `Sales` in the **Value Label** property. Then, in the **Color Source** property, choose **Green Above Target**.

10. In the **CHART PROPERTIES** panel, click on **Target** and enter `Target` in the **Value Label** property. Then, in the **Value** property, enter `120000000`.

11. For the summary number, we want to show the attainment percentage. To do this, we need a new beast mode. Click on **ADD CALCULATED FIELD** and enter `Attainment %` in the **DETAILS Calculated Field Name** box. Then, enter the following formula:

    ```
    sum(`Amount`)/120000000
    ```

12. Click on **SAVE & CLOSE**.

13. Click on the summary number and change the **Label** property to **Attainment**. In **Column**, select **Attainment %**. Click on **Show Formatting Options** and select **Percentage** as **Display as**.

14. Click on **SAVE**. Finally, click on the **X** icon to close Analyzer.

Now the **Sales** dashboard will look similar to *Figure 6.7*:

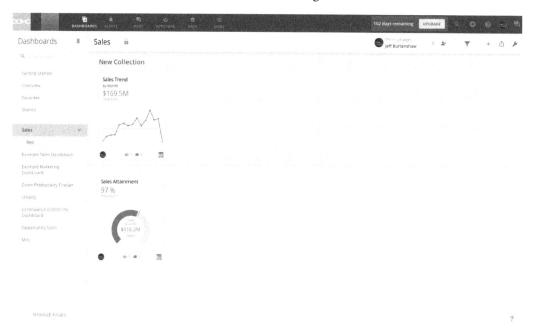

Figure 6.7 – The Sales dashboard with gauge

So, you have enhanced your dashboard with a nice gauge that indicates progress.

Next, let's show how to create the sales breakdown by lead source.

Creating a sales period over period breakdown card

The period over period breakdown card shows the actual change and the percentage change from one period to another. Let's walk through how to create a card by showing the change in sales from a year ago for each lead source. Perform the following steps:

1. In the **Sales** dashboard header, click on + and select + **Create new Card**.

2. Click on **SELECT** underneath **Visualization**. Click on **SELECT** under **Existing data** and select the **Opportunity CH4** dataset. Then, click on **CHOOSE DATASET**.

3. In Analyzer, rename the chart by double-clicking on **Opportunity CH4** in the chart preview area and replacing it with `Change in Sales by Leadsource`.

4. In the **DATA** panel, enter `stage` in the filter search box, and drag and drop the **Stage** beast mode into the **FILTERS** panel. Check **5 Closed Won** and, click on **Apply**.

5. In the **CHART TYPES** panel, select **Vertical Bar** and click on **Symbol + stacked bar**.

6. Click on the date range selector, and in **Date Range**, pick **All Time**.

7. Remove the default fields. Click on the field in **X AXIS**, and click on **REMOVE**. Click on the field in **Y AXIS**, and click on **REMOVE**. Click on the field in **SERIES**, and click on **REMOVE**.

    Now we need to create beast mode calculations for the last 12 months of sales along with the year-ago sales totals. Since the test data covers 2019 and 2020, our formulas will reflect that.

8. Click on **ADD CALCULATED FIELD** and enter `L12M Sales` as the **Calculated Field Name** value. Then, enter the following formula:

```
-- L12M sales
sum(case when `CloseDate`<= date('12/31/2020')
-- on live data use CURRENT_DATE()
   and `CloseDate`> DATE_SUB(date('12/31/2020'),interval
12 Month)
   then `Amount` else 0 end)
```

    This formula sums the sales amount for the last 12 months. To see the data from the example, we might need to change the formula to more than 12 months to see results, as the underlying data is static.

9. Check **Share Calculation on Dataset** and then click on **SAVE & CLOSE**.

10. Click on **ADD CALCULATED FIELD** and enter `Year Ago Sales` as the **Calculated Field Name** value. Then, enter the following formula:

```
-- Year ago sales
sum(case when `CloseDate` <= DATE_SUB(date('12/31/2020'),
interval 12 Month) and `CloseDate` > DATE_
SUB(date('12/31/2020'),interval 24 Month) then `Amount`
else 0 end)
```

    The preceding formula calculates the year-ago sales from December 31, 2020. Again, due to the static dates on the dataset, we might need to adjust the lookback beyond 12 months to see any data.

11. Check **Share Calculation on Dataset** and then click on **SAVE & CLOSE**.

12. Click on **ADD CALCULATED FIELD** and enter `Sales $ Change` as the **Calculated Field Name** value. Then, enter the following formula:

```
-- Sales $ Change on live data use CURRENT_DATE() inplace
of 12/31/2020
-- L12M sales
sum(case when `CloseDate`<= date('12/31/2020') and
`CloseDate`> DATE_SUB(date('12/31/2020'),interval 12
Month) then `Amount` else 0 end)

-
-- Year ago sales
sum(case when `CloseDate` <= DATE_SUB(date('12/31/2020'),
interval 12 Month) and `CloseDate` > DATE_
SUB(date('12/31/2020'),interval 24 Month) then `Amount`
else 0 end)
```

The preceding formula yields the dollar change in sales amount during the last 12 months from the year-ago sales.

13. Check **Share Calculation on Dataset** and then click on **SAVE & CLOSE**.

14. Click on **ADD CALCULATED FIELD** and enter `Sales % Change` as the **Calculated Field Name** value. Then, enter the following formula:

```
-- Sales % Change from Year Ago on real data use CURRENT_
DATE()
-- L12M sales
Round((sum(case when `CloseDate`<= date('12/31/2020')
and `CloseDate`> DATE_SUB(date('12/31/2020'),interval 12
Month) then `Amount` else 0 end)

-
-- Year ago sales
sum(case when `CloseDate` <= DATE_SUB(date('12/31/2020'),
interval 12 Month) and `CloseDate` > DATE_
SUB(date('12/31/2020'),interval 24 Month) then `Amount`
else 0 end))
/
-- Year ago sales
```

```
sum(case when `CloseDate` <= DATE_SUB(date('12/31/2020'),
interval 12 Month) and `CloseDate` > DATE_
SUB(date('12/31/2020'),interval 24 Month) then `Amount`
else 0 end) *100,1)
```

The preceding formula yields the percentage change in sales amount during the last 12 months from the year-ago sales.

15. Check **Share Calculation on Dataset** and then click on **SAVE & CLOSE**.

16. Click on **ADD CALCULATED FIELD** and enter `Sales Change Summary` as the **Calculated Field Name** value. Then, enter the following formula:

```
-- Sales Change Summary on real data use CURRENT_DATE()
in place of 12/31/2020
```
```
-- L12M sales
```
```
concat('Change from Year Ago ', round((sum(case when
`CloseDate`<= date('12/31/2020') and `CloseDate`> DATE_
SUB(date('12/31/2020'),interval 12 Month) then `Amount`
else 0 end) -
```
```
-- Year ago sales
```
```
sum(case when `CloseDate` <= DATE_SUB(date('12/31/2020'),
interval 12 Month) and `CloseDate` > DATE_
SUB(date('12/31/2020'),interval 24 Month) then `Amount`
else 0 end)) /
```
```
-- L12M Sales
```
```
  sum(case when `CloseDate` <= DATE_
SUB(date('12/31/2020'), interval 12 Month) and
`CloseDate` > DATE_SUB(date('12/31/2020'),interval 24
Month) then `Amount` else 0 end)*100,1),'% | $ ',
```
```
round((
```
```
-- L12M sales
```
```
sum(case when `CloseDate`<= date('12/31/2020') and
`CloseDate`>DATE_SUB(date('12/31/2020'),interval 12
Month) then `Amount` else 0 end) -
```
```
-- Year ago sales
```
```
sum(case when `CloseDate` <= DATE_SUB(date('12/31/2020'),
interval 12 Month) and `CloseDate` > DATE_
SUB(date('12/31/2020'),interval 24 Month) then `Amount`
else 0 end)) /1000000,1),'M')
```

The preceding formula uses the `concat()` function to create a phrase presenting the percentage change and dollar change in sales from a year ago as a summary. Because there is no way to reference other beast modes from within a beast mode, there is a lot of repetition to get the `Year Ago` and `L12M` sales. The high-level formula is `Change from Year Ago % = (L12M Sales - Year Ago Sales)/Year Ago Sales` and `Change from Year Ago $ = L12M Sales - Year Ago Sales`.

17. Check **Share Calculation on Dataset** and then click on **SAVE & CLOSE**.

18. Drag and drop the **Lead Source Standardized** beast mode field into the **X AXIS** area, drag and drop the **Sales % Change** beast mode field into the **Y AXIS** area, and drag and drop the **Sales $ Change** beast mode field into the **SERIES** area.

19. In the **CHART PROPERTIES** panel, click on **Value Scale (Symbols)** and change **Label Format** to **Percentage**.

20. In the **CHART PROPERTIES** panel, click on **Value Scale (Bar)** and change **Label Format** to **Currency**.

21. Drag and drop the **Sales $ Change** beast mode field into the **SORTING** panel. Now, click on the **Sales $ Change** field in the **SORTING** panel and select **Descending**. Then, click on **Apply** to see the lead sources in descending order of actual dollar change in sales from a year ago.

22. Click on the summary number and choose **Sales Change Summary** in **Column**. In **Label**, clear the field and leave it blank.

23. From the data, it appears as though some lead source types are empty, and there are still some spelling mistakes. Let's go back and edit the beast mode to clean these errors up. Enter `lead` in the **DATA Filter** search, then double-click on the **Lead Source Standardized** beast mode field. Then, enter the following updated formula:

```
-- Lead Source Standardization
case
when `LeadSource` = 'Patnes' then 'Partners'
when `LeadSource` = 'Diect' then 'Direct'
when `LeadSource` = 'Sales Ceated' then 'Sales Created'
when `LeadSource` = 'efeal' then 'Referral'
when `LeadSource` = 'Self-Souced' then 'Self-Sourced'
when `LeadSource` = 'Maketing Outbound' then 'Marketing
Outbound'
when `LeadSource` = 'Stategic Accounts Maketing' then
'Strategic Accounts Marketing'
```

```
when `LeadSource` = 'Patneing' then 'Partners'
when `LeadSource` = 'enewal' then 'Renewal' -- new
when `LeadSource` is null then 'Unknown' -- new
else `LeadSource` -- If not mapped above then keep as is
end
```

The lines commented as --new are the added lines. The *x* axis on the chart looks much better.

Let's set some colors to demarcate the changes:

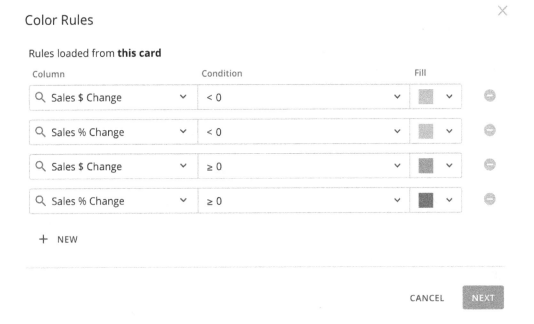

Figure 6.8 – The color property settings

1. Click on **Colors** in the **Chart Properties** panel. Set **Color Rules** up, as shown in *Figure 6.8*. Click on the **NEXT** button and then click on **APPLY TO ONLY THIS CARD**.

2. Click on **SAVE & CLOSE**.

The **Sales** dashboard will now look similar to *Figure 6.9*:

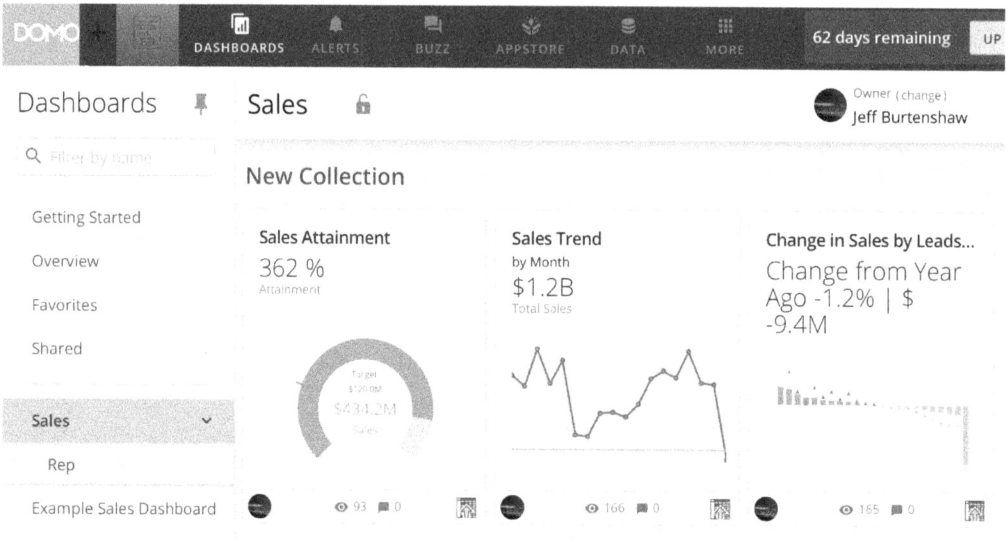

Figure 6.9 – The Sales dashboard with the Changes in Sales card

Your monitoring dashboard is starting to take shape as you add more cards.

Next, let's add a simple regression-based forecast prediction to the sales trend card.

Adding a forecast to a card

Domo provides a simple way, based on regression analysis, to add forecast values to card trendlines. Let's walk through an example:

1. Hover over the **Sales Trend** card and click on the wrench icon in the upper-right corner of the card. Then, click on **Edit in Analyzer**.

2. In **CHART PROPERTIES**, select **Multi-Period Projection** and enter 3 in the **Projection Period Count** property. Choose **Linear Regression** as the **Project Using** property value.

The dashed lines in *Figure 6.10* indicate the forecast values for the next 3 months along with the current month, as this, usually, will not be complete. The darker shaded area of the graph highlights the current period gap between the actual values and the forecasted values:

Figure 6.10 – Sales forecast

3. Click on the down arrow tip to the right-hand side of **SAVE** and select **Save and Close** to close the card. Then, click on **SALES** in the **SALES > SALES TREND** breadcrumb to return from the card detail view back to the dashboard.

Nice! Now we can see the predicted values for the next few periods on the chart.

Great! Next, we'll create a card to show the sales pipeline activity.

Creating a pipeline running total card

A **Running Total** chart shows the running total of a desired value over time. The value will rise and fall as new items come and go. In this case, let's go through an example of how the total sales in the sales pipeline are growing or shrinking over time. Perform the following steps:

1. In the **Sales** dashboard header, click on + and select + **Create new Card**.

2. Click on **SELECT** under **Visualization**, and click on **SELECT** under **Existing data**. Select the **Opportunity Burn Up** dataset and click on **CHOOSE DATASET**.

3. In Analyzer, rename the chart by double-clicking on **Opportunity Burn Up** in the chart preview area and replacing it with `Pipeline Burnup`.

4. In the **CHART TYPES** panel, select **Vertical Bar** and click on the **Line + Grouped and stacked bar** option.

5. Remove the default fields. Click on the field in **X AXIS** and then click on **REMOVE**. Click on the field in **Y AXIS** and then click on **REMOVE**. Click on the fields in **SERIES** and then click on **REMOVE**.

6. In the **DATA** panel, drag and drop **Day**, in the **DIMENSIONS** area, into **X AXIS**.

7. Click on **ADD CALCULATED FIELD** to create a new beast mode to show the open opportunities total for the last day of each month. Enter `Open Last Day of Month` in the **Calculated Field Name** box. Enter the following into **FORMULA**:

```
sum(case when
Month(DATE_ADD(`Day`,Interval 1 Day)) <> Month(`Day`)
then `Open` else 0 end)
```

When the date is the last day of the month and you add one day, then the date will be the next month, so the months will not match. Since the `Open` field is a running total, this is necessary to only get month-end totals that we can show on the chart.

8. Check **Share Calculation on DataSet** and click on **SAVE & CLOSE**.

9. In the **DATA** panel, drag and drop **Open (Last Day of Month)** in the **MEASURES** area into **Y AXIS**.

10. In the **DATA** panel, drag and drop **DayNet** in the **MEASURES** area into **SERIES**.

11. In the **DATA** panel, drag and drop **New** in the **MEASURES** area into **SERIES**.

12. In the **DATA** panel, drag and drop **Won** in the **MEASURES** area into **SERIES**.

13. In the **DATA** panel, drag and drop **Lost** in the **MEASURES** area into **SERIES**.

14. Click on the date range selector and check **Hide Date on Card Details**. In **Date Range**, pick **Month** in **Graph by**.

15. In **CHART PROPERTIES**, click on **General** and enter 1 in **Series on Left Scale**; check the **Sync Values Scales**, **Hide Second Scale**, and **Sync Zero Lines** properties. Click on **Bar Settings** and enter 1 inside the **First Bar Series Count** property. Click on **Trellis | Tiered Date Settings**, and choose **Tiered Dates** in the **Show As** property. Click on **Data Table** and check **Show Data Table**.

16. Click on the down arrow tip on the right-hand side of **SAVE** and select **Save and Close** to close the card. Then, click on **SALES** in the **SALES | SALES TREND** breadcrumb to return from the card detail view back to the dashboard.

That's it! You have a great burn-up view of the sales pipeline activity that will look similar to *Figure 6.11*:

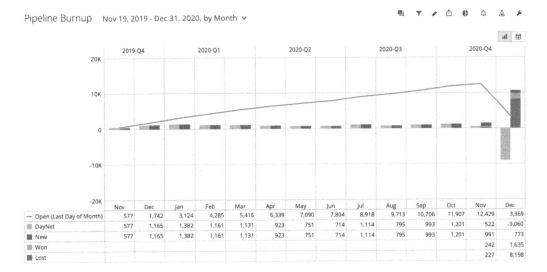

Figure 6.11 – The running total chart

Now, let's try a different way to visualize the pipeline, using a waterfall chart or a bridge chart.

Creating a waterfall card

A **waterfall** chart, also known as a walk chart or a bridge chart, shows the progression from a starting value. Then, it shows the increases and decreases categorized and netted to an ending value. For example, the sales pipeline started at X, increased A because of B, decreased C due to D, and ended at Z. The waterfall chart requires a specific data format to work. So, before we create the card, we will use **Magic ETL (Extract, Transform, and Load)** to sculpt the data.

> **Important Note**
>
> **Magic ETL BETA features**: At the time of writing, the **PIVOT** and **UTILITY** tile panel sections are only available as part of the ETL BETA program. Contact Domo Sales or ask your Domo Account Manager to get access to the Beta features.

Sculpting the data

Let's walk through the steps to make a waterfall chart in Domo, starting with data sculpting. Perform the following steps:

1. Click on **DATA** in the main menu, and then click on **ETL** in the **MAGIC TRANSFORM** area.

2. Under **DATASETS**, in the tile panel, click and drag the **Input Dataset** tile onto the canvas. Click on **SELECT DATASET** and choose **Opportunity Burn Up**. Recall that the **Opportunity Burn Up** dataset is the output of the **Opportunity Aggregated** ETL job we created in *Chapter 4, Sculpting Data*.

3. Click on **PIVOT** in the tile panel, and drag and drop the **Unpivot** tile onto the canvas. This step stacks the data from the **Opportunity Burn Up** dataset.

4. Click on the circle on the right-hand side of the **Opportunity Burn Up** tile, and drag the line to the triangle on the center-left side of the **Unpivot** tile.

5. In the **Unpivot** tile configuration panel, enter Item in the **New column name** box under **Name the column that will hold the column labels**.

6. In the **Unpivot** tile configuration panel, enter Value in the **New column name** box under **Name the column that will hold the values**.

7. In the **Unpivot** tile configuration panel, select **Open** in the **Select column** box under **Select columns to collapse**.

8. Click on **ADD COLUMN** and select **New** in the **Select column** box. Repeat the **ADD COLUMN** step for the **Won, Lost, Closed**, and **DayNet** columns.

9. Click on **DONE** to close the configuration panel, as shown in *Figure 6.12*:

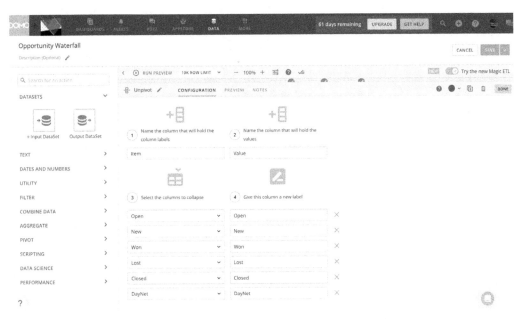

Figure 6.12 – The Opportunity Waterfall ETL Unpivot configuration

10. For reference, a preview of the output of the **Unpivot** step can be seen in *Figure 6.13*:

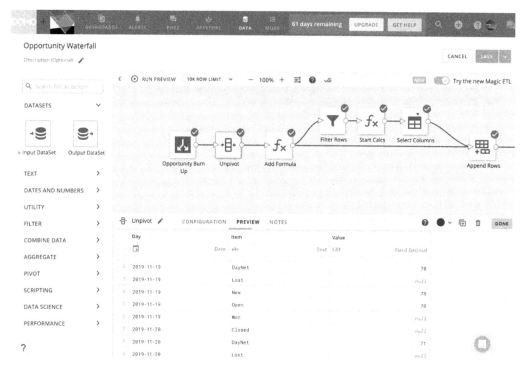

Figure 6.13 – The Opportunity Waterfall ETL Unpivot step preview

11. Next, we need to add a column for the starting value of each month based on the ending value from the prior month, as the starting value for a month is the same as the ending value of the prior month. This new column will be used downstream, in the ETL job steps, to append rows with the opening value period. Click on **UTILITY** in the tile panel, and drag and drop the **Add Formula** tile onto the canvas.

12. Click on the circle on the right-hand side of the **Unpivot** tile. Then, drag the line toward the triangle on the center-left side of the **Add Formula** tile.

13. Click on the **Add Formula** tile on the canvas to open the tile configuration panel, and click on the **CONFIGURATION** tab.

14. Enter `Start Month` in the **Enter column name** box. Then, enter the following formula into the **Write formula** box, as shown in *Figure 6.14*:

Figure 6.14 – The Opportunity Waterfall ETL Add Formula configuration

Let's take a look at the formula that we just entered:

```
CASE
WHEN MONTH(`Day`) <> MONTH(DATE_ADD(`Day`,INTERVAL 1
Day))
AND `Item` = 'Open'
THEN DATE_ADD(`Day`, INTERVAL 1 Day)
ELSE null END
```

Since the starting number of open opportunities at the beginning of the month is the same as the end of the previous month, this formula adds a new column to the dataset that also marks the last day of each month as the start of the next month, as shown in *Figure 6.15*:

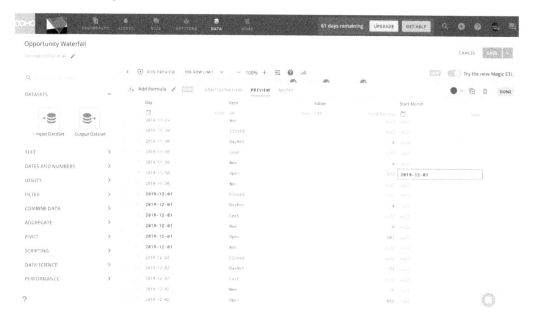

Figure 6.15 – The Opportunity Waterfall ETL Add Formula preview

This will be used to show the start value in the waterfall chart.

15. Click on **DONE**.

Now that we have the start and end dates for the periods nicely set up in the dataset, the next step is to append new rows for the opening values to the dataset. This will take four new tile steps to the dataflow to accomplish adding rows with the number of opportunities in the sales pipeline at the start of each month.

16. Click on **FILTER** in the tile panel, and drag and drop the **Filter Rows** tile into the canvas. For this branch of ETL steps, we are only going to filter out the opening value rows.

17. Click on the circle on the right-hand side of the **Add Formula** tile, and drag the line toward the triangle on the center-left side of the **Filter Rows** tile.

18. Click on the **Filter Rows** tile on the canvas to open the tile configuration panel. Then, click on the **CONFIGURATION** tab.

19. Click on **ADD FILTER RULE** and select **Start Month** as the **Select a column to filter on** choice.

20. Click on **Select operation** and choose **is not null**, as shown in *Figure 6.16*:

Figure 6.16 – The Opportunity Waterfall ETL Filter Rows configuration

21. Now we only have the opening value rows in this step, as shown in *Figure 6.17*. Let's continue refining the format of the columns for appending back into the dataset:

Figure 6.17 – The Opportunity Waterfall ETL Filter Rows preview

22. Let's add in the constant value for the starting period values. Click on **UTILITY** in the tile panel, and drag and drop the **Add Formula** tile onto the canvas.

23. Click on the circle on the right-hand side of the **Filter Rows** tile, and drag the line toward the triangle on the center-left side of the **Add Formula 1** tile.

24. Click on **Add Formula 1** and rename it to Start Calcs in the tile configuration panel.

25. Enter Day in the **Enter column name** box, and enter `Start Month` in the **Enter formula** box. This will replace the last day of the month with the first day of the next month.

26. Click on **ADD FORMULA** and enter Item in the **Enter column name** box. Then, enter 'Start' in the **Enter formula** box. This sets up the constant value for the item value in the dataset, as shown in *Figure 6.18*:

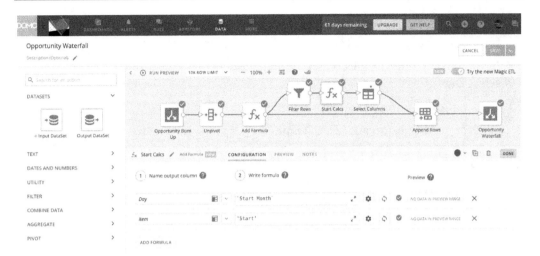

Figure 6.18 – The Opportunity Waterfall ETL Start Calcs configuration

Here is the preview, as shown in *Figure 6.19*:

Figure 6.19 – The Opportunity Waterfall ETL Start Calcs preview

27. Next, let's set up the final columns for the operation to append start period rows. From the **UTILITY** section of the tile pane, drag and drop the **Select Columns** tile onto the canvas.

28. Click on the circle on the right-hand side of the **Start Calcs** tile, and drag the line toward the triangle on the center-left side of the **Select Columns** tile.

29. In the **Select Columns** tile configuration panel, click on **Add Column** and then select **Day**.

30. Click on **Add Column** and select **Item**.

31. Click on **Add Column** and select **Value**.

32. The configuration of **Select Columns** can be seen in *Figure 6.20*:

Figure 6.20 – The Opportunity Waterfall ETL Select Columns configuration

Now that we have the beginning of month values in the correct format, as shown in *Figure 6.21*, we need to append them together with the pivoted burn-up data:

Figure 6.21 – The Opportunity Waterfall ETL Select Columns preview

33. From the **COMBINE DATA** section of the tile pane, drag and drop the **Append Rows** tile onto the canvas.

34. Click on the circle on the right-hand side of the **Select Columns** tile, and drag the line toward the triangle on the center-left side of the **Append Rows** tile.

35. Click on the circle on the right-hand side of the **Add Formula** tile, and drag the line toward the triangle on the center-left side of the **Append Rows** tile.

36. In the **Append Rows** tile configuration panel, select **Only include shared columns** under the **Which columns should be included in this append?** area, as shown in *Figure 16.22*:

Figure 16.22 – The Opportunity Waterfall ETL Append Rows configuration

We are almost there. All that remains is to write out the data sculpted for the waterfall chart format into a new dataset.

37. In the tile panel, click and drag the **Output Dataset** tile onto the canvas.

38. Click on the circle on the right-center side of the **Append Rows** tile, and drag the line toward the triangle on the center-left side of the **Output DataSet** tile.

39. Rename the **Output DataSet** name to `Opportunity Waterfall` in the **CONFIGURATION** tab.

40. Click on **Run Preview** and wait for the dataflow to run. When complete, the **PREVIEW** tab will appear in the tile configuration panel. Click on the **PREVIEW** tab. You can refer to *Figure 6.23*:

Figure 6.23 – The Opportunity Waterfall ETL Output DataSet preview

41. Click on the settings icon and choose **Only When DataSets are updated** for **When should this DataFlow run?**, and check **Opportunity Burn Up**. Click on **APPLY**. That will configure this dataflow to run after each completion of the **Opportunity Burn Up** dataflow updates the **Opportunity Burn Up** dataset, as shown in *Figure 6.24*:

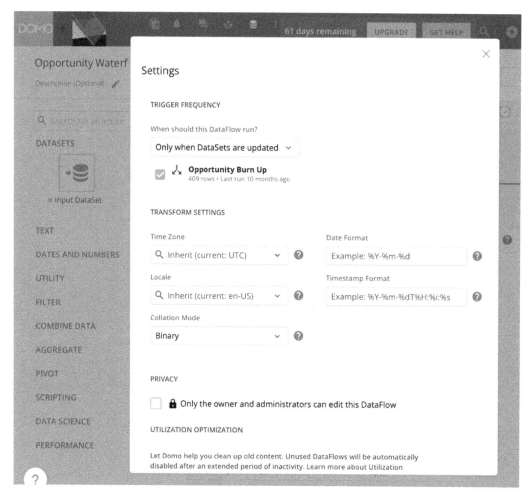

Figure 6.24 – The Opportunity Waterfall ETL settings

42. Click on the down arrow tip on the right-hand side of **SAVE** and select **Save and Run**. Enter `Waterfall` as the **Version Description (Optional)** value. Click on **SAVE and RUN**.

That concludes the steps to sculpt the data for the waterfall chart. This period snapshotting of data over time is a very useful skill to have a pattern for.

Creating the waterfall card

Now that we have the data sculpted into a new dataset, let's create the card. Perform the following steps:

1. Go back to the **Sales** dashboard by clicking on **DASHBOARDS** in the main menu.

2. In the **Sales** dashboard header, click on + and select + **Create new Card**.

3. Click on **SELECT** under **Visualization** and click on **SELECT** under **Existing data** and. Select the **Opportunity Waterfall** dataset, and click on **CHOOSE DATASET**.

4. In **Analyzer**, rename the chart by double-clicking on **Opportunity Waterfall** in the chart preview area. Replace it with `Pipeline Bridge`.

5. In the **CHART TYPES** panel, select **Vertical Bar** and click on **Waterfall**.

6. Remove the default fields. Click on the field in **ITEM NAMES** and click on **REMOVE**. Click on the field in **VALUE**, and click on **REMOVE**. Click on the fields in **SUMMARY GROUP**, and click on **REMOVE**.

7. In the **DATA** panel **DIMENSIONS** area, click and drag **Item** onto the **ITEM NAMES** area.

   Now, let's add a beast mode that changes the values to negative for items that take opportunities out of the funnel so that they can be displayed properly.

8. Click on **ADD CALCULATED FIELD**. In the **DETAILS** panel **Calculated Field Name** value, enter `Value +/-`. Then, enter the following into **FORMULA**:

```
SUM(CASE
WHEN `Item` in ('Won','Lost','Closed') THEN `Value` * -1
ELSE `Value`
END)
```

9. Check **Share Calculation on DataSet** and click on **SAVE & CLOSE**.

10. In the **DATA** panel **BEAST MODES** area, click and drag **Value +/-** onto the **VALUE** area.

    Now we need to control the order of the items that have been sorted for the waterfall chart. For that, let's create another beast mode.

11. Click on **ADD CALCULATED FIELD**. Then, in the **DETAILS** panel **Calculated Field Name** value, enter `Sort`. Enter the following into **FORMULA**:

```
CASE
WHEN `Item` = 'Start' then 1
WHEN `Item` = 'New' then 2
```

```
WHEN `Item` = 'Won' then 3
WHEN `Item` = 'Lost' then 4
ELSE `Item`
END
```

12. Check **Share Calculation on DataSet** and click on **SAVE & CLOSE**.

13. In the **DATA** panel **BEAST MODES** area, click and drag **Sort** onto the **SORTING** area.

14. Select **No aggregation** under **Aggregation**, and click on **APPLY**.

    Now we need to filter out the items that are not needed in the chart.

15. In the **DATA DIMENSIONS** area, click and drag the **Item** field onto the **FILTERS** area. Check **Start**, **New**, **Won**, and **Lost**.

16. Click on **Apply**.

    Next, let's add in a quick filter so that it is simple to pick which period to view on the card. To do this, we need to—you guessed it—add a beast mode for the period.

17. Click on **ADD CALCULATED FIELD**, and in the **DETAILS** panel **Calculated Field Name** value, enter Period. Then, enter the following into **FORMULA**:

```
CASE
WHEN YEAR(`Day`) = YEAR(CURRENT_DATE)
AND MONTH(`Day`) = MONTH(CURRENT_DATE)
THEN ' This Month'
ELSE DATE_FORMAT(`Day`,'%Y-%m')
END
```

The preceding formula flags rows with a day matching the current month to 'This Month' and all others to the actual year and month value.

18. Check **Share Calculation on DataSet** and click on **SAVE & CLOSE**.

19. Click and drag the **Period** field from the **BEAST MODES** area onto the **FILTERS** area.

20. Turn on the **Display as Quick Filter** switch and click on **Apply**.

21. In the quick filters panel, inside the **Period** area, uncheck the **Select All** row.

22. In the **Period** quick filter area, click on the gear icon and choose **Single select**. Select **2020-12** in the quick filter as the default setting. If you have current data instead of test data, you can select **This Month** to always see the current month waterfall.

We don't need a summary number for this card, so turn it off.

23. Click on the summary number, and choose **No Summary Number**.

Next, let's set some of the chart properties.

24. In the **CHART PROPERTIES** area, click on **General** and enter End in the **Final Summary Text** property. Also, check the **Bridging Lines** property.

25. In the **CHART PROPERTIES** area, click on **Value Scale (Y)** and enter Opportunities in Pipeline into the **Title** property.

26. The card configuration can be seen in *Figure 6.25*:

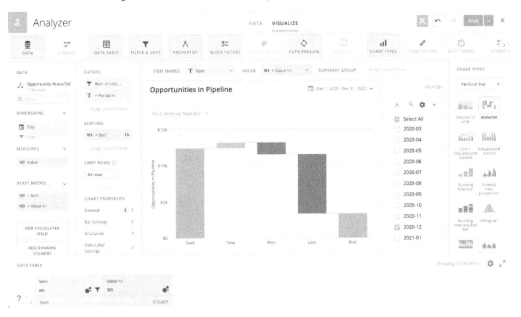

Figure 6.25 – The card configuration

27. Click on the down arrow point next to **SAVE**, and click on **Save and Close**. This takes you to the card detail page:

Figure 6.26 – The waterfall card detail page

And there you have it, a beautiful waterfall chart, as shown in *Figure 6.26*.

It's time to summarize what we have accomplished.

Summary

In this chapter, we learned how to create dashboard pages and sub-pages. We took a detailed tour of the Analyzer feature set and used Analyzer to create a few new cards with some cool new beast modes and quick filters. The trend and forecast, period-over-period, burn-up, and waterfall charts are heavily used patterns for many use cases beyond sales reporting. We learned that creating a card often involves iteratively going back to data sculpting and beast mode creation activities to get the data in the right shape for card building. Additionally, we got a feel for using properties to fine-tune the visuals on a card. The result was a sales dashboard. It's pretty amazing how quickly we can create fully functional dashboards using the Domo platform.

In the next chapter, we will cover many features for interacting with dashboards and cards.

Further reading

Here are related articles of interest for you to explore:

- *A refresher on Regression Analysis*: `https://hbr.org/2015/11/a-refresher-on-regression-analysis`.

- An introduction to the SQL Interval command: `https://www.mysqltutorial.org/mysql-interval/`.

- Background on the SQL Case statement: `https://www.w3schools.com/sql/sql_case.asp`.

- For reference material on creating waterfall charts, take a look at `https://knowledge.domo.com/Visualize/Adding_Cards_to_Domo/KPI_Cards/Building_Each_Chart_Type/Waterfall_Graph`.

7
Working with Drill Pathways

Often when viewing cards, users will want to drill down to see more details. Every card in Domo, by default, is enabled to drill down to the raw dataset. However, as a card designer, you may want to pre-define the drill path with custom visualizations that communicate more effectively than a raw table view. Creating these drill paths in Domo couldn't be simpler—actually, you just create another chart as a drill path under the main chart and Domo automatically figures out how to enable the drill-down linkages! There is even the ability to drill down to a completely different dataset than the parent card's dataset, provided the column names are the same. This ability to drill to a different dataset is particularly useful on large datasets where the parent dataset is aggregated and enables drilling to the larger dataset details.

In this chapter, we will specifically cover the following:

- Creating card drill paths
- Drilling down to a different dataset
- Linking cards
- Comparing parts to the whole with segments

Technical requirements

To follow along with this chapter, you will need the following:

- Internet access

- Your Domo instance and login details

- The ability to download example files from GitHub

- The cards created in *Chapter 6*, *Creating Dashboards*

> **Important Note**
>
> If you don't have a Domo instance, get a free trial instance here: `https://www.domo.com/start/free`.

All of the code found in this chapter can be downloaded from GitHub: `https://github.com/PacktPublishing/Data-Democratization-with-Domo`.

Creating card drill paths

Let's go through some scenarios where we use card drill paths to better understand sales performance. As is normally the case, before we get to cards, we need to do some data sculpting work. The following steps will update the opportunity data to include regional data:

1. We need to update the `Opportunity CH4` dataset to include a **Region** column and some additional date dimension columns. Download the `Opportunity CH7.xlsx` worksheet as the data update source from `https://github.com/PacktPublishing/Data-Democratization-with-Domo/blob/1a56847225e2226ceb6caaa8da339e621f6b26e1/Opportunity%20CH7.xlsx`.

2. On the **Sales** dashboard, double-click to open the **Sales Attainment** card. Scroll down to see the card footer and click on the **Opportunity CH4** link at the lower left of the screen.

3. Click on the **SETTINGS** tab, then click **browse your files** in the **Details** panel, and select the **Opportunity CH7.xlsx** worksheet from where you downloaded it.

4. In the **Select Tables** panel, select **OPPORTUNITY** as the **SHEET** option to use.

5. Scroll down in the **Select Tables** panel and click the **SAVE** button. This will import the data from the spreadsheet, replacing the existing data in the dataset with **Replace** as the **Update method** setting in the **Update Mode** panel.

6. While we are on the dataset view, let's also change the name of the dataset to reflect that it is now sourced from the `Opportunity CH7.xlsx` worksheet. Click the three stacked dots icon at the upper right and select **Edit Name & Description**. Replace 4 in the dataset name with 7, and click **SAVE**.

> **Important Note**
>
> Surrogate object identifiers: You may be wondering what happened to the cards that were connected to the dataset. Well, Domo uses surrogate keys internally for all objects, so name changes won't impact any existing relationships.

Now that the opportunity data is in a dataset, let's also bring in the sales quota data, which resides in the **Quota** sheet in the same `Opportunity CH7.xlsx` spreadsheet, to create a different dataset.

7. Click on the **DATA** icon in the main menu bar. Then, click on the **DataSets** icon on the left side.

8. Click on the **File** icon on the **DataSets** page header, click the **Browse your files** link, and select `Opportunity CH7.xlsx` from its downloaded location.

9. In the **Select Tables** panel, select **Budget** as the **SHEET** option to use and click **Next >**.

10. Enter `Quota` as the **DATASET NAME** and click **SAVE AND RUN**.

Excellent, we just created the aggregated **Quota** dataset.

Next, let's open the **Sales Attainment** card and create the drill path to regions.

Creating a drill-down to a region from the Sales Attainment card

To drill down from the gauge to a regional view, you will first define and save the drill path for the card. Then, to drill down from the card details view, click on the chart visual, which triggers a link to the defined drill pathway. You might be wondering how will the card defined as the drill-down path know what to filter on? That is a fair question. When a user clicks on the chart object, Domo automatically senses what fields are involved and their specific values based on the click location and then uses those as the filters on the drill-down card. OK, ready to create your first drill-down path? Let's go and create a drill-down to see the regional performance:

1. Double-click on the **Sales Attainment** card on the **Sales** dashboard.

2. Click the *wrench* icon at the upper right and select the **Edit Drill Path** option.

3. Click the **+ Add a View** link to go to the **Analyzer** card creation page to configure the drill-down chart.

4. Change the chart type to a **Pie** chart using the **CHART TYPES** panel.

5. Click and drag the **Region** field from the **DIMENSIONS** area into the **PIE NAME** area.

6. Click and drag the **Amount** field to the **PIE VALUE** area.

7. In the chart title, enter Region. This is the name that will show in the drill-down breadcrumb navigation as well.

The **Analyzer** page will look like *Figure 7.1*:

Figure 7.1 – Region drill path pie chart definition

8. Click **SAVE and CLOSE**.

9. Click the < **DETAILS** breadcrumb at the upper left of the screen.

10. Click on the chart gauge to drill down and see the regional breakout in the pie chart. The filter settings from the top level are automatically pushed down to the pie chart as well.

11. Click on the **East** pie slice to see the detailed tabular data for the region, as shown in *Figure 7.2*:

SALES ATTAINMENT > REGION > DATA

Data

Region	Amount	AccountId	CampaignId	CloseDate	ConnectionReceivedId	ConnectionSentId	CreatedDat
East	475,119	8b4248f76e7947959f4833f9a60c2...		7/31/2020			7/31/2020
East	605,757	9a020b45e91640d4ad6e6a7b10a...		9/1/2020			7/8/2020 6
East	508,711	cfb141f850f14728a95c4d8ca41a4...		7/3/2020			5/14/2020
East	525,804	e14a4dab2ac044608b74e074c23e...		6/30/2020			1/30/2020
East	339,858	5aef5ae686884674a0acf3581457...		4/21/2020			4/15/2020
East	45,582	5ab7812d0f3d4193a853abcc46f5...		4/4/2020			3/17/2020
East	726,462	ab2c62b2f9544ce19f5389627352...		2/19/2020			11/26/2019
East	138,507	168a3004a7894a1ebcfdfd2fcda4e...	aad28c4133684b2592c243ee2007...	12/25/2019			11/26/2019
East	398,983	695daaec6d1444f9a9b593c28fa9...		7/21/2020			7/14/2020 1
East	395,211	695daaec6d1444f9a9b593c28fa9...		7/3/2020			6/17/2020 1
East	248,969	695daaec6d1444f9a9b593c28fa9...		11/26/2020			11/25/2020
East	69,477	73ff20f979434713bd31f1aab6bf7...		11/4/2020			1/30/2020
East	364,528	f8c327dc612642d79892ce4cc638...		10/2/2020			9/29/2020

Figure 7.2 – Drill down detail view

Note that the breadcrumb now shows the full drill pathway. Wow, that was easy, right?

However, some of you might be wondering, *Well, what about the target data? This is just showing the actual sales.* We'll cover that next.

Creating drill-downs on sales actuals and targets

It is very typical to have *sales actual data* in one dataset and the *sales quota data* in another dataset. We all know how useful it is to show the actuals versus the quota. So, we could use data sculpting to add columns for the target in the actuals data, but they are usually at different grains, so that gets messy. A simpler method is to create an aggregate dataset (using sculpting tools or getting a targets spreadsheet from the finance team) that has both the budget and actual values at the same grain, and then enable a drill down from the aggregate to the raw actuals data. For convenience, I have already created this aggregate data, which you downloaded in the *Creating Card Drill Paths* section into the **Quota** dataset. Next, let's convert the **Sales Attainment** card to use the **Quota** dataset:

1. First, to adjust the **Sales Attainment** card to use the **Quota** dataset, click on the **Analyzer** beaker icon on the card's details page.

2. Click on **Opportunity CH7** in the **DATA** panel. Search for and select **Quota** in the dialog.

3. Click **CHOOSE DATASET** to change the chart to use the **Quota** dataset.

4. You will see a warning saying **Field names do not match**. That is OK in this case because the **Quota** dataset doesn't have an amount field; instead, it has both budget and actual fields that we will use. Click **Switch DataSet**.

5. Click and drag the **Actual** field from the **MEASURES** area into the **GAUGE VALUES** area. Click on the **SUM of Actual** field and click **Format**. Choose **Currency** in the **Display as** box, and **1** for the **Decimals** precision.

6. Click and drag the **Budget** field from the **MEASURES** area into the **TARGET VALUES** area. Click on the **SUM of Budget** field and click **Format**. Choose **Currency** in the **Display as** box, and **1** for the **Decimals** precision.

7. Click the **Target** item in the **CHART PROPERTIES** panel and clear the number in the **Value** property so this hardcoded value doesn't override the dynamic value from the dataset.

8. Click the **General** item in the **CHART PROPERTIES** panel and enter 10 in the **Ring Width Percent** property.

9. Edit the beast mode for **Attainment %** by hovering on the field and clicking the pencil icon in the **BEAST MODES** area of the **DATA** panel. Change the formula to the following:

```
sum(`Actual`) / sum(`Budget`)
```

Remember to check **Share Calculation on Dataset**.

10. In Summary Number, select **Use All Values** and choose the **Attainment %** column. Click **Show Formatting options…** and select **Format**, **Display as**, **Percentage**, **0**, and **Decimals**.

11. Add a **Quick Filter** instance by dragging the **Period** field into the **FILTERS** panel and turning on the **Display as Quick Filter** option, and clicking **Apply**.

12. Uncheck **Select All** in the **Period** panel to the right and check **2020-12**.

The **Analyzer** view for the card will look like *Figure 7.3*:

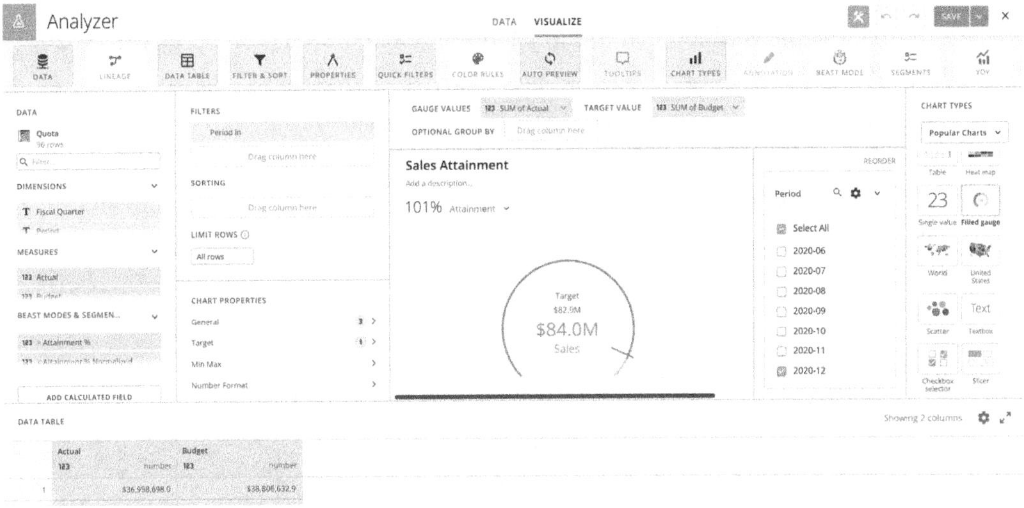

Figure 7.3 – Analyzer view of the Sales Attainment card using the quota dataset

13. Click **Save and Close**.

Great, we have the **Sales Attainment** card now using the Quota dataset.

Next, let's re-create the regional drill-down complete with a bullet bar chart showing **Actuals** and **Targets**:

1. Click on the *wrench* icon at the upper right of the **Sales Attainment** card detail page and select **Edit Drill Path**.

2. Click the *trashcan* icon to the right on the **Region** drill path row to delete the drill path and click **REMOVE**. Who likes pie charts anyway?

> **Important Note**
>
> When changing the parent card's dataset, it is best to delete all previous drill paths on the card as they are all still tied to the original dataset. Changing the primary dataset does not cascade down to the drill path cards.

3. Click the + **Add a view** link.

4. Let's use a bullet bar chart type to show the budget and actuals. Click **Horizontal bar** in the **CHART TYPES** panel, and select the **Bullet** chart type.

5. Click and drag the **Region** field from the **DIMENSIONS** area into the **NAMES** area.

6. Click and drag the **Actual** field to the **ACTUAL VALUE** area. Format as **Currency** with **0** decimals.

7. Click and drag the **Budget** field to the **TARGET VALUE** area. Format as **Currency** with **0** decimals.

8. Change the chart title to Region. This is the name that will show in the drill-down breadcrumb navigation as well.

9. Click **SAVE and CLOSE**, which will take you to the **Drill Path** page, as shown in *Figure 7.4*:

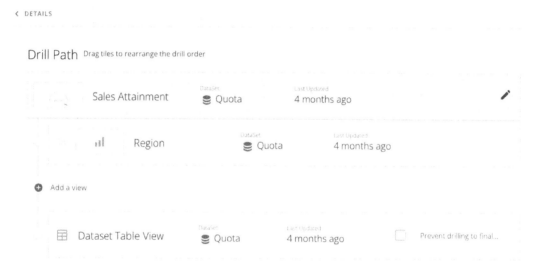

Figure 7.4 – Edit Drill Path page

If you want to prevent the users from drilling down to the raw dataset, check the **Prevent drilling to final** box.

Next, we will add another drill-down layer that drills into the **Opportunity CH7** details dataset from the aggregate **Quota** dataset.

Creating a drill-down to a different dataset

Continuing the example, we will create a drill-down card that starts from an aggregate dataset but drills to a detailed dataset, in this example, from the **Region** drill-down card to a **Type** drill-down card. When drilling from one dataset to another, the same context-sensitive field and value click detection are used but the filtering on the destination dataset is based on field names. So, be sure that field names to filter on from the aggregate dataset match the field names on the drill-down dataset. Let's give it a go:

1. From the **Sales Attainment** wrench icon **Edit Drill Path** view, click the + **Add a view** link to add the next drill-down level under the **Region** path.

2. Change the dataset to drill down to the detail data by clicking **Quota** in the **DATA** panel, selecting the **Opportunity CH7** dataset, and clicking **CHOOSE DATASET**.

3. Ignore the warning and click **Switch DataSet**.

4. Change the chart type to **Bar** in the **Horizontal bar** chart types.

5. Drag and drop the **Type** field into the **Y AXIS** area.

6. Drag and drop the **Amount** field into the **X AXIS** area.

7. Change the summary number to the sum of the **Amount** field, set **Format** to **Currency** with **1** decimal place, and check the **Abbreviate** property.

8. Drag and drop the **IsWon** field into the **FILTERS** panel and select **true**; then, click **Apply** to filter the card to only won opportunities.

9. Change the card title to Type from Region.

10. Click **SAVE and CLOSE** (see *Figure 7.5*):

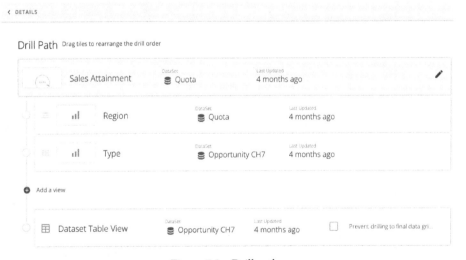

Figure 7.5 – Drill paths

> **Important Note**
>
> On occasion, the card owner may not wish to allow users to drill down to the raw dataset. In that case, the designer would check the **Prevent drilling to final data grid** box.

Now, let's navigate back up to the top-level card using the breadcrumb and test the full drill-down path:

1. Click the **< DETAILS** breadcrumb at the upper left of the screen.
2. Click on the chart gauge in the **Sales Attainment** card to drill down and see the regional breakout in the pie chart, as seen in *Figure 7.6*:

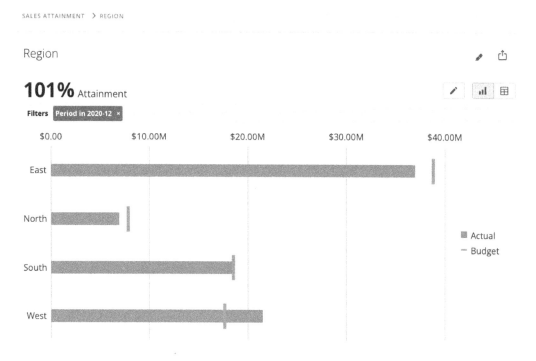

Figure 7.6 – Regional drill path with Budget and Actual bullet bar

3. Click on the **East** region bar to drill into the **East** region's sales by **Type**, as seen in *Figure 7.7*:

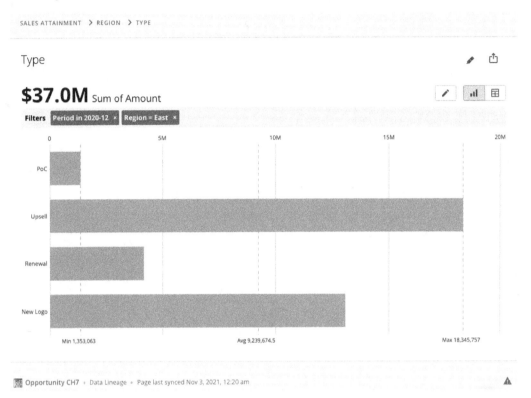

Figure 7.7 – Type drill path bar chart

4. Click on the **Renewal** bar to see the renewal sales, as seen in *Figure 7.8*:

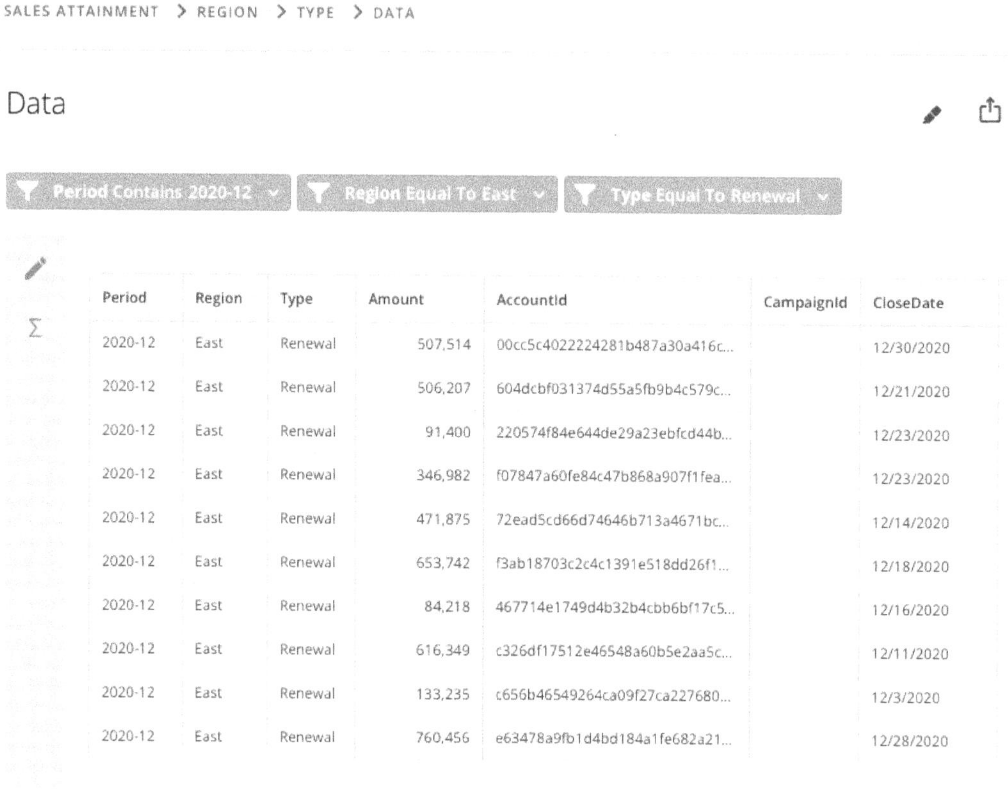

SALES ATTAINMENT ❯ REGION ❯ TYPE ❯ DATA

Data

Period Contains 2020-12 ⌄   Region Equal To East ⌄   Type Equal To Renewal ⌄

Period	Region	Type	Amount	AccountId	CampaignId	CloseDate
2020-12	East	Renewal	507,514	00cc5c4022224281b487a30a416c...		12/30/2020
2020-12	East	Renewal	506,207	604dcbf031374d55a5fb9b4c579c...		12/21/2020
2020-12	East	Renewal	91,400	220574f84e644de29a23ebfcd44b...		12/23/2020
2020-12	East	Renewal	346,982	f07847a60fe84c47b868a907f1fea...		12/23/2020
2020-12	East	Renewal	471,875	72ead5cd66d74646b713a4671bc...		12/14/2020
2020-12	East	Renewal	653,742	f3ab18703c2c4c1391e518dd26f1...		12/18/2020
2020-12	East	Renewal	84,218	467714e1749d4b32b4cbb6bf17c5...		12/16/2020
2020-12	East	Renewal	616,349	c326df17512e46548a60b5e2aa5c...		12/11/2020
2020-12	East	Renewal	133,235	c656b46549264ca09f27ca227680...		12/3/2020
2020-12	East	Renewal	760,456	e63478a9fb1d4bd184a1fe682a21...		12/28/2020

Figure 7.8 – Renewal sales grid

Notice at this **Data** level, the filters just above the table are active.

That was awesome. Remember, when you have larger datasets for performance, it is a good practice to create an aggregate dataset for top-level cards to use for speed, but then enable the user to drill to the supporting detailed larger dataset.

Next, we will show how to create a guided analytics path using card linking.

Linking cards

Users will often want to see data related to the information they are viewing, whether it is a different cut of the same information or a completely different but related dataset. Domo enables this **guided analytics** journey using the **Link Related Card** feature.

Let's walk through linking a related cut of the same information to the **Sales Attainment** card; namely, a quarterly trend of the sales data:

1. Obviously, to link to a card, the card must exist, so let's create a **Quarterly Sales Trend** card.

2. From the **Sales** dashboard, click + on the **Sales** header, and select + **Create new Card**.

3. Click **SELECT** under **Visualization**, then click **SELECT** under **Existing**.

4. Select the **Quota** dataset, then click **CHOOSE DATASET**.

5. Change the title to Quarterly Sales Trend.

6. In the **CHART TYPES** panel, select **Vertical bar** and click **Nested bar**.

7. In the **DATA** panel, drag and drop **FiscalYear** to the **X AXIS** area.

8. Drag and drop **Actual** to the **Y AXIS** area.

9. Drag and drop **Fiscal Quarter** to the **SERIES** area.

10. Drag and drop **FiscalYear** to the **SORTING** area.

11. Select **No Summary Number**.

12. Click **SAVE and CLOSE**.

13. In the **Quarterly Sales Trend** card detail view, scroll down and click **Select Card** in the **Link Related Card** footer area. Enter `Sales`, and select **Sales Attainment** (see *Figure 7.9*):

Figure 7.9 – Linked card

14. Click on the **Sales Attainment** thumbnail image to go to the **Sales Attainment** card details page. Notice that the **Quarterly Sales Trend** card is also linked there, so you can toggle back and forth between linked cards.

Now let's link in different but relevant information, specifically, the sales pipeline.

15. From the **Sales Attainment** card detail view, scroll to the footer and enter `Pipeline` in the **Link Related Card | Select Card** box. Select the **Pipeline Bridge** card.

Now, the **Sales Attainment** card will have two cards linked to guide users.

Next, let's learn how to use the segments feature.

Comparing parts to the whole with segments

Segments in Domo are a way to create specific filters on a dataset and use them as a series in cards. The advantage of segments over other filters is that they show up as a series in the chart while also preserving the context of the whole. So, it is helpful to quickly see the whole in relation to the segment. For instance, you could see the East region's sales in relation to total sales on a line chart. You could also do this with custom beast modes, but segments are a shortcut. Segments work at the dataset level, so once a segment is created, it can be used on any card consuming the dataset.

Let's get a feel for how segments work with an example of creating regional sales segments and applying them to the **Sales Trend** card:

1. From the **Sales** dashboard, click on the **Sales Trend** card.
2. In the **Sales Trend** card detail view, open **Analyzer** by clicking the beaker icon.
3. Click **SEGMENTS** on the **Analyzer** toolbar.
4. Enter `East Region Sales` in the **Segment Name** box.
5. Click + **Add columns to filter**, select **Region**, check **East**, and click **APPLY**.
6. Click + **Add columns to filter** again, select **IsWon**, check **true**, and click **APPLY**.

7.   Click **SAVE & CLOSE** (see *Figure 7.10*):

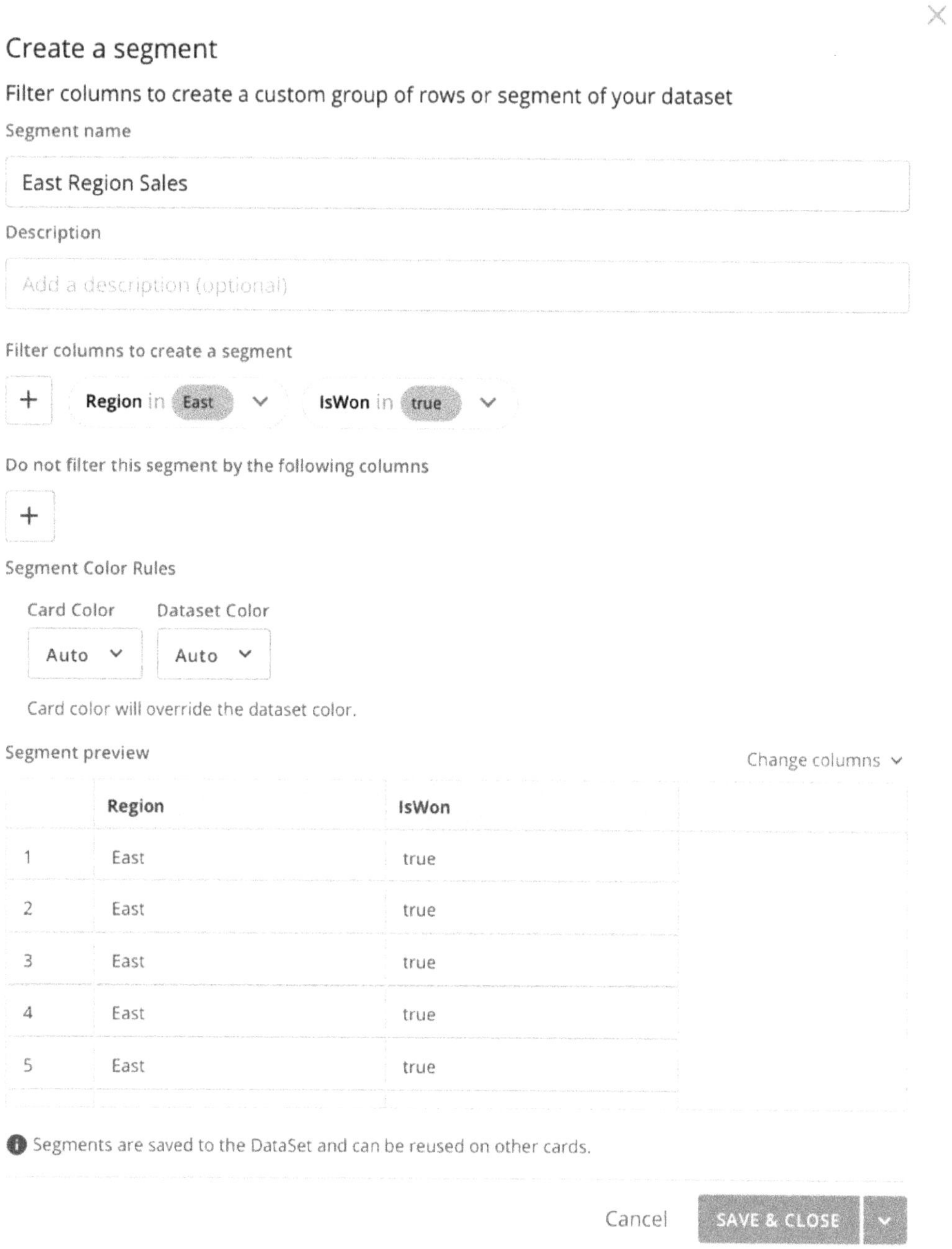

Create a segment

Filter columns to create a custom group of rows or segment of your dataset

Segment name

East Region Sales

Description

Add a description (optional)

Filter columns to create a segment

| + | **Region** in East ∨ | **IsWon** in true ∨ |

Do not filter this segment by the following columns

| + |

Segment Color Rules

Card Color   Dataset Color

Auto ∨   Auto ∨

Card color will override the dataset color.

Segment preview   Change columns ∨

	Region	IsWon
1	East	true
2	East	true
3	East	true
4	East	true
5	East	true

ⓘ Segments are saved to the DataSet and can be reused on other cards.

Cancel   SAVE & CLOSE ∨

Figure 7.10 – Creating a segment

8. Click **SEGMENTS** on the **Analyzer** toolbar.

9. Enter New Logo in the **Segment Name** box.

10. Click + **Add columns to filter**, select **Type**, check **New Logo**, and click **APPLY**.

11. Click + **Add columns to filter** again, select **IsWon**, check **true**, and click **APPLY**.

12. Click **SAVE & CLOSE**.

    Once created, segments reside in the **DATA** panel under the **BEAST MODES & SEGMENTS** section and can be edited from there.

13. Click **Save and Close** on the **Analyzer** toolbar to return to the card detail view.

14. Click on the **SEGMENTS** tab in the right-hand panel. Check **East Region Sales** and **New Logo** to see the segments on the chart along with the total sales. See the chart legend in *Figure 7.11*:

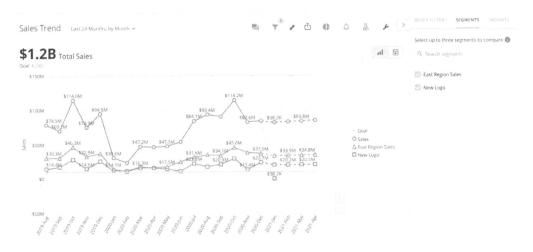

Figure 7.11 – Segments in a chart

Summary

In this chapter, we learned how to use card drill-downs, linked cards, and segments features to provide guided analytical paths. The ability to drill from an aggregated dataset to a detailed dataset was showcased. These features are straightforward to leverage and provide a robust framework for answering the next level of questions from any starting point card.

In the next chapter, we will examine interactive dashboard features.

Further reading

Here are related articles of interest:

- For discussion material on segments, visit the following link: `https://dojo.domo.com/discussion/comment/53862#Comment_53862`.

- For further reference information on adding drill paths, go here: `https://domohelp.domo.com/hc/en-us/articles/360042924094-Adding-a-Drill-Path-to-Your-Chart`.

8
Interacting with Dashboards

One of the useful things a Domo dashboard can do is allow the user to start from a default dashboard view and then interact with its content, exploring the data in real time. The **Collections** feature organizes the layout of cards on the page into logical sections, almost like organizing slides in a PowerPoint. There are also page filters so that you can filter all the cards on a page at once on any field, including beast modes in the dataset. Once set, filters can be named and saved as a group for reuse. Another advanced feature is **interactive card filtering, which**, when enabled, sets the page filters for all cards based on what is clicked on in a particular card. In other words, you can use one card to filter the contents of other cards.

In this chapter, we will cover the following topics:

- Describing dashboard page interactions
- Working with collections
- Using page filters

Technical requirements

To complete this chapter, you will need the following:

- Internet access

- Your Domo instance and login details

- The ability to download this chapter's example files from GitHub

- The dashboard and cards you created in *Chapter 6, Creating Dashboards*, and *Chapter 7, Working with Drill Pathways*

> **Important Note**
> If you don't have a Domo instance, then you can get a free trial instance at `https://www.domo.com/start/free`.

The code for this chapter can be downloaded from this book's GitHub repository at `https://github.com/PacktPublishing/Data-Democratization-with-Domo`.

Describing dashboard page interactions

A great dashboard allows the user to interact with it. This typically involves organizing how the content is presented and enabling the user to filter the data. Domo dashboard pages come with a rich set of features for users to interact with. Many of the interactive features are dynamic and do not affect the default entry state of the dashboards or cards. For example, clicking the browser page's refresh button will reset the dashboard back to its default clean-slate state.

Let's review the high-level interaction features that are shown in the following screenshot:

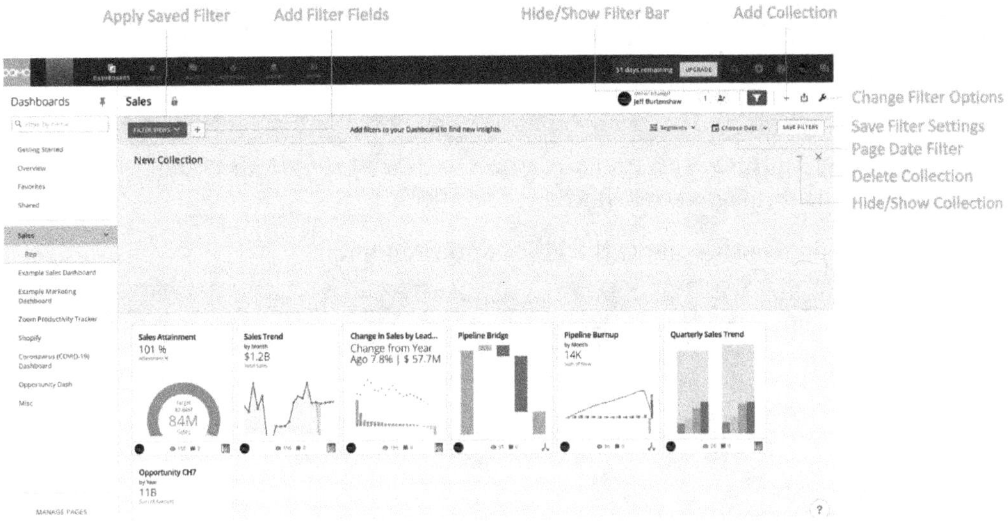

Figure 8.1 – Page interaction features

The following are some brief descriptions of the features that were highlighted in the preceding screenshot:

- **Apply Saved Filter**: This applies the page filter settings from a previously saved filter.

- **Add Filter Fields**: This adds additional fields from the datasets to be filtered on the page. The field values are combined from fields sharing the same name across all datasets that are used in the page's cards. Additionally, all the filter fields **cascade**, which means that the values in one field are pared down to the corresponding value occurrences in the other fields. This avoids non-sensical, non-occurring choices from being presented. It is also a useful tool for viewing valid data-value relationships across the filtered fields.

- **Hide/Show Filter Bar**: This toggles the page filter bar.

- **Add Collection**: This adds a collection section. Essentially, this is a page section divider.

- **Save Filter Settings**: This saves the current field filter settings as a reusable filter in **Apply Save Filter**.

- **Page Date Filter**: At the time of writing, this is a beta feature that filters all the date fields that are used in card visuals on the page.

- **Delete Collection**: This removes a collection from the page. Don't worry – all the cards in the deleted collection are not deleted with the collection; the cards in the collection are automatically placed at the bottom of the page and must be deleted or repositioned independently from there.

- **Hide Show Collection**: Here, each collection can be collapsed or expanded.

- **Change Filter Options**: This provides access to the **Filter options** popup, which can be accessed via the page's wrench icon.

The following screenshot shows the **Filter options** popup:

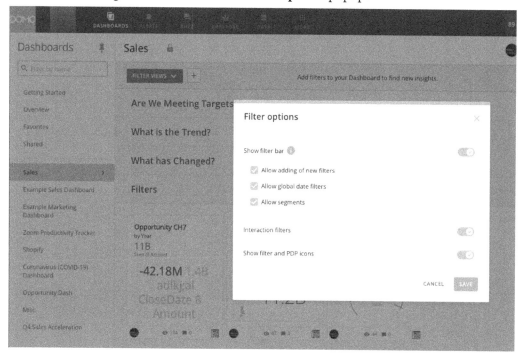

Figure 8.2 – The Filter options popup

The following are brief descriptions of the items that are available via the **Filter options** popup. These options are page-specific:

- **Show filter bar**: This allows page owners and admins to determine whether the page filter bar is visible or not.

- **Allow adding of new filters**: This determines whether users can add new filters and filtered columns for the page.

- **Allow global date filters**: This sets whether the page-level date filter feature is visible on the page.

- **Allow segments**: This controls whether segments have been enabled on the page.

- **Interaction filters**: This allows you to click on values in one card to automatically filter other cards on the page with the indicated value.

- **Show filter and PDP icons**: This controls the appearance of the filter icons (a light blue circle icon with a funnel inside; it appears to the left of the card's title when the card is being filtered) and **Personal Data Privacy** (**PDP**) icons (a gray shield with a person inside appears in the card's footer when PDP privileges are applied) on the cards.

Now, let's try out some of these features, starting with collections in the next section.

Working with collections

Page **collections** let you break up a page into sub-sections, with each collection holding cards under the collection's name on the page. A page can have none-to-many collections and the collections can be collapsed and provide flow to a dashboard, similar to slides in a presentation deck.

For our first use of collections, let's implement a monitoring dashboard pattern that can be applied to almost any function. A great monitoring dashboard answers questions such as the following:

- For this function, where are we now, and how does that compare to our goals?

- How did we get there?

- What are the trends?

- What has changed?

- Where are we going to be?

Not surprisingly, these questions also form the basis of an outline for a high-performance monitoring dashboard pattern.

Let's learn how to use **collections** to implement a monitoring dashboard pattern for the Sales Department:

1. In the **Sales** dashboard header, click the + icon and select **New Collection**.

2. Enter `Are we Meeting Targets?` in the collection's **Title** box.

3. Click (-) on the right-hand side of the collection's title banner to collapse the collection.

4. Repeat *Steps 1* to *3* for three more collections: What is the Trend?, What has Changed?, and Filters.

The dashboard, with all its collections collapsed, will look as follows:

Figure 8.3 – Collapsed monitoring dashboard collections

Note that a **collection** can be collapsed and expanded either by double-clicking anywhere on the title banner or by clicking on the (-) or + toggle on the right-hand side of the title banner. The **collection** name can be edited by double-clicking on it as well. Finally, the order of a **collection** can be changed by clicking and dragging on the title banner and dropping it into a new position on the page. Collections can be deleted by clicking on the **X** icon on the far right of the collection title banner.

> **Important Note**
> If you delete a collection with cards in it, those cards won't be deleted. Rather, those cards will be automatically moved out of the collection to the bottom of the page before the collection is removed.

Great! Now, the page has been organized neatly into our monitoring dashboard pattern.

Now, we can move the cards into the appropriate collections.

Moving cards into collections

Now that the collections have been set up, for improved readability, let's move each card into the relevant collection. Let's get started:

1. Before you can drop a card into a collection, you need to expand the collection by clicking the + icon on the collection's title banner.

2. Click, drag, and drop the **Sales Attainment** card into the **Are We Meeting Targets?** collection area.

3. Click, drag, and drop the **Quarterly Sales Trend** card into the **Are We Meeting Targets?** collection area.

4. Click, drag, and drop the **Sales Trend** card into the **What is the Trend?** collection area.

5. Click, drag, and drop the **Pipeline Bridge** and **Pipeline Burnup** cards into the **What is the Trend?** collection area.

6. Click, drag, and drop the **Change in Sales by Leadsource** card into the **What has Changed?** collection area.

Now, your **Sales** dashboard should look as follows:

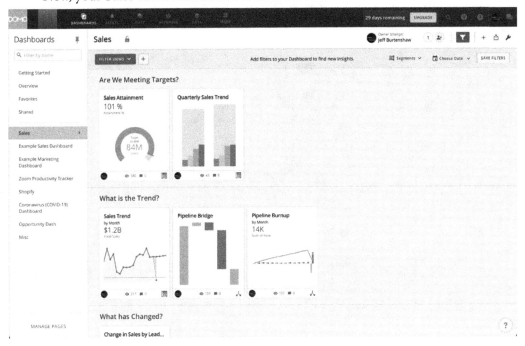

Figure 8.4 – Page collections with cards

That looks much more organized instead of just a bunch of random cards. Well done!

Next, let's practice using page filters to see different slices of the data.

Using page filters

Mastering the use of page filters is a core skill if you wish to slice and dice the contents of a dashboard to answer questions. Let's try adding filters to the **Type** and **Region** fields and save those together as a new **Saved Page Filter** choice:

1. Click the + button on the filter bar just to the right of the **FILTER VIEWS** combo box.

2. Select **Type** from the drop-down list, check **New Logo**, and click **Apply**.

    Notice that the cards that are using the **Opportunity CH7** dataset are the only cards that have been filtered on the page:

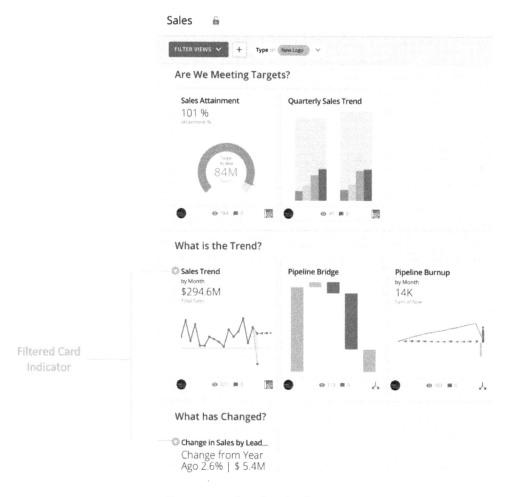

Figure 8.5 – Filtered card indicator

3. Hover over the **Filtered Card Indicator** icon on each filtered card to see a description of the filter that's been applied.

> **Important Note**
> Domo supports multiple datasets on a single dashboard. By default, filter fields will filter all the datasets with the same field names.

In this case, cards that are connected to other datasets that don't have a **Type** column in the dataset are not filtered. However, if the datasets have the same field name, then the filter options will show the unique values from one dataset with the same field name and then apply the filter across all the datasets and cards on the page. Let's look at the **Region** field to see this in action.

4. Click the + button on the filter bar again and click on the > icon in the **Region** field. The > icon indicates that multiple datasets contain the **Region** field. If you just select **Region**, it will automatically apply the filter to all the datasets with the field, but the > icon gives you the option to control which underlying datasets the filter is applied to. In this case, check **Select All** and click the **SELECT** button. Then, check the **East** region. To set the primary dataset that will be used to populate the filter values, click the database icon in the **Filter Region** popup and click **Opportunity CH7**. This option only affects the unique values in the filter choices and not which underlying datasets to filter cards; it also shows which underlying dataset will be filtered.

5. Click **Apply**.

6. Fantastic! Now, let's save these filter savings so that we can use them again with a click.

7. Click the **SAVE FILTERS** button on the filter bar, enter `Sales-East Region & New Logo` as its name, and click **SAVE**.

8. Click the down arrow in the **Filter Views** combo box, hover over **Sales-East Region & New Logo**, and click the three stacked dots button to the right. In the popup, turn on the **Share with everyone** option and click **SAVE**.

> **Important Note**
>
> **Dataset Dimensionality**: A note about card dataset design: when a dashboard uses multiple datasets, it is very common to see some cards filtered down while others are not. This can be confusing to users as information on the dashboard may not appear consistent. For that reason, wherever possible, keep the dimensionality of filters consistent across all datasets. For instance, the **Pipeline** cards on the **Sales** dashboard don't have a **Region** column in their datasets. This could create confusion if you're filtering on **Region** as the pipeline cards would show the total across all regions, even with the **Region** filter applied to other datasets.

Now that we know how to create, save, and reuse page filters, let's look at a built-in feature that we can use to filter a page by date.

Using the page date filter

There are times when it may be helpful to filter all the cards on a page by a particular date range. Let's learn how to do this:

1. Reset the page filter by selecting **None** from the **FILTER VIEWS** combo box on the filter bar.

2. Click the **Choose Date** combo box on the filter bar, choose **Between** for **Date Range**, and enter `01/01/2020` and `12/31/2020`.

    It's as simple as that! Remember that the page date filter only filters dates that are used in the card visual.

Now, let's learn how to use click selections on one card to set page filters.

Using interactive card filtering

Sometimes, it is useful to be able to have what is selected on one card drive the filtering on the other cards – for example, having a pie chart or treemap filter a sales trend. As it so happens, this is exactly the kind of interaction that **Interactive Filter Mode** enables. Essentially, in this mode, Domo detects where we click on a card and sets the appropriate page filters automatically. Let's take a look:

1. On the **Sales** page, create a pie chart card called `Regional Sales` using the **Region** and **Amount** fields on the **Opportunity CH7** dataset.

2. On the **Sales** page header, click the wrench icon and select **Filter options**. Then, turn on the **Interaction filters** option and click **SAVE**.

3. Navigate to the **Regional Sales** card and click on the **East** region pie slice so that it is highlighted. This will create a page filter on the **East** region that is also visible in the page filter bar.

To reset the interaction filter on the card, click the crossed-out filter icon on the card, as shown in the following screenshot:

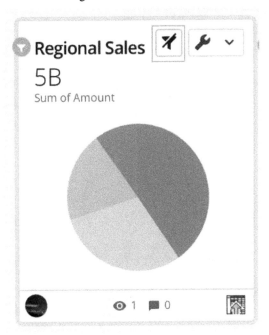

Figure 8.6 – Interaction filter reset

Next, let's try using the **Filter Card** type to filter the page.

Adding a filter card to a page

Often, as dashboard designers, we may want to make the page filter more apparent to the user rather than having it hidden in the filter bar. So, the filter card type is used to add a filtering visual card right to the dashboard. Let's take a look:

1. Create a new card by clicking + on the **Sales** page header and selecting **+ Create new Card**.

2. Click **SELECT** under **Visualization** and click **SELECT** under **Existing data.**

3. Select the **Opportunity CH7** dataset and click **CHOOSE DATASET**.

4. Name the card `Region Filter`.

5. In the **CHART TYPES** panel, select either the **Checkbox selector** or **Slicer** chart type based on your style preference.

6. Drag and drop the **Region** field into the **ITEMS** and **OPTIONAL GROUP BY** areas. Regardless of the underlying card dataset you choose, the filter card will apply the filter to all of the datasets on the page with the same field name.

7. Choose the **No Summary Number** option.

The **Analyzer** card's design will look as follows:

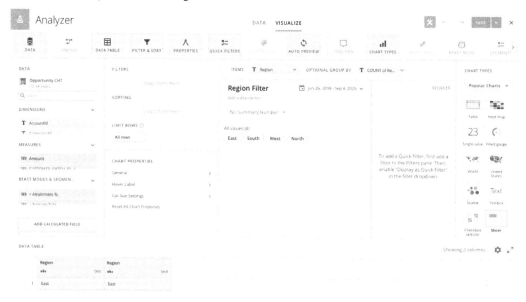

Figure 8.7 – The Analyzer card's design settings

8. Click **SAVE**.

9. Drag and drop the **Region Filter** card into the **Filters** collection.

10. Click on the **East** option in the **Region Filter** card on the **Sales** page.

Good job! Now, we can easily create a page filter by selecting the values from the card.

Let's summarize what we have learned in this chapter.

Summary

In this chapter, we saw that there are powerful features we can use to interact with cards on pages. By using collections to organize the layout, as well as interactive card clicks to create page filters, date filters, and specialized cards to create interactive filters, it is easy to get to the slice of data you desire. Filters can be saved as **Filter Views** for reuse and shared with other users. Context is always available by looking at the filter bar and by hovering over filtered card indicators. Recall that you need to provide consistent dataset dimension field naming when you're designing the dataset field when multiple datasets are being used on a page. This helps ensure that the applied filters won't show inconsistent information across the datasets that are being used on the page.

In the next chapter, we will go a level deeper and learn about card interactions.

Further reading

- For reference material on applying saved page-level filters, go to `https://domohelp.domo.com/hc/en-us/articles/360042923914-Applying-Page-Level-Filters-with-Filter-Views`.

9
Interacting with Cards

Card interactions give us fine-grained control of what we can see in a card. These controls encompass changing the chart type, zooming in and out, filtering the data, exporting the data to Excel, and even utilizing pivot table functionality. All of these interactions are temporary in nature, as a browser refresh will reset the card to the default state set by the designer. These features are located on the card details page in Domo. Additionally, there is a lot of information about when the card data was last updated, who has access to it, and who has viewed the card. This information provides context, and the interactivity is helpful for users when they need to do data discovery on the fly. Also, these features enable the user to slice the information up, experiment with different visuals, or make permanent annotations of significant events that impact the data. There is even a feature to bring attention to any issues with the card to the card's owner.

In this chapter, we will cover the following topics:

- Describing card interactions
- Using card interactions
- Using pivot tables
- Exporting data to Excel

Technical requirements

To follow along with this chapter, you will need the following:

- Internet access
- Your Domo instance and login details
- The ability to download example files from GitHub
- The dashboard and cards that were created in *Chapter 6, Creating Dashboards*, and *Chapter 7, Working with Drill Pathways*

> **Important Note**
>
> If you don't have a Domo instance, you can get a free trial instance at `https://www.domo.com/start/free`.

All of the code found in this chapter can be downloaded from GitHub at `https://github.com/PacktPublishing/Data-Democratization-with-Domo`.

Describing card interactions

In this section, we will introduce the card-level interactions that are available to users to filter and change visuals on the fly. We will work through exercises on how users can interact with cards without being a card designer. These card interactions are accessed via the card details page. Let's walk through a quick overview of the features, as shown in *Figure 9.1*:

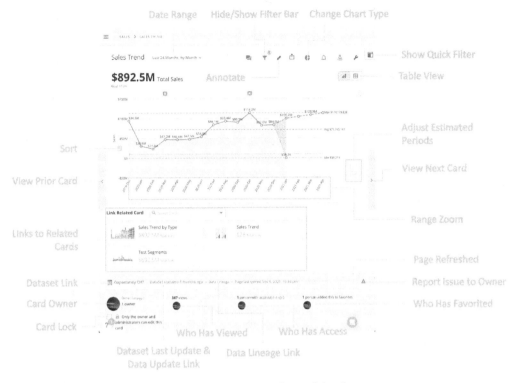

Figure 9.1 – Interactions on the card details page

Here is a brief description of each of these features:

- **Date Range**: This shows the toolbar for managing the date range and grain on the charts.

- **Hide/Show Filter Bar**: This toggles the card filter bar.

- **Annotate**: This captures a comment about a specific date on the chart and saves the comment with the card. It is helpful to capture significant events for context.

- **Change Chart Type**: This shows a palette of possible alternative chart types for the data.

- **Show Quick Filter**: This opens the **Quick Filter** panel.

- **Table View**: This shows a tabular view of the chart data. It is very useful for cut and paste operations.

- **Sort**: This allows you to cycle through the numeric and alpha sort options. It only appears when you hover over it.

- **View Prior Card**: This allows you to navigate directly to the previous card's details page.

- **View Next Card**: This allows you to navigate directly to the next card's details page.

- **Links to Related Cards**: When a card has been linked, a thumbnail of the linked card appears at the bottom of the card. Clicking on the thumbnail image will take you to the linked card.

- **Adjust Estimated Periods**: In time series charts, this feature allows you to increase or decrease the estimated values.

- **Dataset Link**: This allows you to link to the dataset details page for the card.

- **Card Owner**: This shows the owner of the card.

- **Card Lock**: This locks the card so that only the owner or the admins can change the design in Analyzer.

- **Dataset Last Update & Data Update Link**: In addition to showing the last time the dataset was updated, this serves as a link to a popup to change or refresh the dataset.

- **Data Lineage Link**: This allows you to link to the dataset's lineage page.

- **Who Has Viewed**: This is the area that shows who has viewed the card details page and a count of the number of times viewed.

- **Who Has Access**: This is the area that displays who has access to the card.

- **Who Has Favorited**: This is the area that displays who has flagged the card as a favorite and how many times it has been favorited.

- **Page Refreshed**: This item presents a timestamp of when the page was last refreshed.

- **Report Issue to Owner**: This enables a user to send a message to the card owner with a description of the issue they are seeing.

- **Range Zoom**: Click and drag inside this area to zoom in on a specific date range.

- **Legend Filter**: For any chart with a legend, clicking on specific legend items will exclude/include that series on the chart.

  That's the rundown of card interactions.

Now, let's try out some of these features in the next section.

Using card interactions

Let's get a feel for filtering, zooming, annotating, pivoting, and exporting cards by trying examples of these frequently used features.

Working with card filters

Card filters are similar to page filters, but their scope is limited to the active card. All card filters are temporary and will not affect the card design settings. So, let's add a few filter fields to a card to get a feel for how to use this powerful feature. Perform the following steps:

1. From the **Sales** dashboard, open the **Sales Trend** card in the **What is the Trend** collection.

2. On the **Sales Trend** details page, click on the funnel icon button from the page header toolbar. Notice that the funnel icon has an indicator number on the icon. This indicates the number of columns that are being filtered on the card. As you can see in the filter bar, by design, this card is filtered on **Stage = 5 Closed Won**.

3. Next, let's add a filter to just see the new logo wins by clicking on the + button in the filter bar, selecting the **Type** field, and checking **New Logo** in the **Filter Type** dialog. Then, click on **Apply**.

4. Notice that the funnel's filtered field indicator increases to **2**, and the trend line now shows **New Logo** wins only.

5. But what if we wanted to also see all of the wins except new logos? Well, we could uncheck **New Logo** and check all the other options, or we can just click on the **Type** filter in the filter bar, change the **In** value to **Not In** in the selector, and click on **Apply**.

> **Important Note**
>
> Built-in filters only use AND logic across fields; I can hear you asking – "what if I need to do OR logic instead of AND when using multiple columns?" Well, that is where **Beast Mode** comes in. Simply create a beast mode formula by applying the more complex logic and use that beast mode field in the filter; for example, `Case when field1=x or field2=y then 'My Special Choice' else 'Everything Else' end`.

Very good; now we understand how to filter cards.

In the next section, we will look at how to use the interactive **Quick Filters**.

Using quick filters

Quick Filters are established by the card designer and are for convenience and anticipated frequent use. They are given a permanent placement in the card details right-hand side tool drawer. Perform the following steps:

1. Continuing with the **Sales Trend** card details page, make sure the card details page drawer is expanded to show the **QUICK FILTERS** tab by clicking on the **Expand Quick Filters** icon in the page header toolbar.

2. Then, in the **QUICK FILTERS** tab, uncheck **Closed Won** and check **Negotiate** to see the trend in opportunities that are currently being negotiated. Take a look at *Figure 9.2*:

Figure 9.2 – QUICK FILTERS

Quick filters are a handy way to enable users to filter the data on a card with just a click.

In the next section, we will learn how to use the **Date Range Selector**.

Using the date range selector

As users, we frequently want to change the date range or data granularity that is presented. Conveniently, the card detail page provides a **Date Range Selector** to do this. Perform the following steps:

1. Staying with the **Sales Trend** card, click on the **Date Range** selector on the right-hand side of the card title. It will say something similar to **Last 24 Months, by Month**.

2. This will open the **Date Range Shelf** section. Click on the first combo box, select **Between**, and enter **03/01/2020** and **04/30/2020** as the date ranges, as shown in *Figure 9.3*.

3. Since we are looking at a smaller range of dates, let's also change the grain to daily by selecting **Day** in the **Graph by** combo box.

Figure 9.3 – Using the Date Range Shelf section

Perfect; you can see how simple it is to change the date range and granularity!

4. Next, let's look at the zoom capability to hone in on a smaller range on a chart.

Using range zoom

Sometimes, we just need a way to quickly zoom in on a particular date range on a chart. In Domo, you can click and drag to highlight the axis values, and when you release the mouse button, the chart zooms into the highlighted area. Let's walk through how to do this:

1. Refresh the **Sales Trend** card by clicking on the browser's refresh icon. This will reset all the filters and bring the card back to its default state.

2. In the **Sales Trend** card just above the chart date axis in the chart area, click and drag from **2020-Oct** to **2020-Dec** and release the mouse button, as shown in *Figure 9.4*. This date range will be highlighted on the chart, and upon releasing the mouse, the chart will zoom in to just the date range highlighted.

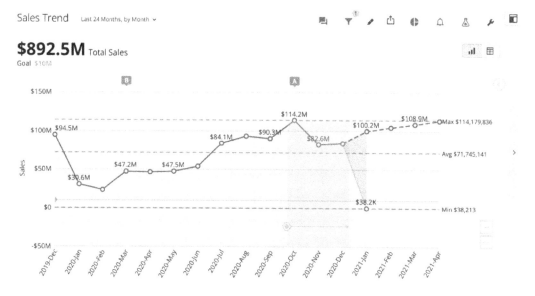

Figure 9.4 – Click, drag, and release to zoom in

3. To undo the zoom, click on the back arrow icon in the upper-right corner of the chart area, as shown in *Figure 9.5*:

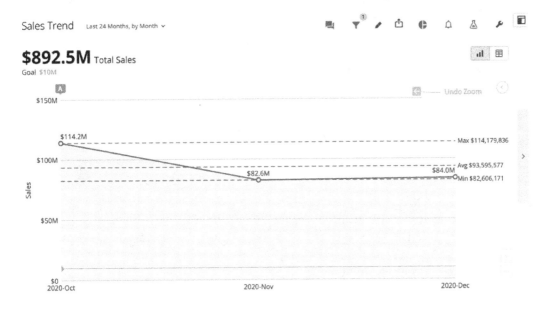

Figure 9.5 – Undo Zoom

It's pretty slick being able to zoom in and out on the card!

Next, let's see how to make a permanent comment on a chart.

Creating chart annotations

Have you ever looked at a chart and wondered when that new product was launched, or when that big event occurred? At the time, you think you will never forget, but if you are anything like me, you have learned that having the ability to put a quick note on the chart will ensure you never forget any important information.

Let's see how to do this in Domo. Perform the following steps:

1. Let's start by refreshing the page using the browser's page refresh function. This will reset all the filters and bring the card back to its default state.

2. Now, let's make an annotation for when the COVID pandemic started in March 2020 by clicking on the **Annotation** pencil icon in the page header toolbar.

3. Click on the date range drop-down menu under the **Sales Trend** chart title and select **All Time** from the **Date Range** drop-down list.

4. Click on **Create Chart Annotation** and move the mouse pointer so that the dropline is over **March 2020** and click again. Enter `Covid Pandemic Hit` in the **Describe your annotation** box and click on **SAVE**.

5. Make another annotation, called `Fall Sales Campaign`, in **October 2020**.

6. Your chart will now have the annotation indicators on the chart along with the annotation drawer on the right-hand side of the chart. See *Figure 9.6*:

Figure 9.6 – Card annotation

The annotation tool is very handy for marking relevant events on a chart permanently—if you use this feature, never again will you have to go back and look up that big event in your email or calendar!

Next, let's investigate how to use chart legends to filter.

Using the chart legend to filter

As the heading implies, we can use the chart legend to choose what items to include or exclude from the chart. Perform the following steps:

1. Open the **Quarterly Sales Trend** card from the **Sales** dashboard.

2. Find the legend on the right-hand side of the chart area, and click on **Q1** and **Q3** to deselect those periods from the chart. This will leave you with a comparison of the **Q2** and **Q4** sales. See *Figure 9.7*:

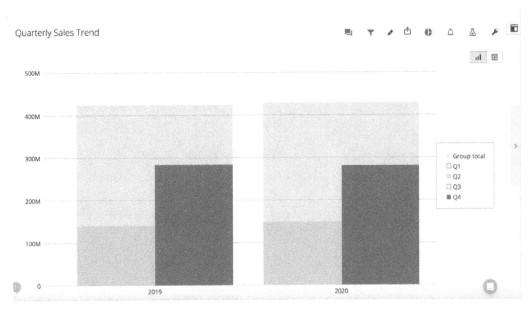

Figure 9.7 – Legend filtering

Wow, how convenient! The legend also serves as a filter.

Okay! Next, let's dive into using **Pivot Tables** in Domo.

Using pivot tables

Domo's pivot table implementation lets you pivot on any Domo dataset. In Domo, **Pivot Tables** are implemented through something called a **Sumo Card**. The sumo card is a special card type that has been built to pivot table data. So, we need to create a new card using the **Sumo Table** card type. Let's go through an example:

1. From the **Sales** dashboard, click on the + icon in the page header and then click on **+ Create new card**.

2. Click on the **VIEW MORE** button, and then click on the **SELECT** button under **Sumo Table**. Then, choose the dataset by selecting **Opportunity CH7** in the **DataSet:** box, and click on **CHOOSE DATASET**.

3. Let's pivot on the sales amount by type and date. Click on the pivot table icon just underneath the pencil icon on the left-hand side. Click, drag, and drop the **Type** field under the **CATEGORIES** section to the rows area that is directly above **CATEGORIES**.

4. Under the **VALUES** list, click, drag, and drop the **Amount** field into the **DRAG VALUES HERE** box.

5. Now, let's add formatting and subtotals. Click on **AMOUNT TOTAL** on the column heading and select **Formatting**. Choose **Currency** in the **Display as** box, and select **0** in the **Decimals** box. Check the **Abbreviate (1.23k)** item, and then click on **APPLY**. Now your screen should look like *Figure 9.8*:

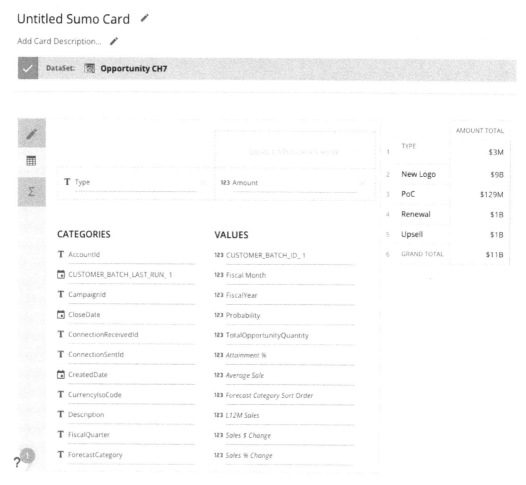

Figure 9.8 – Sumo pivoting on type and amount

6. Click, drag, and drop the **CloseDate** field into the **DRAG CATEGORIES HERE** area. In the resulting table header, click on **CLOSEDATE**, select **Date Grain**, and then click on **Year**.

7. To add another level to rows, click, drag, and drop the **Lead Source Standardized** beast mode field below **Type** in the row area. Here, **Lead Source Standardized** will be toward the bottom of the **CATEGORIES** list in italics, indicating that it is a beast mode field.

8. To add subtotals, click on the show subtotals icon on the left-hand side, which is directly underneath the Σ icon. The pivot table will now have subtotals, as shown in *Figure 9.9*:

Figure 9.9 – The sumo card in Analyzer

9. Enter Opportunity Sumo as the card title and then click on **SAVE**.

10. Scroll down the dashboard to see the **Opportunity Sumo** card, and click the card to go into the details page view.

11. Click on the pencil icon to change the layout of the pivot.

Isn't it great to know how to create pivot tables in Domo! Well done!

In the next section, we will go over how to export the data into **Excel**.

Exporting data to Excel

One of the most sought-after features in any BI tool is the ability to export the data to an Excel worksheet. Although with Domo, you will find less of a need to do this, at times, it is still helpful to extract the data into **Excel** for further analysis.

Let's go through the steps to do an export from a card into **Excel**:

1. On the **Quarterly Sales Trend** card, click on the rectangle with the up arrow icon and select the **Send / Export** option to bring up the choices, as shown in *Figure 9.10*:

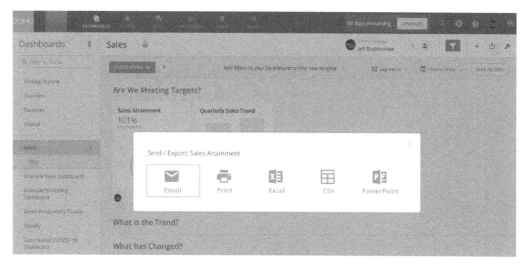

Figure 9.10 – The Send / Export options

2. Click on **Excel**. This will immediately download an Excel file to the **Downloads** folder of your machine's hard drive. The file will be named after the title of the card, which, in this case, is Quarterly Sales Trend.xlsx.

And that's it—just a couple of clicks to get the data into Excel.

Important Note

Being able to readily export card data from Domo reflects a philosophy that Domo is the steward of customer data assets. Additionally, the customer has unfettered access and simple ways in which to leverage their data exported from Domo.

Well, that's it for this section! Let's move on to a quick recap of what we have learned in this chapter.

Summary

In this chapter, we reviewed many ways in which to interact with cards in Domo, including filtering, zooming, selecting date ranges, entering notes, pivoting data, and getting data into Excel. All of these interactions enable the user to interactively work with their data of interest in the cards without needing to change the underlying card design or structure. And if the Domo interactions aren't enough, the option to export to Excel is only a few clicks away.

In the next chapter, we will review a standard approach for telling business stories.

Further reading

Here is a list of related articles of interest:

- For more detailed information about all the table chart types, take a look at `https://domohelp.domo.com/hc/en-us/articles/360043429573-Table-Chart`.

- For reference material on sumo cards, navigate to `https://domohelp.domo.com/hc/en-us/articles/360042925674-Adding-a-Sumo-Card-2`.

Section 3: Communicating to Win

Everyone has a need to tell stories and they don't have to be boring.

The person who communicates the story in a presentation usually gets the lion share of the credit or blame for the ideas. Having great card content in a dashboard is just a starting point for a story. Telling a story with the right data communicated in an effective way is where we win the hearts and minds of our target audiences. This section will show you a framework for identifying relevant stories and for presenting those stories effectively. Then you will learn the many options for distributing your content with Domo. This includes data driven alerts and context provided instant messaging.

If you know your stuff and can talk about it, but struggle to put it down into an effective story, Domo Stories and best practices shared here will help get your ideas implemented with you as the driving force. And hence, you may reap the rewards of your efforts and success.

This section comprises the following chapters:

- *Chapter 10, Telling Relevant Stories*
- *Chapter 11, Distributing Stories*
- *Chapter 12, Alerting*
- *Chapter 13, Buzzing*

10
Telling Relevant Stories

Being relevant with data means bringing information to light that is important to current issues at hand. When deciding what story to tell, relevance to the audience should always be paramount. Telling relevant stories also requires context or a frame of reference. In a business context, a business model provides a framework for storytelling. Leadership audiences expect concise arguments both for and against the proposal, while peer and subordinate audiences may need more details to help them understand the full picture and to facilitate execution. We will learn about deciding what the relevant parts of the business are and organizing the story presentation using a pattern. A pattern for relevant storytelling seeks to explain the journey of where the story began, what happened, where things are now, and where things could be in the future. A story is not meant to merely present facts and figures but should also advocate for a position. Stories that don't decide what to advocate for or against may be interesting but are irrelevant and not actionable.

We will learn how to use a business framework to focus on a relevant story, then create a specific statement for the story to cover, and then apply a storytelling pattern in the Domo Stories tool.

In this chapter, we will specifically cover the following topics:

- Using a business framework to be relevant
- Deciding on a story to tell
- Using a monitoring dashboard to refine story statements
- Learning about storytelling patterns
- Applying a story pattern using Domo Stories

Technical requirements

To follow along with this chapter, here's what you'll need:

- Internet access
- Your Domo instance and login
- Ability to download example files from GitHub
- Dashboard and cards created in *Chapter 6*, *Creating Dashboards*, and *Chapter 7*, *Working with Drill Pathways*

> **Important Note**
> If you don't have a Domo instance, get a free trial instance here:
>
> `https://www.domo.com/start/free`

- All of the code found in this chapter can be downloaded from GitHub at this link: `https://github.com/PacktPublishing/Data-Democratization-with-Domo`

Using a business model framework to be relevant

Deciding what to focus on in business performance can be a challenge. Business model frameworks are used to simplify real-world complexities, facilitating our ability to think about and conceive ideas without an overburdening amount of detail. Many different business models could be used. For the purposes of data storytelling, we will use a **business model** with five **framework areas**—namely, **Resources**, **Operations**, **Value Creation**, **Customer Experience**, and **Risk**—that will guide us in our pursuit of relevant story content.

The following diagram illustrates **business model framework areas** for identifying relevant story content:

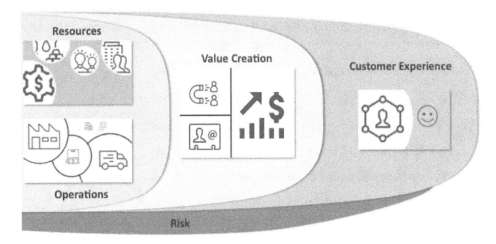

Figure 10.1 – Business model framework areas for relevant story content

We'll now look at detailed descriptions of the five framework areas seen in *Figure 10.1.* Each area description includes several questions that are typically relevant to a business when answered.

Resources

Resources are inputs to business operations for generating products and services of value to the customer. A few examples of resources are cash, labor, raw materials, equipment, facilities, suppliers, people, and intellectual property. Resources are predominantly costs consumed by the business, and hence are measured in terms of efficiency and scarcity.

Here are some typical questions on resources:

- What are the resource costs and how have they changed?
- Are there shortages or surpluses? Where? How long is this likely to be for?
- What is the resource utilization and how has it changed?
- In terms of people as resources, how satisfied are they?

Knowing how resources are being used to contribute value to an organization is an important aspect of success.

Next, let's discuss the Operations framework area.

Operations

Operations are business processes executed to create and deliver products and services of value to the customer. Examples of operations are research and development, design, manufacturing, distribution, professional services, finance, **human resources** (**HR**), and legal. Operations are measured through cost, quality, quantity, and efficiency.

Here are some common questions relevant to operations:

- Are the operations profitable?

- What are the operational costs and how have they changed?

- What is the operational quality and how has it changed?

- What is the volume and how has it changed?

- Are there additional efficiencies to be gained?

- Are resource constraints impacting operations?

- Are operations conducted in an ethical manner?

Measuring operational excellence is critical for improving business results.

Now, let's understand the Value Creation framework area.

Value Creation

Value creation encompasses the marketing and selling activities a business engages in. Ultimately, these activities are related to influencing customers' willingness to pay for products and services.

Some relevant questions are listed here:

- Is our customer acquisition pipeline healthy?

- Are we growing or losing customers? By segment?

- How has our conversion rate from marketing qualified lead to sales qualified lead changed?

- Which campaigns have the best performance?

- Which sales representatives and channels are performing the best?

- How has our average order size changed?

- Are total sales growing? By product and service line?

- What is the **return on investment** (**ROI**)?

Measuring the performance of sales and marketing activity is a core discipline to enhance growth.

Next, let's talk about the Customer Experience framework area.

Customer Experience

Customer experience focuses on understanding how the customer experiences the product and services delivered.

Relevant questions are listed here:

- How have our **net promoter scores** (**NPS**) changed? By product/service?
- How are customer satisfaction scores trending? By channel?
- What are the trends in complaint and praise volumes?
- Which types of things do customers complain about or praise the most?
- How is our retention rate trending?
- How much, how frequently, and how recently have our customers purchased?
- What are the trends in customer referrals?
- Which are the leaders and laggards in terms of our products and services?

Quantifying customer experiences across customer touchpoints is becoming table stakes to be competitive in many industries.

Next, let's discuss how the Risk framework area plays into potential stories.

Risk

Risk focuses on understanding what can go wrong in all framework areas of the business model.

Sample relevant questions are listed here:

- Which resource risks are there?
- What are the operational risks?
- What are the value creation risks?
- What are the customer experience risks?
- Can risks be quantified monetarily?

Anticipating and quantifying risks are important activities to avoid being blindsided by unexpected, high-impact risk events.

Now that we have a broad business model and framework areas covering the current relevant issues in a business, let's turn our attention to how to focus on a specific story to tell.

Deciding on a story to tell

In the previous section, we learned how to apply a business model to determine the relevant framework area(s) to cover. In this section, we will dive deeper into narrowing our focus down from a framework area of interest by creating a story statement. A **story statement** is a single sentence that encapsulates the story to be told and the context in which it is to be told. A story statement is created by making choices under the following **story categories**: **Narrator Voice**, **Presentation Method**, **Audience**, **Framework Area**, **Action**, and **Results**. The following screenshot shows a **story statement guide** with category options that will help you create a basic story statement to guide your story creation:

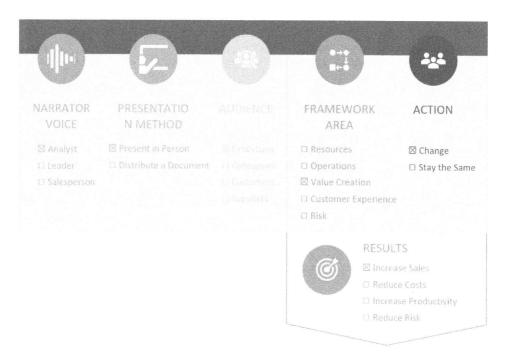

Figure 10.2 – Story statement guide

Typically, you are the story narrator, and the challenge is to craft a storyline so that it resonates with the audience. Another important consideration is the method of presentation—in-person live or a remote presentation distribution. Audiences can be grouped into management, colleagues, suppliers, or customers. Often, you are either presenting an idea to executives, communicating an initiative to your colleagues, or persuading suppliers or customers. When we tell a story, we should also communicate recommended actions and expected results.

Here is a story statement template that you could use:

```
As a [Narrator Voice] I want to [Presentation Method]
to [Audience] presenting evidence of why there was a
[Framework Topic] issue and what needs to [Action] to
[Result].
```

Let's work through an example of using the story statement guide shown previously, as follows:

1. Using the story statement guide in *Figure 10.2*, select items to substitute into each category.

2. Then, insert the choices made in the story statement guide into a **story statement template**. For example, create your story statement by applying the checked boxes in *Figure 10.2* to the story statement template, as seen here:

```
As an Analyst I want to Present in Person to Executives
presenting evidence of why there was a Value Creation
issue and what needs to Change to Increase Sales.
```

Here's another story statement example:

```
As a Leader I want to Distribute a Document to Colleagues
presenting evidence of why there was an Operations issue
and what needs to Change to Increase Productivity.
```

And here's another example:

```
As a Salesperson I want to Present in Person to Customers
presenting evidence of why there was an Operations issue
and what needs to Change to Reduce Costs.
```

Granted—these statements are broad, but they are still a great way to focus our efforts on where to begin the relevant story. Who would have thought that deciding which story to tell could be so easy? Keep your story statement close as a guidepost to your story creation activity.

In the next section, let's go through an exercise using a monitoring dashboard to refine story statements.

Using a monitoring dashboard to refine story statements

A good **monitoring dashboard** will enable us to see anomalies and issues in a few moments that when addressed will drive improvements in performance. These are perfect trailheads to build a **story statement** around. Let's enhance our **Sales** dashboard and make some permanent card additions as a by-product of our analysis to support our story statement.

We can see on the **Sales Attainment** card shown in the following screenshot that we were at **96%** target attainment for the year 2020:

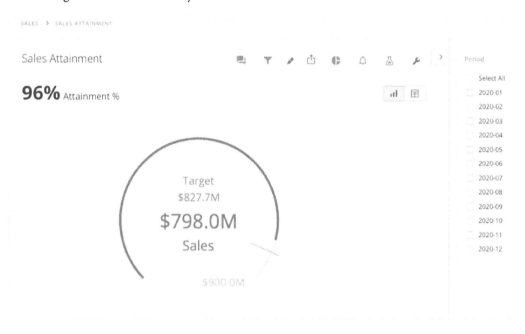

Figure 10.3 – Sales Attainment card

> **Note**
> Make sure the card is filtered to all of 2020; we may need to check all the months in 2020.

Based on this observation of the 4% miss in sales versus quota, we will refine the story statement by changing the Value Creation category to be more specific, as follows:

```
As an Analyst I want to Present in Person to Executives,
evidence why there was a 4% miss in the 2020 Sales Target issue
and what needs to change to increase sales.
```

Creating a quarterly target variance trend card

Let's enhance our monitoring dashboard beyond a gauge card, which shows a snapshot of our performance, and make a card to show the trend of quota to actual performance by quarter, as follows:

1. A quarterly trend will show the timing of when we fell short and by how much. To do that, we need to create a card that makes it simple to see variances to targets for each quarter.

2. Click on the **+** icon on the **Sales** dashboard menu bar and pick **+ Create New Card**. Click **SELECT** under **Visualization** and then click **SELECT** under **Existing data**.

3. Click on the **Quota** dataset and then click the **CHOOSE DATASET** button.

4. Now, in the **Analyzer** view, rename the card `Target Variance Trend`.

5. In the **CHART TYPES** section, pick **Vertical Bar**, and then click on the **Bullet** chart-type thumbnail.

6. Create a beast mode formula to rename the **Fiscal Quarter** field so that the new name will match the **QTR** field in the `Opportunity CH7` dataset by clicking on **ADD CALCULATED FIELD** and entering `Qtr` as the **Calculated Field Name** value. Then, enter the following as the **FORMULA** value:

```
-- Rename to use the same name here as in the Opportunity
dataset
-- for page level filtering matching
`Fiscal Quarter`
```

7. Check **Share Calculation on DataSet** and then click **SAVE & CLOSE**.

8. Drag and drop **Qtr** from **BEAST MODES** into the **NAMES** area and to the **SORTING** area.

9. Drag and drop **Actual** from **MEASURES** into the **ACTUAL VALUE** area.

10. Drag and drop **Budget** from **MEASURES** into the **TARGET VALUE** area.

11. Add a beast mode called `Variance $` with the following **FORMULA** value:

```
sum(`Actual`)-sum(`Budget`)
```

12. Check the **Share Calculation on Dataset** option and click **SAVE & CLOSE**.

13. Drag and drop `Variance $` from **BEAST MODES** into the **RANGE1 VALUE** area.

14. Add a beast mode called `Variance %` with the following **FORMULA** value:

```
(sum(`Actual`)-sum(`Budget`)) / sum(`Budget`)
```

15. Check the **Share Calculation on Dataset** option and click **SAVE & CLOSE**.

16. Click **Summary Number** and choose **Use All Values**, select `Variance %` as the **Column** value, enter `Variance` as the **Label** value, and format as a **Percentage** value.

17. Add a beast mode called `Year` with the following **FORMULA** value:

```
left(`Period`,4)
```

18. Check the **Share Calculation on Dataset** option and click **SAVE & CLOSE**.

19. Drag and drop the **Year** filter from the **BEAST MODES** area into the **FILTERS** area and click **Selection** in the filter-type combobox, and then turn on the **Display as Quick Filter** slider. Then, click **Apply**, as seen in the following screenshot:

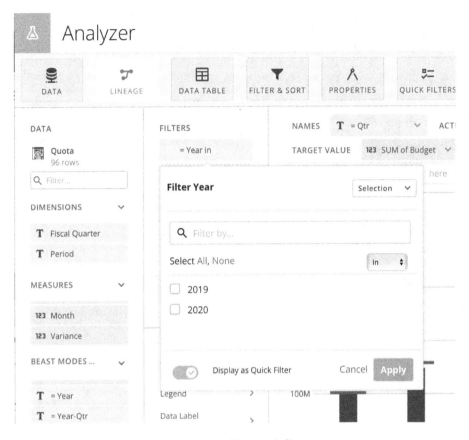

Figure 10.4 – Year quick filter

20. Next, in the **QUICK FILTERS** pane, uncheck **2019** so that only **2020** is checked.

21. Let's set colors for the chart. In **CHART PROPERTIES**, click **Colors** and select **Actual** as the column and choose blue as the **Fill** color.

22. Click **+ NEW** and change the **Column** value to **Budget** on the new row and choose dark gray in **Fill**.

23. Click **+ NEW** to add another row and change the **Column** value to `Variance $`. Then, click on the **Condition** box on the row and choose is **greater than or equal to** 0, and click in the **Fill** box and select green.

24. Click **+ NEW** to add another row and keep the **Column** value as `Variance $`. Then, click on the **Condition** box on the row and choose **is less than** 0, and click in the **Fill** box and select red.

25. Repeat *Step 21* to *Step 24* to add custom colors for Variance % too.

26. The **Color Rules** values will look like this:

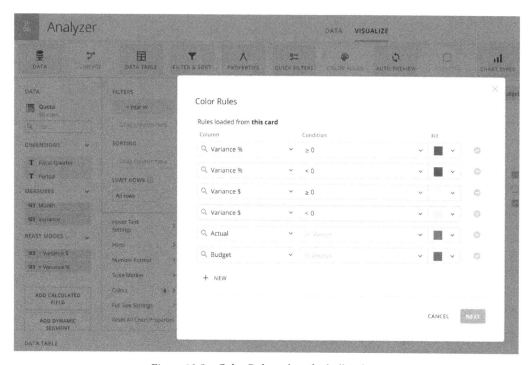

Figure 10.5 – Color Rules values for bullet chart

27. Click **NEXT** and then click **APPLY TO ALL CARDS**. The result is seen in the following screenshot:

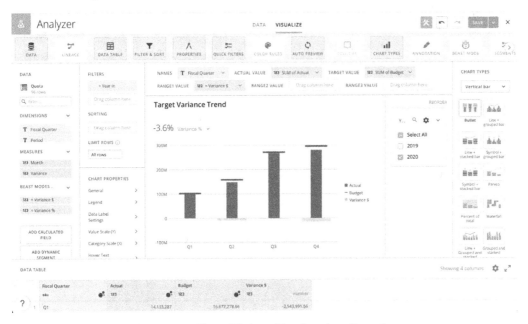

Figure 10.6 – Target Variance Trend card configuration

The blue vertical bars are the actual sales, and the horizontal gray marker lines are the target values. The position of the marker line indicates the over-/underrelation of sales to target. When the marker line is on the bar, then actuals are over target, and when the marker line is above the bar, then actuals are under target. Sales variance to target is represented by the wider red and green shaded bars. This chart gives a quick visual of target achievement trends.

28. Click **Save and Close**.

29. Create a collection called **Are We Meeting Targets?** by clicking + on the **Sales** dashboard header and selecting **New Collection**.

30. Drag and drop the **Target Variance Trend** card into the **Are We Meeting Targets?** collection.

31. Great! Now that we have created the card, click into the **Target Variance Trend** card and hover over the `Variance $` legend item to see that in **2020-Q4**, we had the largest target miss, as seen in the following screenshot:

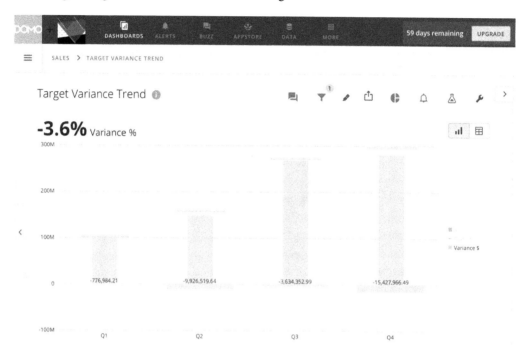

Figure 10.7 – Quarterly sales variances

Now, we can tune the story statement further, as follows:

> As an **Analyst** I want to **Present in Person** to **Executives** evidence why there was a **$15.4 million miss in the 2020-Q4 Sales Target** issue and what needs to **Change** to **Increase Sales**.

Important Note

For an extra challenge, create a drill down to a bullet chart by month. See *Chapter 7, Working with Drill Pathways,* as needed.

Creating a regional target variance card

Seeing that the most relevant issue in terms of overall impact is in **2020-Q4**, we are curious to see how each region performed against targets as well, specifically in **Quarter 4 (Q4)**. We could use page or card filters on **Region** and **QTR**, but that takes a few filtering clicks where we tend to lose comparison context. So, let's create a card that shows the trend for every region on one card since this will be a commonly used diagnostic. Proceed as follows:

1. Click on the + icon on the **Sales** dashboard menu bar and pick + **Create New Card**. Click **SELECT** under **Visualization** and then click **SELECT** under **Existing data**.

2. Click on the **Quota** dataset, and then click the **CHOOSE DATASET** button.

3. Now, in **Analyzer**, rename the card `Regional Target Variance`.

4. In the **CHART TYPES** section, pick **Horizontal Bar** and then click on the **Symbol + grouped bar** chart-type thumbnail.

5. Drag and drop the region **DIMENSION** value into the **X AXIS** area and then to the **SORTING** area.

6. Next, we need to create beast modes for variances by amount and percent. Click on **ADD CALCULATED FIELD** and enter `Variance $` as the **Calculated Field Name** value. Then, enter the following as the **FORMULA** value:

   ```
   sum(`Actual`)-sum(`Budget`)
   ```

7. Check **Share Calculation on DataSet** and then click **SAVE & CLOSE**.

8. Drag and drop `Variance $` from **BEAST MODES** into the **SERIES** area and `Variance %` to the **Y AXIS** area.

9. Click on the `Variance %` field in the **Y AXIS** area and format it as a **Percentage** value.

10. Drag and drop **Year** from the **BEAST MODES** area into the **FILTERS** area and click **Selection** in the filter-type combobox, and turn on the **Display as Quick Filter** slider. Then, click **Apply**, as shown in *Figure 10.4*.

11. Next, in the **QUICK FILTERS** pane, uncheck **2019** so that only **2020** is checked.

12. Click on the **General** section of **CHART PROPERTIES** and check the **Sync Zero Lines** property.

13. The **Analyzer** configuration for the card is seen in the following screenshot:

Figure 10.8 – Regional Target Variance card configuration

The actual and percentage variances from targets are easy to see. The red vertical bars are the sales variances from targets, and the triangle symbols are the variance percentages. This is a dual-axis chart with a percentage scale on the left and a dollar scale on the right.

14. Click **Save and Close**.

15. Drag and drop the **Regional Target Variance** card into the **Are We Meeting Targets?** collection, as seen in the following screenshot:

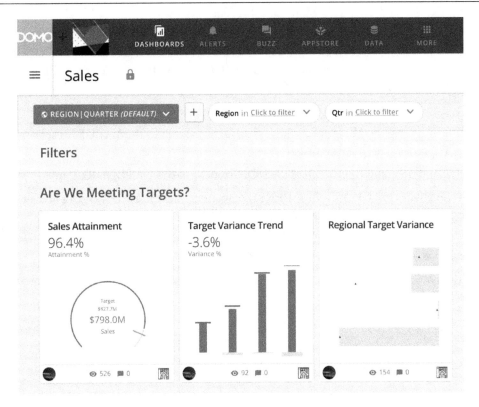

Figure 10.9 – Sales dashboard Are We Meeting Targets? collection

Great! Now that we have created a **Regional Target Variance** card, click on it, and observe that the **West** region has the highest variances on both an actual and percentage basis—an interesting observation, but one that doesn't give us a quarterly view on the timing of when the variances happened. Let's keep this card as a permanent part of our region monitoring; however, recall that our story statement is about **2020-Q4** sales, so we need another card with both region and timing to be relevant.

Creating a regional quarterly target variance trend card

To get a better insight as to what is going on both regionally and quarterly simultaneously, we need to combine these dimensions into a multidimensional visualization using a **heatmap**. Here's how to do this:

1. Click into the **Regional Target Variance** card and then click on the wrench menu icon, then click **Save as** and change the **Title** value to **Regional Target Variance Heatmap**.

2. Open **Analyzer** by clicking the beaker icon on the **Regional Target Variance Heatmap** card details page.

3. In the **CHART TYPES** section, pick **Other Charts** and then click on the **Heat map** chart-type thumbnail.

4. Drag and drop the **Qtr DIMENSION** value into the **CATEGORY 1** area.

5. Drag and drop the **Variance % BEAST MODE** value into the **VALUE** area, and set the **Format** value as **Percentage**.

6. Drag and drop the **Region DIMENSION** value into the **CATEGORY 2** area.

7. In the **Diverging CHART PROPERTIES** section, click **Show Diverging** and then choose **Zero** in the **Midpoint Value Type** property.

8. In the **Data Label Settings CHART PROPERTIES** section, select `%_Value` in the **Text** box and check **Auto Abbreviate Values**.

9. Add a quick filter on **Year** and check the Year **2020** only, as shown in the following screenshot:

Figure 10.10 – Regional Target Variance heatmap

10. Click **Save and Close**.

We see in *Figure 10.10* that the **West** region has been under the target all year; it's interesting to note that **Q4** loss was light compared to the losses in the **West** region in **Q2** and **Q3**. So, we can conclude that **Q4** performance overall was not just a **West** region issue. In fact, we see that the **South** region was the furthest percentage from the target in **Q4** and that perhaps sales were frontloaded in the **South** region. Let's see if this might be a forecasting issue by looking at actual region sales variances to prior-year sales.

Creating a regional actual sales variance card

A regional actual sales variance card will give us perspective on how actual sales are changing on a year-over-year basis. This will be very useful in providing color to the forecast variances. To create a regional actual sales variance card, proceed as follows:

1. Copy the **Regional Target Variance** card by clicking on the **Save as** option in the **Card** wrench menu. Name the card `Regional Actual Sales Variance`, check the **Take me to the new card when I'm done** option, then click **SAVE**.

2. In the **Regional Actual Sales Variance** card, open **Analyzer**, click **Quota** in the **DATA** area, and select the **Opportunity CH7** dataset; then, click **CHOOSE DATASET**.

3. Drag the **Stage** filter from **BEAST MODES** and drop it into the **FILTERS** area, check **5 Closed Won**, and click **Apply** to filter only actual sales.

4. Fix the card by dragging `PYV%` from **BEAST MODES** and dropping it into the **Y AXIS** area. Set the **Format** value as **Percentage**.

5. Then, drag `PYV$` from **BEAST MODES** and drop it into the **SERIES** area, and then set **Format** as **Currency Decimals 0**.

6. Reset the **Year** quick filter so that **2019** and **2020** are the only years checked.

7. The configuration can be seen in the following screenshot:

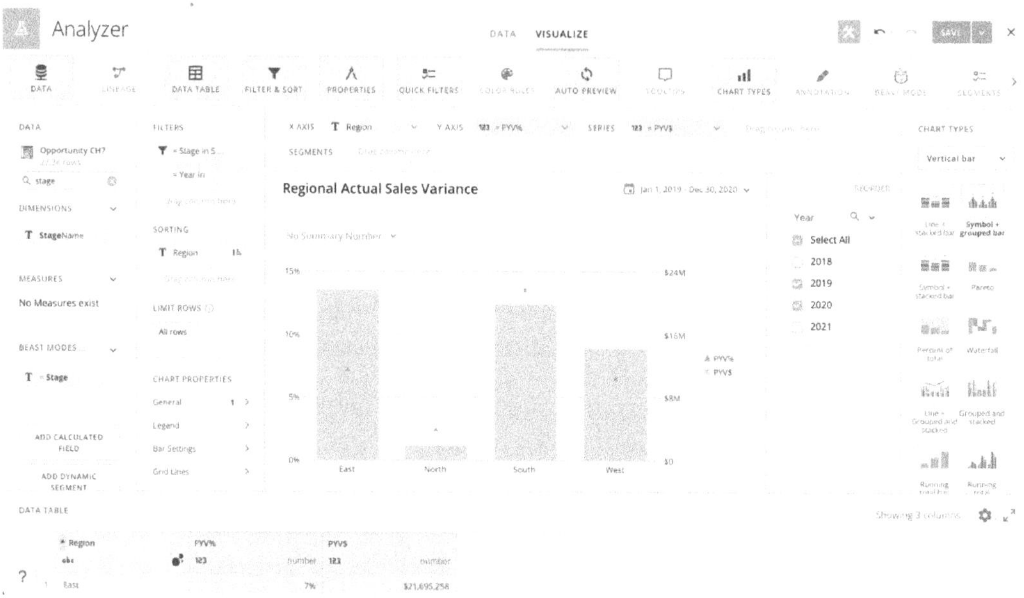

Figure 10.11 – Regional Sales Variance card configuration

8. Click **SAVE and CLOSE**.

9. Drag and drop this card into the **What has Changed?** collection.

Wow! All the regions increased sales over last year, so that puts us in a strong position and points to shortages being around missed expectations to plan. Let's look at the quarterly trend on actual sales next.

Creating an actual sales prior year variance card

The timing of the actual sales variance by quarter compared to the prior year will give us insights into the timing of sales acceleration/deceleration. Let's create a card that shows the quarterly actual sales and actual sales variance timing. Proceed as follows:

1. Copy the **Regional Actual Sales Variance** card by clicking on the **Save as** option in the **Card** wrench menu. Name the card `Actual Sales Prior Year Variance`, check the **Take me to the new card when I'm done** option, then click **SAVE**.

2. In the **CHART TYPE** area, select **Vertical Bar** and click the **Bullet** chart type.

3. Add a beast mode called `TY$` with the following **FORMULA** value:

```
-- This Year Amount
-- use year(CURRENT_DATE()) with live data
sum(case when year(`CloseDate`) = year('1/1/2020') then
`Amount` end )
```

4. Check the **Share Calculation on Dataset** option and click **SAVE & CLOSE**.

5. Add a beast mode called `PY$` with the following **FORMULA** value:

```
-- Previous Year Amount
-- use year(CURRENT_DATE())-1 with live data
sum(case when year(`CloseDate`) = year('1/1/2020')-1 then
`Amount` end )
```

6. Check the **Share Calculation on Dataset** option and click **SAVE & CLOSE**.

7. Add a beast mode called `PYV$` with the following **FORMULA** value:

```
-- Prior Year Variance
-- use year(CURDATE()) in place of year('1/1/2020') when
live
sum(case when year(`CloseDate`) = year('1/1/2020') then
`Amount` else 0 end)

-
sum(case when year(`CloseDate`) = year('1/1/2020')-1 then
`Amount` else 0 end)
```

8. Check the **Share Calculation on Dataset** option and click **SAVE & CLOSE**.

9. Add a beast mode called `PYV%` with the following **FORMULA** value:

```
-- Prior Year Variance %
-- use year(CURDATE()) when live
(sum(case when year(`CloseDate`) = year('1/1/2020') then
`Amount` else 0 end)
/
sum(case when year(`CloseDate`) = year('1/1/2020')-1 then
`Amount` else 0 end))
-1
```

10. Check the **Share Calculation on Dataset** option and click **SAVE & CLOSE**.

11. Drag QTR from the **BEAST MODES** area and drop it into the **NAMES** area.

12. Drag QTR from the **BEAST MODES** area and drop it into the **SORTING** area.

13. Drag TY$ from the **BEAST MODES** area and drop it into the **ACTUAL VALUE** area.

14. Drag PY$ from the **BEAST MODES** area and drop it into the **TARGET VALUE** area.

15. Drag PYV$ from the **BEAST MODES** area and drop it into the **RANGE VALUE 1** area.

16. The card configuration can be seen in the following screenshot:

Figure 10.12 – Actual Sales Prior Year Variance Trend card configuration

This view confirms a deceleration in actual sales in **Q4**. So, which region(s) is decelerating actual sales in **Q4**?

Creating a regional year-over-year quarterly actual sales variance trend card

Now we know overall actual sales are up across the board **year over year** (**YoY**) but decelerated in **Q4**, looking at the regional actuals trend on a quarterly basis will give us the ability to visually compare this card to the target variance card and hone in on improvement areas. For course correction, fixing the actuals variance issues may be more important than fixing the forecasting issues. So, let's see what's going on with the YoY quarterly sales trend.

To create a **Regional YoY Quarterly Sales Variance Heatmap** card, proceed as follows:

1. Copy the **Regional Target Variance Heatmap** card by clicking on the **Save as** option in the **Card** wrench menu. Name the card `Regional YoY Actual Variance Heatmap`.

2. Open the **Regional YoY Actual Variance Heatmap** card in **Analyzer**, click **Quota** in the **DATA** area, and select the **Opportunity CH7** dataset; then, click **CHOOSE DATASET** and then click **Switch DataSet**. Ignore the warning about missing columns in the new dataset as we are going to remap the columns used in the chart.

3. Drag and drop the **BEAST MODE PYV$** (prior-year variance amount) into the **VALUE** area.

4. Drag and drop **Stage** from **BEAST MODES** to the **FILTERS** area and select **5 Closed Won** to get only wins.

5. Drag and drop the **Year BEAST MODE** into the **FILTERS** area. Replace **Range** with **Selection** in the filter-type combo-box and then slide on the **Display as Quick Filter** option and click **APPLY**.

6. In the **Quick Filter** panel **Year** area, click **Select All**, and then click **2019** and **2020**.

7. Set the **Data Label Settings Text** property to `%_VALUE` and check **Auto Abbreviate Values**.

8. Click the **Summary Number** field and choose **Use All Values**, select `PYV$` as the **Column** value, enter `Prior Year Variance` as the **Label** value, and format as **Currency**.

9. Compare to the **Analyzer** configuration shown in the following screenshot:

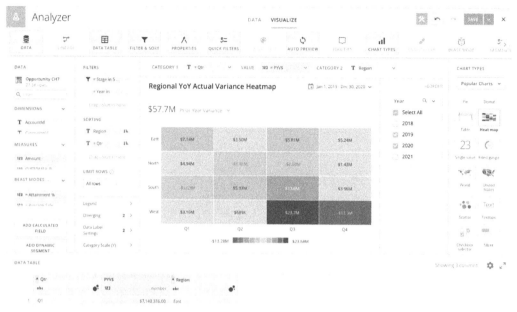

Figure 10.13 – Regional YoY Actual Variance Heatmap configuration

10. Click **Save and Close**.

11. Create a **What has Changed?** collection by clicking + on the dashboard header and selecting **New Collection**.

12. Drag and drop the **Regional YoY Actual Variance Heatmap** card into the **What has Changed?** collection.

13. Click on the **Regional YoY Actual Variance Heatmap** card footer to enter the card details page view, as seen in the following screenshot:

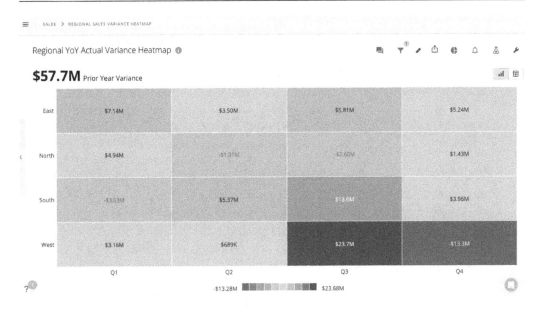

Figure 10.14 – Regional Sales Variances to Prior Year Sales heatmap

As suspected, **South** region actual sales were up from last year, so the forecast variance is an issue of over forecasting, and more so with expectations for the **West** region. However, the **West** region also had a drop in actual **Q4** sales from the prior year, partially explained by a hangover from the large increase in **Q3** over the prior year. Nevertheless, the **West** region is a point of opportunity, so let's see where in **Q4** the **West** region underperformed by looking at additional dimension cuts by **Type**, **Leadsource**, and **Account**.

Creating a leaders and laggards card

A leaders and laggards card clearly highlights areas and the degree of change, making it as simple as reading an ordered list to know where to focus our attention so as to drive better outcomes.

Let's create a leaders and laggards card for regions in **Q4** with a rank ordering of the top percentage changes, as follows:

1. Copy the **Regional Sales Variance Heatmap** card by clicking on the **Save as** option in the **Card** wrench menu. Name the card `Largest Laggards`.

2. Open the **Largest Laggards** card in **Analyzer** and click **Tables and Textboxes** in the **CHART TYPES** area and click on the **Mega Table** chart thumbnail image.

3. Click on the **ADD CALCULATED FIELD** button to create a beast mode titled `Leaders and Laggards` using the following **FORMULA** value:

```
-- substitute year(CURDATE()) for year('1/1/2020') on
live data
case when sum(case when year(`CloseDate`) =
year('1/1/2020')-1 then `Amount` else 0 end) = 0 then
'New'
when (sum(case when year(`CloseDate`) = year('1/1/2020')
then `Amount` else 0 end) /
sum(case when year(`CloseDate`) = year('1/1/2020')-1 then
`Amount` else 0 end)) -1 = -1 then 'No Sales'
when (sum(case when year(`CloseDate`) = year('1/1/2020')
then `Amount` else 0 end) /
sum(case when year(`CloseDate`) = year('1/1/2020')-1 then
`Amount` else 0 end)) -1 < -.7
and (sum(case when year(`CloseDate`) = year('1/1/2020')
then `Amount` else 0 end) /
sum(case when year(`CloseDate`) = year('1/1/2020')-1 then
`Amount` else 0 end)) -1 > -1 then 'Laggard'
when (sum(case when year(`CloseDate`) = year('1/1/2020')
then `Amount` else 0 end) /
sum(case when year(`CloseDate`) = year('1/1/2020')-1 then
`Amount` else 0 end)) -1 > .7 then 'Leader'
else 'In the Middle' end
```

This beast mode creates a dimension with the following values:

- **New**—Items where there were zero prior-year sales
- **No Sales**—Items with sales in the prior year but no sales this year
- **Laggard**—Items where current-year sales were less than 70% of prior-year sales
- **Leader**—Items where current year sales were greater than 70% of prior-year sales
- **In the Middle**—Items where sales variance percentages are between the **Leader** and **Laggard** items

4. Click **Share Calculation on DataSet** and then click **SAVE & CLOSE**.

5. Add `Lead Source Standardized` beast mode with the following **FORMULA** value:

```
-- Lead Source Standardization
case
when `LeadSource` = 'Patnes' then 'Partners'
when `LeadSource` = 'Diect' then 'Direct'
when `LeadSource` = 'Sales Ceated' then 'Sales Created'
when `LeadSource` = 'efeal' then 'Referral'
when `LeadSource` = 'Self-Souced' then 'Self-Sourced'
when `LeadSource` = 'Maketing Outbound' then 'Marketing
Outbound'
when `LeadSource` = 'Stategic Accounts Maketing' then
'Strategic Accounts Marketing'
when `LeadSource` = 'Patneing' then 'Partners'
when `LeadSource` = 'enewal' then 'Renewal'
when `LeadSource` is null then 'Unknown'
else `LeadSource`
end
```

6. Check **Share Calculation on Dataset** and click **SAVE & CLOSE**.

7. Drag and drop the **Region**, **Type**, **Lead Source Standardized**, PVY$, PYV%, and **Leaders and Laggards** fields into the **COLUMNS** area.

8. Drag and drop the PYV$ **BEAST MODE** into the **SORTING** area.

9. Set the **LIMIT ROWS** property to 10.

10. Under **CHART PROPERTIES | General**, check **Financial Style Negatives (100)**.

11. The **Analyzer** configuration is seen in the following screenshot:

Figure 10.15 – Largest Laggards card configuration

12. Click **Save and Close** under **SAVE**.

13. In the card details view of the **Largest Laggards** card, add a card filter by clicking the **Show Filters** funnel icon in the card menu bar and then click + and select **Region**, then check **West** and click **Apply**. Repeat the process to add a filter for **Lead Source Standardized not in Unknown** because a decrease in unknown lead sources is a good thing and not a laggard item.

14. We can see in the following screenshot the 10 largest laggards in the **West** region for **Q4**:

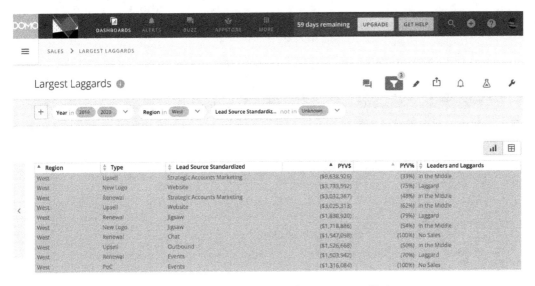

Figure 10.16 – Largest Laggards West region (Q4)

From *Figure 10.16*, it is obvious that the **Strategic Accounts Marketing** channel for both **Renewal** and **Upsell** types, and the **Website** channel for **New Logo** are all large items in terms of dollar variance. These are high-relevance items to investigate ways to improve sales target attainment.

Well, we just worked our way from the target shortfall through the actuals to identify the actual **West** region shortfall. This is typical of what is required to support a story statement. The fantastic thing is that we can do this on our own without having to go to external experts. Also, notice that by building cards to answer questions, we also filled in our monitoring dashboard, as seen in the following screenshot, which will increase the speed of our future diagnostic activity:

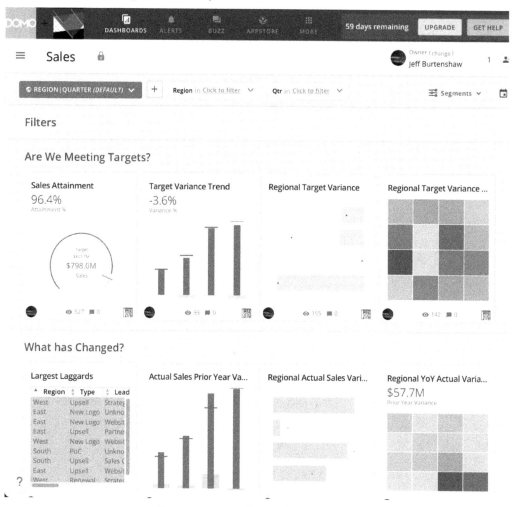

Figure 10.17 – Updated Sales dashboard

Cool stuff, but there is a problem here; looking at the **Sales** monitoring dashboard, the story we just went through around what happened in **Q4** doesn't exactly tell itself, does it? We need a better way than the monitoring dashboard to tell the story. In the next section, will learn about **storytelling patterns**.

Learning about storytelling patterns

Many dashboards are built to organize information, but often, that structure is not telling a story. Rather, these metric container dashboards can be classified as monitoring dashboards and are very functional for discovery. In fact, the monitoring dashboard forms the basis for identifying and supporting a story. But if we stop there, a monitoring approach leads to analysts still creating slide decks to communicate the story. The evolution from monitoring to storytelling in **business intelligence** (**BI**) tools is just beginning, and **Domo Stories** enables better storytelling on a platform. The great news is that a pattern exists for telling business stories, and tools beyond PowerPoint are starting to enable the implementation of the pattern. The following screenshot shows a universal business storytelling pattern:

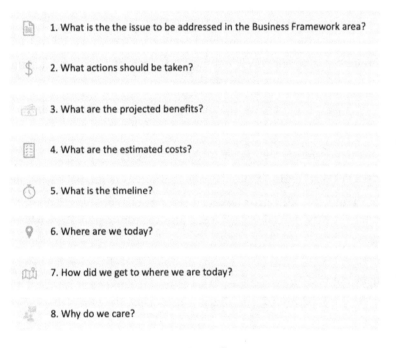

Figure 10.18 – Storytelling pattern

The pattern sequence in *Figure 10.18* is structured for executive audiences who want to get to the ROI and feasibility of the matter as quickly as possible. Executives don't want or need the details upfront—an analogy is *you don't have to know what's going on in the kitchen to eat the meal.* That position is rational; here's why: as time is limited, executives want to assess the overall solution to determine if it is worth investing in. The details (kitchen work) are why they have analysts presenting to them. For an executive, knowing the issue benefits, costs, and recommended actions upfront (eating the meal) enables them to determine if the idea is worth the organization's further attention.

A storytelling pattern is also useful when presenting to colleagues and subordinates but the presentation order in the pattern changes. Often, when presenting to non-executives there is a need to educate first (show them how the meal was made), so the presentation sequence changes to 1,6,7,8,2,3,4,5.

Choose the wrong presentation order of the pattern for the audience and you risk boring the executives or confusing your colleagues/subordinates.

In the next section, we will use our observations from the **Sales** dashboard to flush out opportunities for improvement and apply a storytelling pattern in Domo Stories.

Applying a storytelling pattern using Domo Stories

Time to upgrade our **Sales** monitoring dashboard to use the Domo Stories **user interface (UI)**. This will empower us to effectively tell the story supporting our story statement using a storytelling pattern in a live dashboard.

Next, let's copy the **Sales** dashboard as the starting point for the story dashboard and convert the dashboard page to the Domo Stories UI, as follows:

1. Click on the page wrench menu icon and click **Save as**.

2. Enter **2020-Q4 West Region Sales Miss** as the **New Dashboard Title** value.

3. Check both **Duplicate all Cards** boxes so that any changes we make for the story won't change the original **Sales** dashboard and **Go to the new Page when done** options.

4. Click **SAVE**.

5. On the **2020-Q4 West Region Sales Miss** page, click on the page wrench icon and click **Design Dashboard**, and in the popup, click the **DESIGN DASHBOARD** button.

6. This will open the Stories **EDITING DASHBOARD** interface.

   The **EDITING DASHBOARD** layout controls how the cards on a story dashboard are organized on the page. It is a frame-based paradigm whereby each frame can be sized on the page. However, it is not a freeform layout tool, but rather, a flexible grid layout tool. The idea behind the frames is to simplify design decisions and bring some level of consistency to the presentation.

Let's learn more about the **EDITING DASHBOARD** features, as seen in the following screenshot:

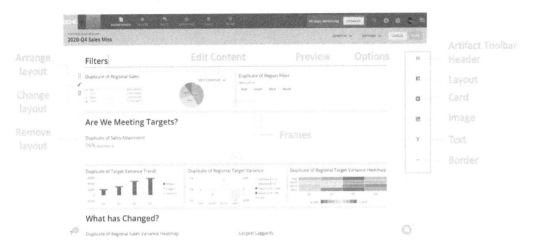

Figure 10.19 – Story editing page features

> **Important Note**
> To make the initial view fit the section, handles were dragged up to make each section shorter so that it showed on screen; otherwise, the **Filters** section would have taken up the whole screen.

Here are descriptions of the main features available on the **EDITING DASHBOARD** page:

- **Arrange layout** is a handle to allow us to drag and reorder the layout sections.

- **Change layout** is a tool that allows us to select canned layouts to use based on the number of frames in the layout.

- **Remove layout** deletes the layout, and the cards in the layout are moved to the **Appendix** section at the bottom of the page.

- **Frames** are layout grid boundaries and can be moved by clicking and dragging them, and the content will adjust.

- **Edit Content** is an options menu for changing the content inside a frame.

- **Preview** is a toggle option to switch the view between **DESKTOP** editing to **MOBILE** preview.

- **Options** is a drop-down menu that lets us change the background **Color fill** or **Image fill** values of the page and change the **Display mode** between **Fixed width** and **Auto width** screen scaling.

- **Artifact Toolbar** contains a pallet of artifacts we drag and drop into the layout frames.

- **Header** is a textual header object.

- **Layout** is a section in the page whose layout can be set independently of other sections.

- **Card** is a container for a Domo Card.

- **Image** is the container for an image.

- **Border** is a horizontal line to separate sections visually.

Not a lot to it, but you will see that simple layout controls can dramatically improve our storytelling ability.

Implementing a storytelling pattern

Now, we can start to implement a storytelling pattern, as seen in *Figure 10.18*:

First, let's remind ourselves of our story statement so that we stay relevant. Here it is:

```
As an Analyst I want to Present in Person to Executives
evidence why the West Region Missed 2020-Q4 Sales Target and
what needs to Change to Increase Sales.
```

Storytelling pattern – Step 1

The first topic in a storytelling pattern is *1. What is the issue to be addressed?*. Let's add a header for this section, as follows:

1. In the **2020-Q4 West Region Sales Miss** editing dashboard, click and drag the **Are We Meeting Targets** header artifact to the top of the page and drop it, and a solid blue line will appear to indicate the drop point.

2. Click on the section layout that was formerly under **Are We Meeting Targets?** and drag and drop under the **Are We Meeting Targets?** header, as seen in the following screenshot:

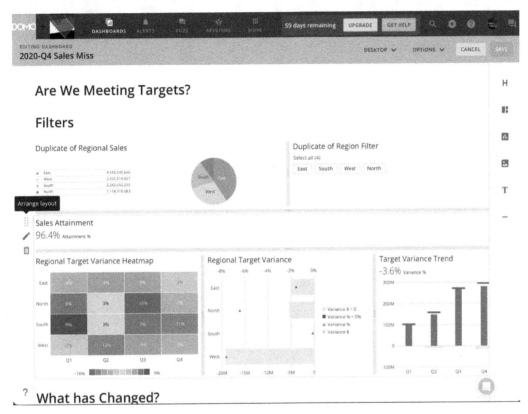

Figure 10.20 – Current layout of the section

3. Click on the **Change layout** pencil, as seen in the following screenshot:

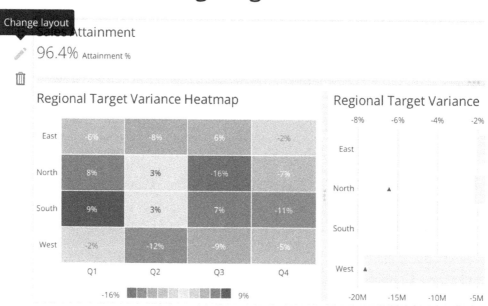

Figure 10.21 – Changing the layout of the section

4. Select the **Hero** category and **3** cards, and then click on the side-by-side vertical **Hero** thumbnail image, as seen in the following screenshot:

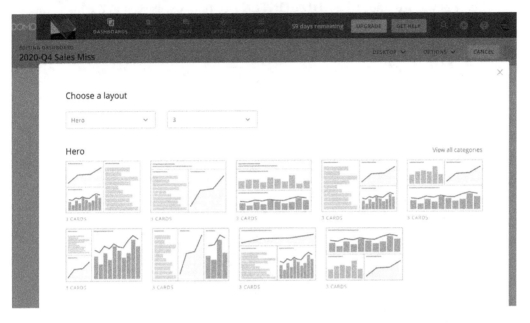

Figure 10.22 – Hero layout selection

> **Important Note**
> **Hero** refers to a card that is intended to get most of the viewer's attention.

5. Click on the **Are We Meeting Targets?** header and rename it What is the 2020-Q4 West Region Sales Issue?.

6. Click on the **Card** icon in the artifact toolbar and drag it above the **Sales Attainment** card until a gray bar above the title highlights, then drop and select + **Create new Card** under the **ADD CONTENT** combobox, as seen in the following screenshot:

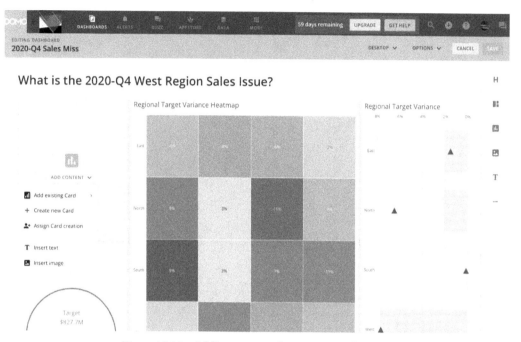

Figure 10.23 – Adding content: Create new Card option

7. Click **SAVE & CONTINUE**, as to go to **Analyzer** to create a new card, we must first save the layout changes to the story.

8. Click **SELECT** under **Visualization**, then click **SELECT** under **Existing data**, find and click the **Quota** dataset, and click **CHOOSE DATASET**.

9. Enter **Banner Phrase** in the card title as the card will display dynamic text that is bound to the dataset and updates in real time.

10. Under **CHART TYPES**, select **Tables and Textboxes**, then click on the **HTML Table** chart thumbnail.

11. Next, we add a beast mode that generates a dynamic formatted text string indicating variances from target and YoY actuals and includes appropriate miss versus beat phrasing and coloring. Click **ADD CALCULATED FIELD**, enter `Performance Phrase` as the title, and get the **FORMULA** value from `Performance Phrase.txt` (`https://github.com/PacktPublishing/Data-Democratization-with-Domo/blob/main/Performance%20Phrase`). Click **Share Calculation on DataSet** and click **SAVE & CLOSE**.

12. Click **Header Row** in **CHART PROPERTIES**, click **Header Font Color**, and enter `FFFFFF` in the color box. This hides the column header values by setting the text color to match the background.

13. For security, Domo supports the rendering of **HyperText Markup Language (HTML) tags** on a limited basis, specifically in the **HTML Table** chart type and in the **Summary Number** card. In this case, we will use the **Summary Number** card to render the beast mode with the formatted text summarizing performance. Click the **Summary Number** dropdown, click **Use All Values**, select **Performance Phrase** as the **Column** value, and clear the **Label** value so that it won't interfere with the phrase text.

14. Drag and drop the **Performance Phrase** beast mode into the **COLUMNS** area and remove any other fields from **COLUMNS**.

15. Drag **Region** into the **FILTERS** area and select the **Display as Quick Filter** option, then click **Apply**.

16. In the **Region** field of the **Quick Filter** panel, click **Select All** and then click **West**.

17. Drag `QTR` into the **FILTERS** area and select the **Display as Quick Filter** option, and then click **Apply**.

18. In the `QTR` field of the **Quick Filter** panel, click **Select All** and then click **Q4**.

19. Compare the card configuration to the one shown here:

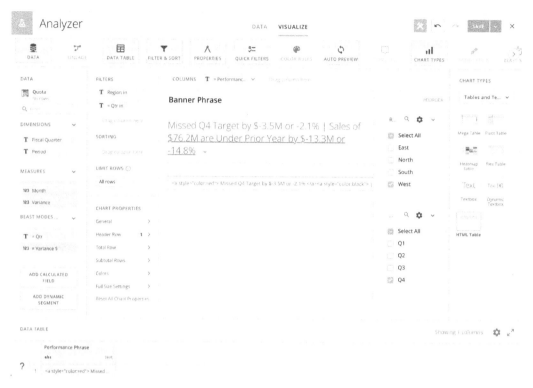

Figure 10.24 – Banner Phrase card configuration

20. Click **SAVE** and choose **Save and Close**, then click on the page wrench icon, and click **Edit Dashboard**.

21. Click **Change layout**, and this time, click the first option under the **With banner** section, as seen in the following screenshot:

Figure 10.25 – Four-card banner layout

22. Click and drag the **Banner Phrase** card over the **Sales Attainment** card until the complete **Sales Attainment** card border is highlighted in blue, then drop to swap their positions.

23. Click in the **Banner Phrase** card, then click the **EDIT CONTENT** dropdown, click **Display settings**, and uncheck **Chart**.

24. Click < **Display settings** and click **Change background**, and then pick a light gray color and click out of the **Color** dialog. Adjust the border of the **Banner Phrase** card and then shorten the section boundary.

25. Remove the layout containing the **Filter** fields by clicking the trashcan icon, and then click **REMOVE LAYOUT**.

26. Click on the **Actual Sales Prior Year Variance** card in the **What has Changed?** section and drag it over the **Regional Target Variance** card and drop it to swap places.

27. Compare your progress to what's shown here:

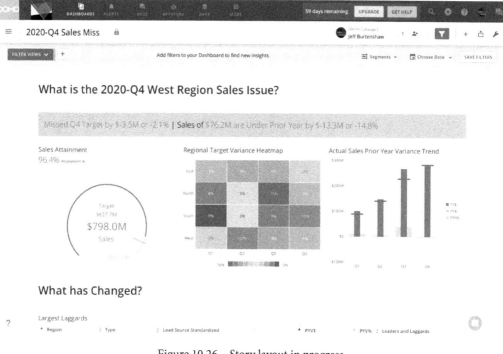

Figure 10.26 – Story layout in progress

28. Click **SAVE**.

29. Notice that the story is showing all regions and quarters, but for our story, we want only the **West** region and **Q4**. To do this, we need to create a **FILTER** view. On the **2020-Q4 Sales Miss** page, click + on the filter toolbar and click **Region**, then **Apply**. Follow those same steps to add a filter on QTR. Then, select **West** in **Region** and **Q4** in QTR, and click **Apply** after each.

30. Click **SAVE FILTERS** on the filter bar and then click **CREATE NEW FILTER VIEW** in the popup. Enter **West Region Q4** as the **New filter view name** value, check the **Set as default for everyone** property, and click **SAVE**.

31. The filter story will look like this:

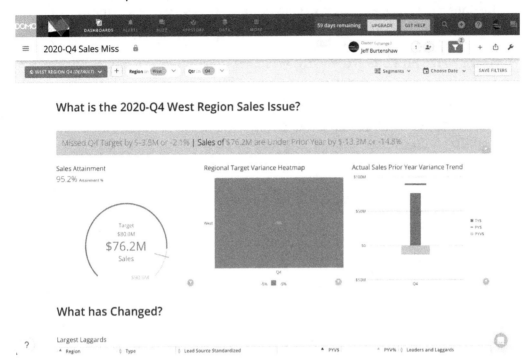

Figure 10.27 – Filtered story in progress

Making progress! For visual appeal, let's add a map with our custom regions.

Adding a map with custom regions

Here are the steps to create a map with custom regions:

1. For the map, we need to download a usa_regions.svg map image file and a Region Mapping.xlsx region mapping table from the GitHub repository.

2. Now, let's create a new custom chart type. From the **2020-Q4 West Region Sales Miss** page, click + and then click + **Create new Card**, click **SELECT** under **Visualization**, click **SELECT** under **Existing data**, select **Opportunity CH7 DataSet**, and click **CHOOSE DATASET**.

3. Under **CHART TYPES**, pick **Custom Charts** from the dropdown, scroll down through the chart types and click on the + thumbnail, click **Upload chart**, drag and drop into the usa_regions.svg file, name the chart USA Regions, and click **SAVE**.

4. Now, in **Card design**, under **Chart Types | Custom Charts**, choose the **USA Regions** chart type, enter **West Region Prior Year Q4 Variance $** as the card title, drag and drop **Region** into the **NAMES** area and PYV$ to the **VALUES** area, drag and drop **Stage** into the **FILTERS** area and select **% Closed Won**, drag and drop **Year** into the **FILTERS** area and change **Range** to **Selection**, and turn on **Display as Quick Filter** and click **Apply**.

5. In the **Quick Filter** area, click **Select All** and click **2019** and **2020**.

6. Under **CHART PROPERTIES**, click **Diverging** and click **Show**, and choose **Zero** as the **Midpoint Value Type** value.

7. Click **Save and Close**.

Now, let's apply our custom regions from the Region Mapping.xlsx file, as follows:

1. From the **DATA** main menu, click **FILE** and drag and drop it into the Region Mapping.xlsx file, select **COLUMN HEADERS** under **Format**, click **Next**, enter **Region Mapping** as the **DATASET NAME** value, and click **SAVE and RUN**.

2. Now, let's bind the **Region Mapping** dataset to the **USA Regions** chart type. Click **ADMIN** on the main menu, click the **Company Settings** icon on the left toolbar, click **Custom Charts**, click **+ ADD CHART**, click **Create regions**, change **Maps** to **Custom Charts** in the dropdown, click the **USA Regions** thumbnail, click **NEXT**, click on the **Region Mapping** dataset, click **NEXT**, choose **Region** in the **Select new custom region** dropdown, and select **Sub-Region** in the **Select sub-region** dropdown. This binds the mapping fields to the usa_regions.svg chart-type image we uploaded. Click **NEXT**, click **NEXT** again, and click **SAVE**. Compare the region boundaries on the **USA Regions** tile to those shown here:

Figure 10.28 – USA Regions custom chart type

3. Click on the page wrench icon, click **Edit Dashboard**, and drag and drop the **Region Map YoY Variance** card from the **Appendix** onto the **Target Variance Trend** card to swap the frame positions of both cards.

4. Click **SAVE**. Compare our results to what's shown here:

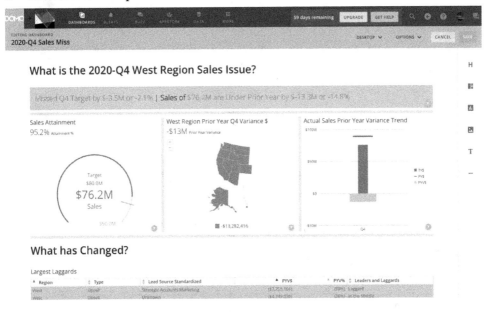

Figure 10.29 – Story page with region map

Nice! We now have a visual map-based representation of sales, and the issue is defined and supported.

Storytelling pattern – Step 2

We are following the *Executive Sequence* path through the *Storytelling pattern*, and the next topic is *2. What actions should be taken?* Let's add a header and layout section for this, as follows:

1. Click and drag the **Header** artifact to the layout area just under the top layout. Enter **What Actions Should Be Taken?** in the header.

2. Click and drag the border artifact above the **What Actions Should Be Taken?** header, and click and drag the **Layout** artifact under the **What Actions Should Be Taken?** header. Select the **Hero** category and **2** cards, and then click on the side-by-side vertical layout **Hero** thumbnail image, as seen in the following screenshot:

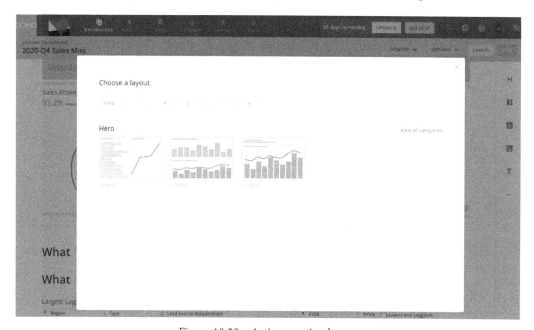

Figure 10.30 – Actions section layout

3. Click **Insert text** in the **ADD CONTENT** dropdown and enter the following:

```
Recommended Actions
  • Review forecasting approach
```

Based on the Largest Laggards:

- Review the Upsell & Renewal sales motions for Strategic Accounts Marketing Channel

- Review the New Logo sales motion from Social Channels

4. From the **What has Changed?** Section, click and drag the **Largest Laggards** card to the right-side frame and drop it.

Terrific! We have created a concise list of recommended actions with supporting evidence! You can see the result here:

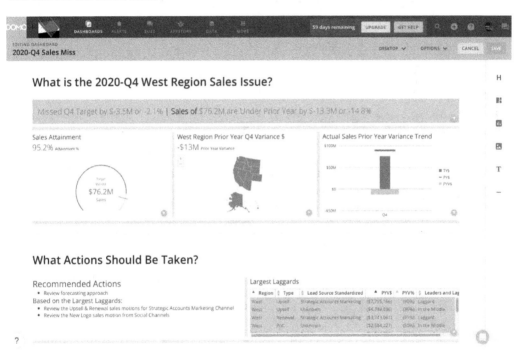

Figure 10.31 – Actions section in progress

Next, we will cover steps *3. What are the projected benefits?* and *4. What are the estimated costs?* of the storytelling pattern.

Storytelling pattern – Steps 3 and 4

The following steps cover the benefits and costs of taking the recommended actions:

1. First, we need to add another layout for **Benefits** and **Costs**, so click and drag the layout artifact under the **Actions** layout. Click **2** in the **Number of Cards** dropdown and click the **Hero** vertical side-by-side layout thumbnail.

2. In the left frame, click **ADD CONTENT**, select **Insert Text**, and enter the following:

```
Benefits
    • Reduced forecast Variances
    • $Increased sales of 3.5M+
```

3. In the right frame, click **ADD CONTENT**, select **Insert Text**, and enter the following:

```
Costs
    • $20K deep-dive analysis
    • Implementation Costs TBD
```

4. Compare the layout to what's shown here:

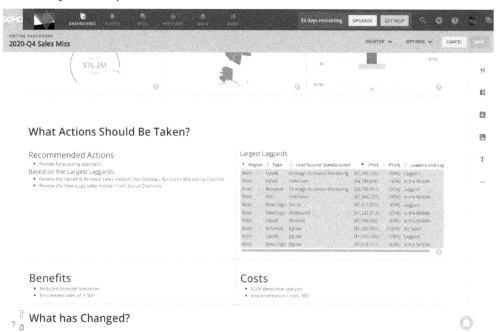

Figure 10.32 – Benefits and costs layout

Next, let's add a timeline section for the recommendations.

Storytelling pattern – Step 5

All actionable proposals address an estimated timeline. Here are the tasks to create a timeline card:

1. Click and drag the **Layout** artifact under the **Benefits and Costs** layout, select **1** in the **Number of Cards** dropdown, and click on the **Hero** card thumbnail.

2. Click **Create new Card** in the **ADD CONTENT** dropdown, click **SAVE & CONTINUE**, click **SELECT** under **Visualization**, and click **CREATE** under **Online data editor**.

3. Enter Timeline in the **Enter DataSet Title** box, enter Item in **New column A**, enter Start in **New column B**, and enter End in **New column C**.

4. Set the field data type to **Date** for **Start** and **End** by clicking on **abc** and choosing **Date**.

5. In the **Item** column, enter Forecast Analysis in row 1, Strategic Marketing Analysis in row 2, Social Channel Analysis in row 3, and Implementation in row 4.

6. In the **Start** column, enter 1/5/2021 in row 1, 1/15/2021 in row 2, 1/15/2021 in row 3, and 2/20/2021 in row 4.

7. In the **End** column, enter 2/2/2021 in row 1, 2/15/2021 in row 2, 2/15/2021 in row 3, and 5/31/2021 in row 4. Compare your output to what's shown in the following screenshot:

Figure 10.33 – Timeline web form

8. Click **SAVE & CONTINUE**.

9.   In **CHART TYPES**, click **Horizontal Bar** and click the Gantt chart thumbnail.

10. In the **Summary number** dropdown, click **No Summary Number**.

11. Drag the **Item** field to the **ITEM** area.

12. Drag the **Start** field to the **START DATE** area.

13. Drag the **End** field to the **END DATE** area.

14. Click **Save and Close**.

15. Scroll down on the story page to see the **Timeline** card, similar to what's shown here:

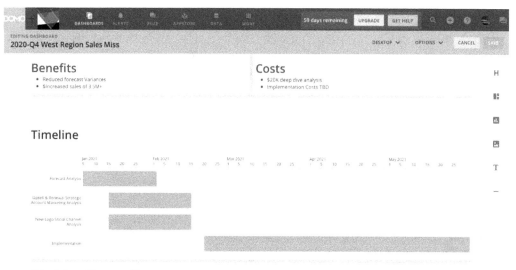

Figure 10.34 – Timeline card

Very nice! We have a compelling story coming through. Next, let's return to the remaining storytelling pattern steps.

Storytelling pattern – Steps 6 and 7

The questions for *Steps 6* and *7* are handled as follows:

1. *Step 6. Where are we today?* is addressed by the **Sales Attainment** card in the banner layout.

2. *Step 7. How did we get to where we are today?* for this story are appendix items with trend cards added from the monitoring dashboard.

Storytelling pattern – Step 8

Our closing statement is made in this section, as follows:

1. *Step 8. Why do we care?* is handled by clicking and dragging a header artifact under the **Timeline** card.

2. Click and drag a **Text** artifact under the header and enter the following rationale for doing the project:

```
Given the underperformance of the West Region in Q4, we
should review the largest underperforming areas compared
to prior year around Upsell and Renewals in the Strategic
Account Marketing channel and New Logo acquisition via
Social Media. We should also review our forecasting
approach for the region. This deeper dive will take a
few months to diagnose and implement corrective actions
so we should start as soon as possible. The stakes are
high going into following quarters and the opportunity to
increase sales is in the millions of dollars while the
cost is the likely just the labor to do the analysis and
implement the changes.
```

Now that we've implemented storytelling patterns, let's clean up the dashboard.

Housekeeping

Lastly, let's remove any extra sections and artifacts from the dashboard, leaving a relevant story, as seen in the following screenshot:

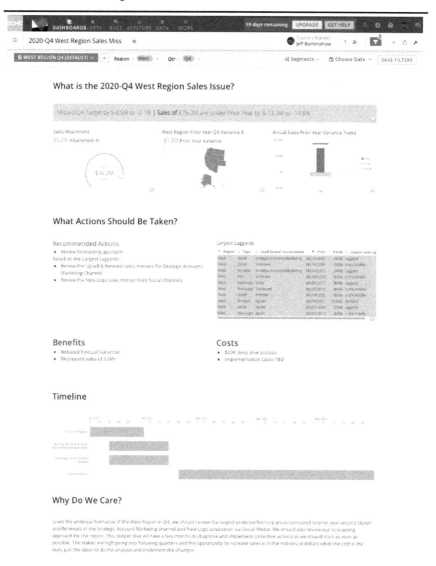

Figure 10.35 – Completed story page

We can see that our story statement has come to life and is much improved by using the storytelling pattern over the raw monitoring dashboard. Our chances of getting our project approved have just increased dramatically.

Well, we covered a ton of ground in this chapter, so it's time for a quick recap.

Summary

In this chapter, we learned about using a **business model** to drive out relevant **framework areas** used to generate **story statements** using a **story statement guide**. Then. we saw how to incrementally refine our monitoring dashboard simply by answering questions to generate evidence for our **story statement**. Then, we shifted to building out the story statement in Domo Stories. We copied the monitoring dashboard as a story starting point and then applied a storytelling pattern. The result is a relevant story that will lead to action being taken. While there was a lot of detail, this chapter is a typical example of how great leaders and analysts use the Domo platform to revolutionize their businesses. The power of democratizing data is on full display here, as one individual with little technical know-how can acquire, sculpt, discover, and present a relevant story in an amazingly fast way.

In the next chapter, we will discuss our options to distribute the fantastic story we just created.

Further reading

The following are related items of interest:

- Suggested reading on business models: *Business Model Generation: A Handbook for Visionaries, Game Changers, and Challengers (The Strategyzer series)*, *Alexander Osterwalder* and *Yves Pigneur* https://www.amazon.com/Business-Model-Generation-Visionaries-Challengers/dp/0470876417

- For variations on period-over-period beast mode calculations, read the following: https://domohelp.domo.com/hc/en-us/articles/360043430133-Sample-Beast-Mode-Calculations-Period-over-Period-Transforms#7.1

- Reference for additional examples on how to use common HTML tags in Domo card summary numbers and HTML tables: https://domohelp.domo.com/hc/en-us/articles/360043430033

- More information on creating custom maps and charts and customizing regions can be found here: https://domohelp.domo.com/hc/en-us/articles/360043428793 and https://domohelp.domo.com/hc/en-us/articles/360042924454-Custom-Charts

11
Distributing Stories

How great would it be, after we have created our story, to have many options to publish and distribute the content with real-time updates included? This chapter will cover the many ways the Domo platform can help you disseminate content. This includes on native iOS and Android devices with no mobile app development required, scheduled emails with attached content, revolving digital wallboard presentations, secured mass distribution via a URL, embedding in PowerPoint, and embedding in bespoke web pages. Get ready – we are going to ensure the message gets out by practicing all the options for communicating through channels that are enabled by the Domo platform.

In this chapter, we will cover the following topics:

- Understanding the content distribution landscape
- Sharing pages
- Sharing cards
- Distributing via email
- Embedding in productivity apps with plugins and add-ons
- Using the mobile app and widget
- Publishing via a URL
- Creating digital wallboards
- Embedding in a web page

Technical requirements

To follow along with this chapter, you will need the following:

- Internet access

- Your Domo instance and login details

- The ability to download the example files from GitHub

- The dashboard and cards we created in *Chapter 6*, *Creating Dashboards*, and *Chapter 7*, *Working with Drill Pathways*

> **Important Note**
> If you don't have a Domo instance, you can get a free trial instance here: `https://www.domo.com/start/free`.

- The code for this chapter can be downloaded from GitHub at `https://github.com/PacktPublishing/Data-Democratization-with-Domo`.

Understanding the content distribution landscape

Content distribution is all about making sure that our stories don't become the greatest stories never told. The Domo platform provides many methods that allow us to directly distribute content. Distribution options range from being as specific as an individual card share to as generic as a mass distribution via a public website. Let's take a moment to understand the distribution options that are shown in the following table:

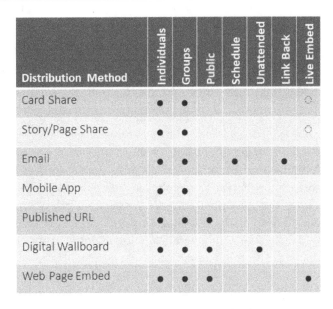

Distribution Method	Individuals	Groups	Public	Schedule	Unattended	Link Back	Live Embed
Card Share	●	●					○
Story/Page Share	●	●					○
Email	●	●		●		●	
Mobile App	●	●					
Published URL	●	●	●				
Digital Wallboard	●	●	●		●		
Web Page Embed	●	●	●				●

Figure 11.1 – Content distribution landscape

The following is a brief synopsis of the **content distribution landscape** in Domo:

- **Card Share** is a way to share a card directly within the Domo application with individuals and groups.

- **Story/Page Share** is like Card Share but works by sharing an entire page.

- **Email** distribution is built into Domo and can be scheduled and links the user back to the Domo interface. The assumption in Domo's design is that the email serves as a nudge to return to the core product.

- **Mobile App** includes apps and widgets for iOS and Android that enable users to access the Domo experience via mobile. Domo aspires to be mobile-first for content consumption.

- **Published URL** is a private or public URL to a Domo real-time slideshow.

- **Digital Wallboard** is a continuously running real-time slideshow that's generated on live Domo content.

- **Web Page Embed** is the Domo embed product for putting cards and stories into a bespoke website application. The content can be authenticated to a Domo login or can be made available to the public.

- **Domo Everywhere Publish** is an instance publish-and-subscribe orchestration tool that can be helpful in situations where third parties such as dealers, distributors, suppliers, corporate subsidiaries, or regions may need a subset copy of a parent publishing instance. The subscribed instance contains the relevant portion of the parent instance's dashboards and data. This provides an extremely scalable and secure way to distribute and maintain a tailored Domo experience to these specialized audiences. This is a paid option.

That was a quick tour of our content distribution options. Now, let's look at some examples.

Sharing pages

Domo makes it very simple to share a page with other Domo users. To share an entire page, go to any page – in this case, **2020-Q4 Sales Miss** for the West region – and click on the share icon on the toolbar, as shown in the following screenshot:

Figure 11.2 – Page sharing icon

This will bring up the page sharing dialog, as shown in the following screenshot:

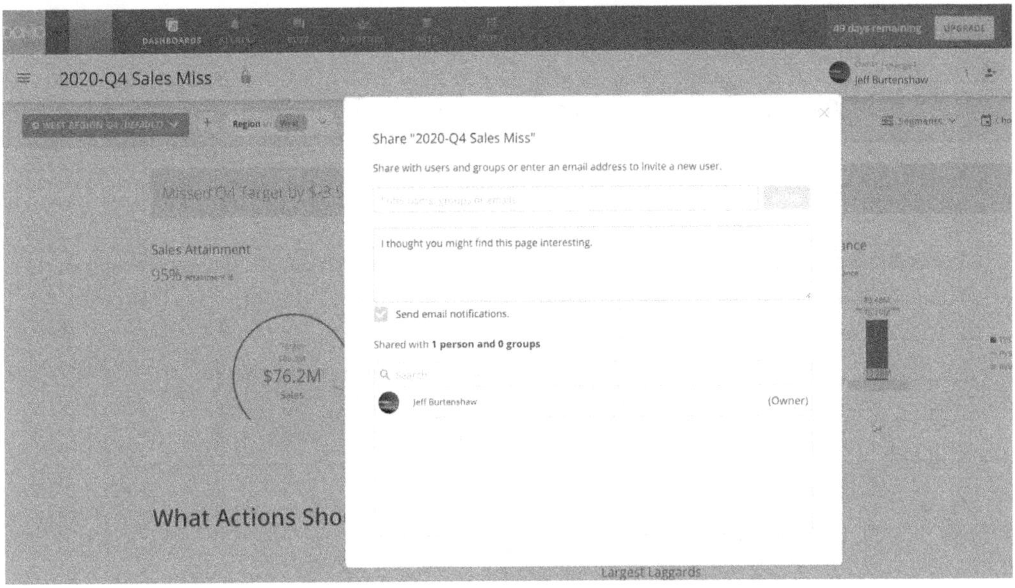

Figure 11.3 – Page sharing dialog

Let's look at the options shown in the preceding screenshot in detail:

- Click inside the **Enter users, groups or emails** box to pick who to share with. You can control whether to send an email notification by toggling the **Send email notifications** property.

- Click **SHARE** and the newly shared entity will appear in the **Shared with** list at the bottom of the dialog.

- The current list of people that the page is being shared with is searchable, so we can see in a long list whether the page has already been shared with a certain person.

- To remove an entity from the share, click the **X** button on their row.

- Click the **X** button in the top-right corner to close the page sharing dialog.

Now, the user that we've shared the page with will be able to see it in their **DASHBOARDS** list.

Next, let's learn how to share individual cards.

Sharing cards

Sharing a card with another Domo user is the most granular way to share content. To share a card, follow these steps:

1. On the **2020-Q4 Sales Miss** page, click on the **Sales Attainment** card, click the rectangle with an up arrow icon, and click **Share**. This will bring up the share dialog with the same functionality that we discussed and showed in *Figure 11.3* just for this card.

2. Once a card has been shared, the users who it was shared with will see the card in their system dashboard, under **Shared**. This can be accessed via the **Dashboards** panel, as shown in the following screenshot:

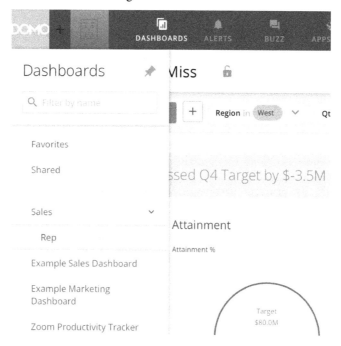

Figure 11.4 – The Dashboards panel's Shared button

A shared card is often moved or copied by the user to another page in the user's instance.

Now, let's learn how to schedule and distribute content via email.

Distributing via email

The Domo user experience is optimized for mobile consumption but since email is still a heavily used delivery channel, let's learn how to send and schedule content via email.

To send an email on a card, on the **2020-Q4 Sales Miss** page, click on the **Sales Attainment** card, and click the wrench icon. You will see three email options in the drop-down menu: **Schedule as Report**, **Send Now**, and **Send / Export**. Let's take a closer look.

Option 1 – Schedule as Report

Follow these steps to schedule a report to be emailed:

1. The first option is to click the **Schedule as Report** option, which opens the **Schedule report** dialog, as shown in the following screenshot:

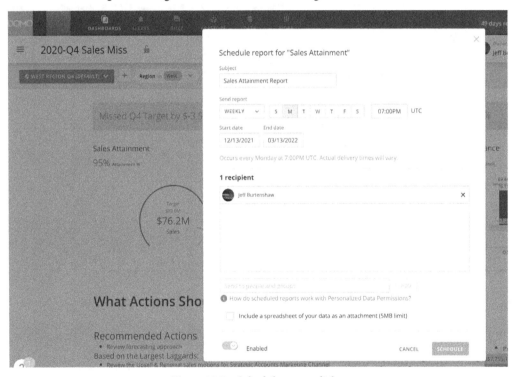

Figure 11.5 – Schedule report dialog

2. Change the email's **Subject** line if desired, set the schedule frequency and time of day in terms of **UTC**, and set a **Start date** and **End date** for the distribution period.

3. Click inside **Send to people and groups**, pick the users and groups you wish to add, and click **ADD** to place them in the **recipient** list. By default, the email will be sent to you. Click the **X** button on the row to remove a name from the distribution list.

4. To attach a spreadsheet of the card data, click the **Include a spreadsheet of your data as an attachment** property.

5. To invite someone that is not a Domo user, simply type their name into the **Send to people and groups** box and click **Invite to Domo**, enter their email address, and click **INVITE**.

6. To see all your scheduled reports, click the **MORE** main menu item and click the **SCHEDULED REPORTS** option. This will take you to the **Scheduled Reports** page.

7. Click the **...** icon on a row to see the scheduling options, as shown in the following screenshot:

Figure 11.6 – The Scheduled Reports page's row options

The following are brief descriptions of the row options that are available on the **Scheduled Reports** page:

- **Edit report** allows you to make changes to the report filters.

- **Edit schedule** allows you to make changes to the distribution participants, schedule frequency, and time.

- **Send Now** sends the report out immediately.

- **View History** shows a log of the scheduled executions and their statuses.

- **Disable** turns off the schedule temporarily and does not remove the schedule.

- **Delete** removes the scheduled report.

Next, let's look at the **Send Now** option.

Option 2 – Send Now

Follow these steps to use the **Send Now** email option:

1. The second option that's available from the card wrench menu dropdown is the **Send Now** option, which brings up the **Send** dialog, as shown in the following screenshot:

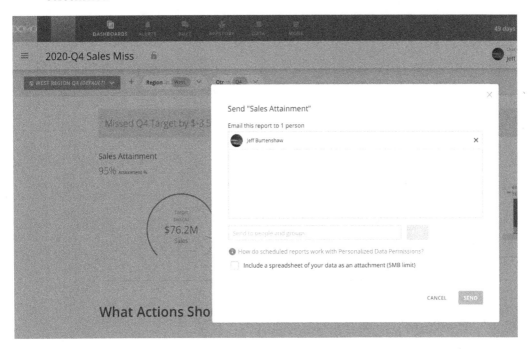

Figure 11.7 – The Send dialog

2. Simply enter the desired people and groups and click **SEND**.

3. The default behavior is for the email to link the user back to Domo based on the emailed content. However, there is also the option to **Include a spreadsheet of your data as an attachment**, albeit with a size limitation of 5 MB.

4. The email will look as follows:

Figure 11.8 – Send Now email example

Next, let's look at the third option for sending emails.

Option 3 – Send / Export

Follow these steps to send/export a report:

1. The third option that's available from the card can be found by clicking the rectangle-with-up-arrow icon and then selecting the **Send / Export** option. This will bring up the **Send / Export** dialog, as shown in the following screenshot:

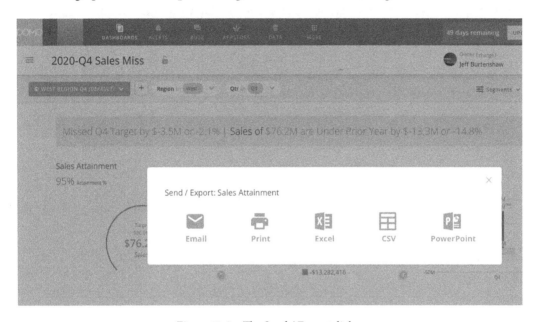

Figure 11.9 – The Send / Export dialog

2. Click **Email** to go to the **Send** dialog.

3. Notice the other distribution options in addition to **Email**. Don't expect much from the **Print** option here since, in this digitally transformed mobile-first world, the **Print** option has not been given much attention.

Now that we understand how to distribute content via an email that contains a link back into Domo, let's learn about embedding the content into other common applications, such as PowerPoint.

Embedding in productivity apps with plugins and add-ons

For those of us that have content consumers that just want to work the same way they have always worked, there's good news as there is a way to embed Domo content into the Microsoft Office PowerPoint, Excel, and Word applications but only on the Windows operating system. Also, there is a Google Sheets add-on that works on Windows and iOS. Let's walk through an example of using the Google Sheets add-on:

1. Download the Google Sheets add-on from the **Tool downloads** page by clicking **MORE** on the main menu, clicking **ADMIN**, and clicking **Tool Downloads** under **… More**, as shown in the following screenshot:

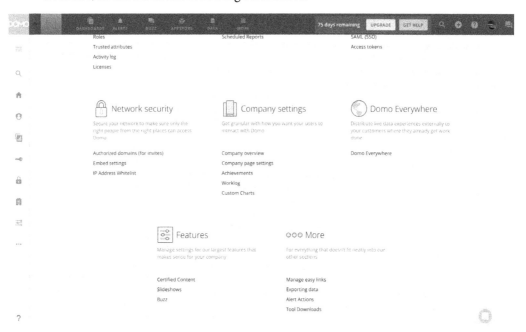

Figure 11.10 – Accessing the Tool Downloads page

2. On the **Tool downloads** page, scroll down to the **Domo Google Sheets add-on** area and click **TRY OUR SHEETS ADD-ON**, as shown in the following screenshot:

Figure 11.11 – Google Sheets add-on

3. Open a new Google Sheet.

4. From the Google Sheet menu, click **Extensions** then **Add-ons**, then **Get add-ons**, as shown in the following screenshot:

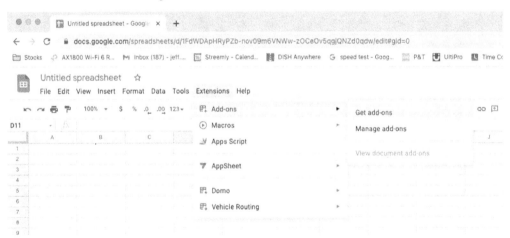

Figure 11.12 – Google Sheets – Extensions for add-ons

5. This will bring up the **Google Workspace Marketplace** dialog. Enter Domo in the search bar and press the *Return* key to return the results, as shown in the following screenshot:

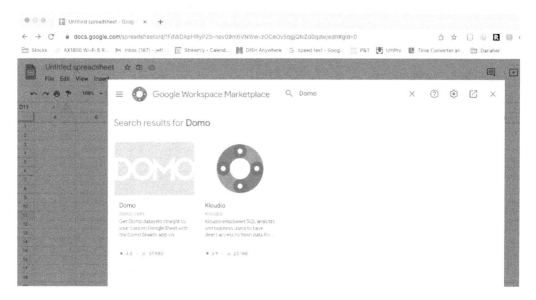

Figure 11.13 – Getting the Domo add-on

6. Click on the **DOMO** tile and then click **Install**, as shown in the following screenshot:

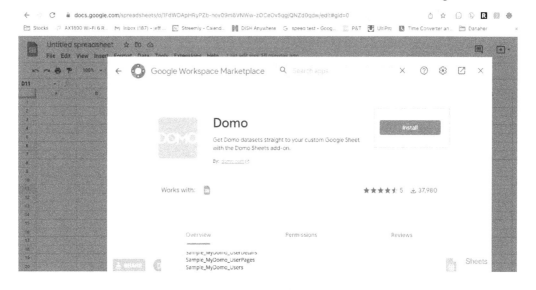

Figure 11.14 – Installing the Domo add-on

7. Click **CONTINUE** to install the Domo add-on, as shown in the following screenshot:

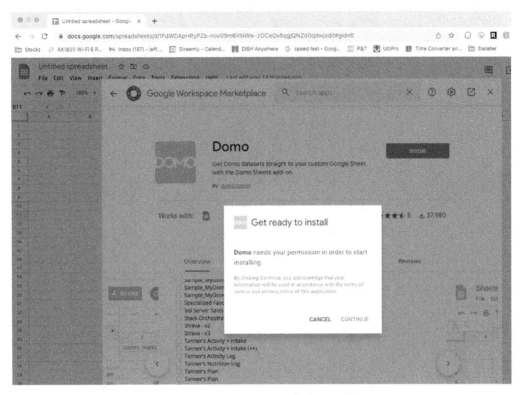

Figure 11.15 – Continuing with the installation

8. Select the Google account that is being used for the Google Sheet when prompted. Then, a message confirming that the add-on has been installed will appear, as shown in the following screenshot:

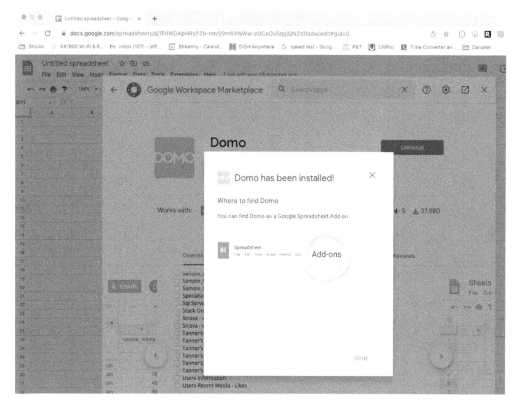

Figure 11.16 – Installation confirmation

9. Click **Done** then click **X** to close **Google Workspace Marketplace**.

10. Refresh your browser and click on **Extensions | Domo | Choose DataSet**, as shown in the following screenshot:

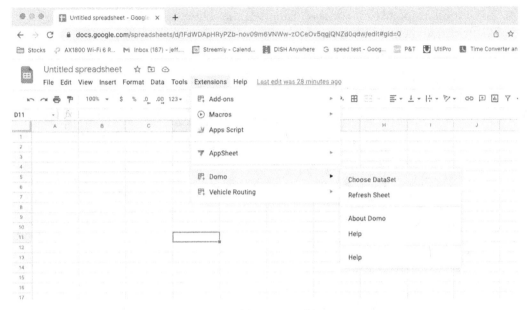

Figure 11.17 – The Domo add-on menu

11. Enter your Domo domain and click **Sign In**, as shown in the following screenshot:

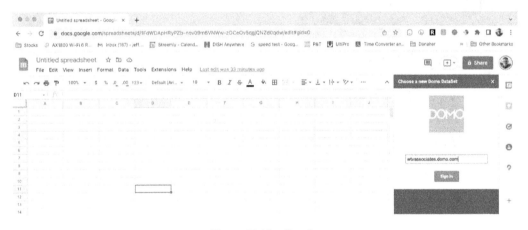

Figure 11.18 – Sign In

12. To connect, respond to the access dialogs. At the authorization end, you will see a confirmation message, as shown in the following screenshot:

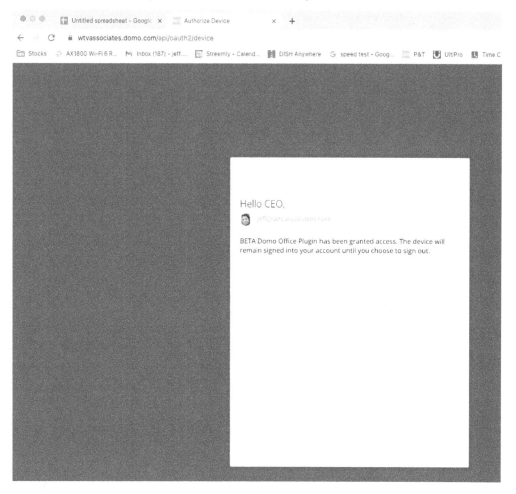

Figure 11.19 – Domo add-on access confirmation

13. Close the confirmation tab and click inside the Google Sheet. Then, click **2. Click to confirm login.**, as shown in the following screenshot:

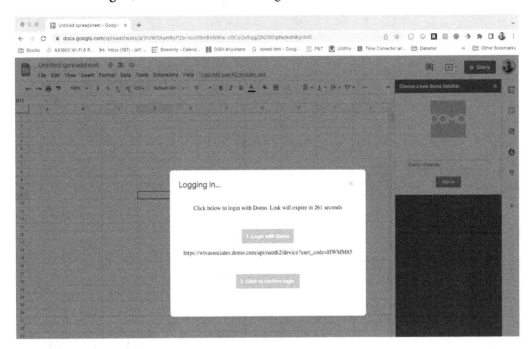

Figure 11.20 – Confirm login

14. This will bring up a list of the instance datasets that you can choose from and embed in the Google Sheet, as shown in the following screenshot:

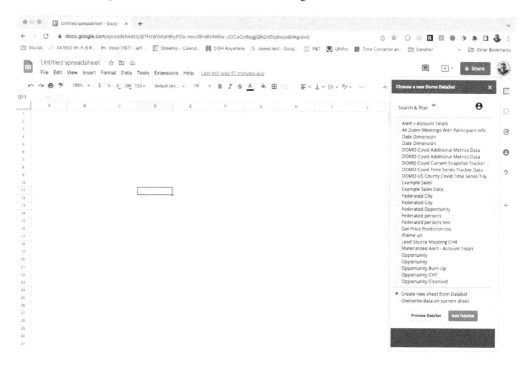

Figure 11.21 – Domo add-on dataset list

15. Select the **Materialized Alert - Account Totals** dataset and click **Add DataSet** to import the data into a new tab in the Google Sheet, as shown in the following screenshot:

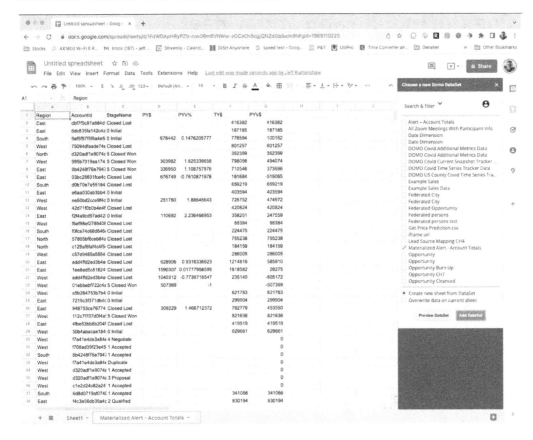

Figure 11.22 – New tab containing the dataset from Domo

16. To refresh the data from the Google Sheet menu, click **Extensions | Domo | Refresh Sheet**. This is a very useful feature for building models that link to the raw data in the sheet that's been pulled in from Domo.

17. On Windows only, there are plugins with similar embedding functionality for PowerPoint, Excel, and Word. Try the respective plugin downloads from the **Tool downloads** page.

Now that we know that we can embed Domo content through plugins and add-ons into common productivity apps, let's learn about Domo's mobile content distribution capabilities.

Using the mobile app and widget

Domo aspires to have us run our enterprises from our phones anywhere via the Domo mobile app, which is available in the Apple App Store for iOS and the Google Play Store for Android by searching for Domo, as shown in the following screenshot:

Apple App Store   Google Play Store

Figure 11.23 – The Domo mobile app in both the Apple App Store and Google Play Store

If you have previously installed the app and connected to another instance, to complete the following exercises, you can skip the installation steps and sign out of the connected instance of the app by clicking the **Menu** option, then touching the gear icon, then clicking **SIGN OUT**. Then sign in to your test instance through the app.

Follow these steps to download the **Domo app** onto your mobile device from the Apple App Store:

1. Open the app and click **SIGN IN**. Enter your Domo instance domain under **Company URL**, click **Remember company URL**, and click **SIGN IN**, as shown in the following screenshot:

Figure 11.24 – The SIGN IN screen

2. Enter your **Email** and **Password** for your Domo account and click **SIGN IN**. This will take you to the home screen, as shown in the following screenshot:

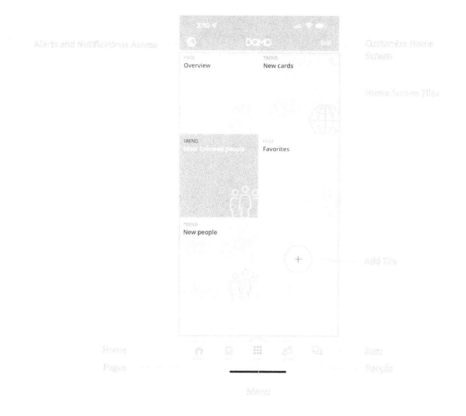

Figure 11.25 – Home screen

The following are brief descriptions of the features that are available from the home screen:

- **Alerts and Notification Access** opens the **Notifications** screen with tabs for **ALERTS** and **BUZZ** messages.

- **Customize Home Screen** opens a screen that allows you to add, delete, and rearrange the tiles from a list of pages.

- **Home Screen Tiles** is the scrollable area of the screen where the page tiles will appear.

- **Home** returns you to the home screen.

- **Pages** takes us to the screen that lists all the pages that are available and allows us to navigate to the desired page.

- **Buzz** opens the Domo messaging screen for receiving and sending **Buzz** messages.

- **People** opens the **People** screen, which allows us to search for Domo users' contact information.

- **Menu** opens the following screen:

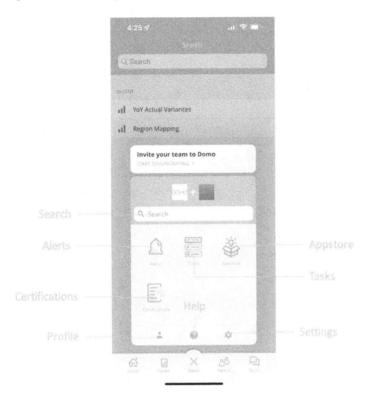

Figure 11.26 – Menu screen options

The following are descriptions of the **Menu** screen's options:

- **Search** allows you to search for **cards, pages**, and **Buzz** conversations, as well as **datasets, people, the Knowledge Base**, and **Dojo** community articles.

- **Alerts** opens a screen that presents alerts information.

- **Appstore** opens a screen that provides a link to the Domo app store.

- **Tasks** opens a screen that allows the user to search for, view, and enter tasks into the Domo **Projects and Tasks** tool.

- **Certifications** opens a screen where you can view content certification workflow requests that have been submitted or are waiting for approval.

- **Profile** opens the user's profile.

- **Help** opens a screen with links to various help resources in the Domo ecosystem.

- **Settings** opens a screen where you can control a variety of settings, as shown in the following screenshot:

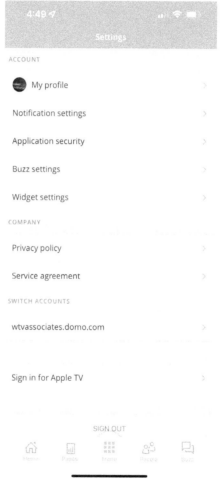

Figure 11.27 – Settings screen

The following settings are available:

- **My profile** accesses the user's profile.

- **Notification settings** controls the personal settings around **BUZZ** and **ALERTS**.

- **Application security** controls the **USE FACE ID** setting.

- **Buzz settings** controls the frequency of message removal and allows you to mark all conversations as read.

- **Widget settings** allows you to drag and drop favorited cards into the **WIDGET SELECTION** area.

- **Privacy policy** links to Domo's privacy policy.

- **Service agreement** links to Domo's service agreement.

- **SWITCH ACCOUNTS** lets you add and change Domo accounts.

- **Sign in for Apple TV** facilitates screen casting to an Apple TV.

- **SIGN OUT** signs the account out of the Domo app.

Now that we know what features are available within the Domo mobile app, let's start customizing.

Important Note

The following steps are for **iOS** devices. If you're using an **Android** device, see *Using mobile widgets* in the *Further reading* section.

Let's customize the home screen:

1. Click **Edit** on the home screen, click the + tile, enter `Sales` in the **Search** box, click + next to **2020-Q4 Sales Miss** so that it changes from a + to a checkmark, and click **Done**.

2. Click and drag **PAGE 2020-Q4 Sales Miss** to the top-left corner, as shown in the following screenshot:

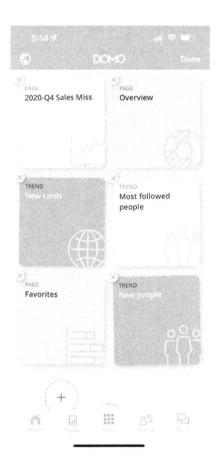

Figure 11.28 – Editing the home screen

3. Click **Done** to exit edit mode.

Next, let's add a card to our app home screen:

1. Click **Home**. In the app's **Search** box, enter `attain` and press **search**. Then, click **Sales Attainment** under the **CARDS** facet.

2. Click on the funnel icon in the lower left to filter, click **+**, and scroll down. Here, click on the **Year** field, click **VALUE**, select **2020**, click **Apply**, and click **X**.

3. Now, let's add this card to our **Favorites** page. To do so, click **Details**, scroll down, and click **ADD TO HOME**. Also, click **ADD TO FAVORITES**, then **<**, and finally **cancel search**.

4. Click **Edit**. Then, drag and drop the **Sales Attainment** tile to the top left, as shown in the following screenshot:

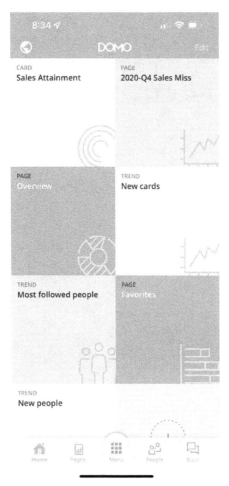

Figure 11.29 – The home screen with a card tile

Next, let's add a **widget** to our mobile device's home screen:

1. Since we have already favorited the **Sales Attainment** card, click **Menu** and then the gear icon to open the **Settings** area. Here, press **Widget settings** and drag and drop the **Sales Attainment** card via its handle into the **WIDGET SELECTION** area where it states **Place a card here**, as shown in the following screenshot. Up to five favorited cards can be added as widgets in the iOS version:

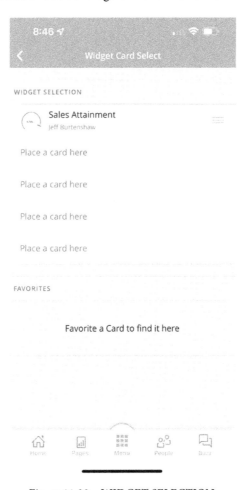

Figure 11.30 – WIDGET SELECTION

2. Click <.

3. Next, we need to add the widget. Swipe right to get to the iOS search page, which also holds the app widgets. Click **Edit**, as shown in the following screenshot:

Figure 11.31 – iOS search screen and editing the widget

4.   Click **Customize**, scroll down to **Domo Favorite Cards**, and press +. Compare what you have to what's shown in the following screenshot:

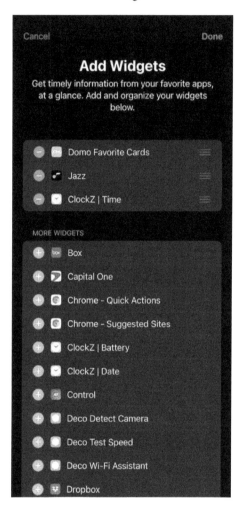

Figure 11.32 – The Add Widgets screen

5.   Click **Done**, and compare your screen's **DOMO FAVORITE CARDS** widget to the one shown in the following screenshot:

Figure 11.33 – Widget added to the iOS search screen

6. Click on **Sales Attainment** in the widget to go back to the Domo app.

> **Important Note**
> If, for some reason, you can't access the mobile app from your mobile device, simply enter your Domo URL into your mobile's browser.

It's great that we can create and distribute all this content via mobile, without knowing any mobile app coding. Using the Domo mobile app and its widgets enables mobile content to be distributed to users with no native device coding. This kind of anywhere, real-time access to information is unprecedented and no longer takes a specialized developer to make it happen. If you are not leveraging the power of the Domo mobile app, you are missing out on a huge part of enabling democratization.

In the next section, we will learn about publishing content via a URL.

Publishing via a URL

Imagine if you could create a public-facing web page that shows a slideshow of Domo cards and publish it publicly or privately, making it accessible to people without a Domo login. This capability comes as a standard feature of the platform. The cards on the web page are not interactive and can't be filtered, so what you see is what you get. However, when used in the right use case, it's an amazing feature. Follow these steps to create and publish a web page with Domo content:

1. Let's say we wanted to publish our sales results to members of the board of directors and large investors. On the **2020-Q4 Sales Miss** page, click the page sharing options icon in the header, as shown in the following screenshot:

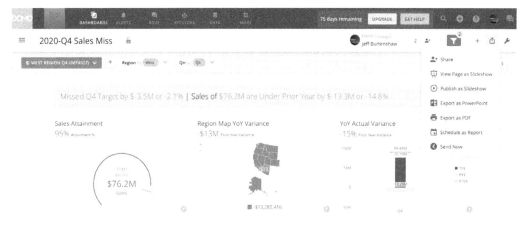

Figure 11.34 – Page sharing options

2. Click **Publish as Slideshow** and enter BOD Sales Report as the title.

3. The left pane shows the cards that are in the slideshow. By default, all the cards from the page are included. We want to limit the information that's published, so we are going to remove all but three cards from the slideshow. To remove a card from the slideshow, hover over the card and click the **X** button next to it. Remove all the cards except for **Sales Attainment**, **West Region Prior Year Q4 Variance $**, and **Actual Sales Prior Year Variance Trend**, as shown in the following screenshot:

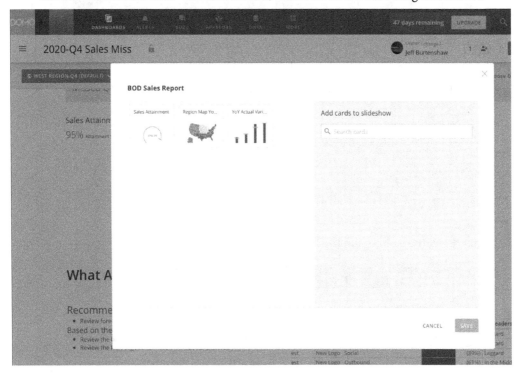

Figure 11.35 – Creating a slideshow

We could add any card from any page in Domo to the slideshow by doing a simple search in the **Add cards to slideshow** section.

4. Click **SAVE**. Then, in the **Congratulations!** dialog, click **Change** under **Current Publish Status**, as shown in the following screenshot:

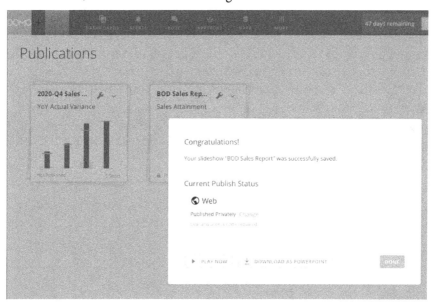

Figure 11.36 – Publication saved dialog

5. Click **Private**, as shown in the following screenshot:

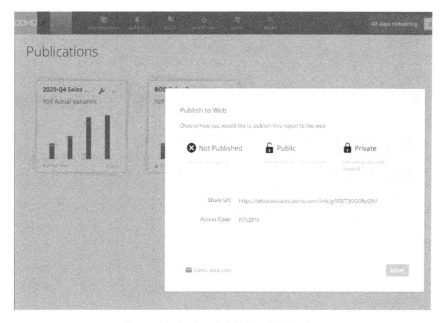

Figure 11.37 – The Publish to Web dialog

6. In the **Publish to Web** dialog, click **EMAIL WEB LINK,** enter the recipients, and send the email.

   To get a public URL to the slideshow, use the **Public** option, which does not have an **access code.**

7. Click **DONE.**

8. If you need to revise the publication later, click **MORE** via the main menu bar and then **PUBLICATIONS,** as shown in the following screenshot:

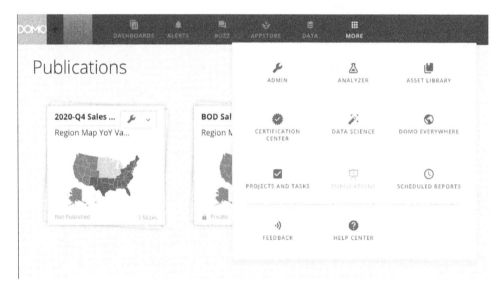

Figure 11.38 – PUBLICATIONS

9. Click on the wrench dropdown in the **BOD Sales Report** publication image and click **Play** for a preview.

10. Click **X** to close the preview.

11. To visit the actual page, click on the wrench dropdown in the **BOD Sales Report** publication image and click **Publish,** copy the URL to your browser, and enter your **access code** when prompted.

    Wow! Now, you are a web developer too... But seriously, this is a powerful way to distribute content to non-Domo users via a web browser.

In the next section, we will discuss how to set up digital wallboards for your office.

Creating digital wallboards

Many large enterprises have digital wallboards – those monitors on the wall running slideshow content. Now, we can too by leveraging the publication we created in the *Publishing via a URL* section in this chapter. There are two ways to do this – via Direct Connect or by streaming. Each wallboard requires a dedicated device that's capable of running a web browser, such as a Raspberry Pi or a notebook. Let's get started:

1. Regarding Direct Connect, open a browser, enter the publication URL, and set the browser to full-screen mode.

2. With the streaming approach (such as using **AirPlay**, **Chromecast**, or **Roku screen mirroring**), open a browser on the dedicated machine that is connected to the streaming/mirroring technology, enter the publication URL, and set the browser to full-screen mode.

   We have just learned how to save thousands of dollars by using Domo's DIY publications to set up cool digital wallboards.

In the next section, we will learn how to embed Domo content in a bespoke web page.

Embedding in a web page

The Domo platform provides a family of products branded Domo Everywhere that aims to make it easy to embed Domo content within other portals and applications. This includes external portals, intranets, custom applications, and multi-tenant Domo installations. These products provide white-labeling so that brand styling is possible. If you need to get to market fast with a new product or extend an existing product, upgrade to a world-class analytic experience through a portal, monetize data through a subscription service, or communicate via an enhanced intranet portal, then embedding Domo content into the solution is a great choice. We can think of Domo as an analytic component library where cards and dashboards can be embedded into other solutions.

Understanding the Domo Everywhere product family

There are three main products in the Domo Everywhere product family. The following table shows each product and a few examples of how the products can be used:

Product	Need	Real-World Example
PaaS	Embed analytics inside a custom application	Retail in-store management analytics
PaaS	Launch a new product that requires analytics	Technology infrastructure provider product performance analytics
Portal Embed	Add analytics to an existing web portal	Manufacturer supplier portal Large energy company PR portal
Portal Embed	Add analytics to an intranet	Financial services company employee intranet analytics
Orchestrator	Distribute data and dashboards from a central repository across Domo tenants securely	Insurance company broker support portal

Figure 11.39 – Domo Everywhere product family summary

Let's look at each product in more detail:

- **Platform as a Service** (**PaaS**) is a complete service stack that organizations can consume from their custom applications to provide white-labeled Domo dashboard and card experiences. Imagine if all we had to do, as a developer, was authenticate and embed dashboards and cards that had already been created in our application. In essence, this is a RESTful service that exposes the core Domo dashboard and card functionality from our custom apps. In a build versus buy analysis of internal build versus Domo Everywhere integration, Domo Everywhere integration wins hands down unless your paycheck depends on lines of code being written and the evaluators suffering from Not Invented Here syndrome.

- **Portal Embed** is a product that allows us to embed Domo cards and dashboards into existing portals such as intranets and external-facing portals. Enterprises with agents, distributors, suppliers, constituencies, boards, shareholders, and more love how fast and easy it is to embed an entire dashboard, story, or cards into an existing portal or website. You may have seen how to embed a Google map into a portal or website by cutting and pasting a snippet of code. This feature provides a similar cut-and-paste embedding functionality for Domo dashboards, stories, and cards.

- **Orchestrator**, also known as **Publish**, was born out of the need for enterprises to monetize information in their Domo data lake with their customer base, which was other businesses. They needed a secure way to distribute dashboards to their customers as a service. Domo created Orchestrator to fill this B2B2B product need. Orchestrator has a publish and subscribe model interface that partitions the data and content designs for each customer instance and keeps it all updated from the master Orchestrator instance. Data updates from clients back to the master Orchestrator instance are also managed.

Full coverage of the Domo Everywhere product family is beyond the scope of this book, but we can get a taste of its capabilities by walking through a simple embedded card example:

1. To embed a card in a web page, from the **2020-Q4 Sales Miss** page, click the sharing options icon, as shown in the following screenshot:

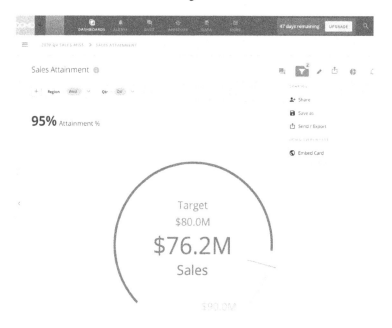

Figure 11.40 – The Embed Card option

2. Click **Embed Card**, then **Private** under **Embed options**:

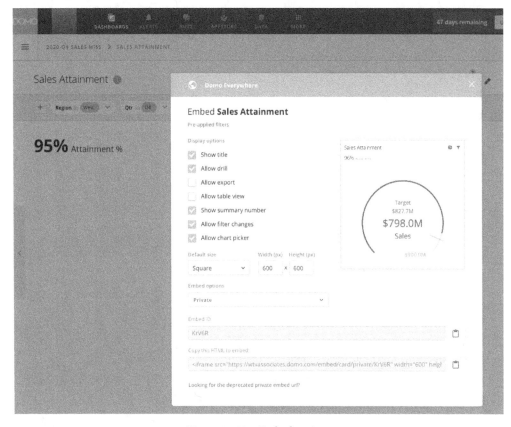

Figure 11.41 – Embed settings

3. Click the clipboard icon in the **Copy this HTML to embed** box. Click **X**.

4. Open any text editor and paste in the text for the frame.

5. Save the text file as `EmbedCard.html`.

6. To see what it will look like on a web page, open the file from a browser or double-click on the filename from the file explorer on your device:

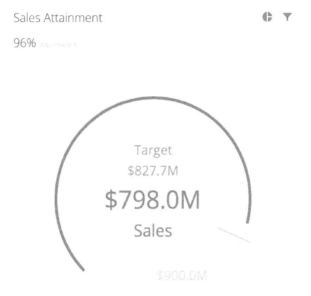

Figure 11.42 – Embedded card iframe rendered from a browser HTML file

The process to embed an entire dashboard is similar but is a paid feature that must be requested from the Domo customer success manager.

With that, we have looked at the many options we have for distributing content using the Domo platform. Now, let's summarize what we've learned.

Summary

In this chapter, we learned that there are many options for distributing Domo dashboards, stories, and cards both inside and outside of Domo. We started by looking at sharing cards, pages, and stories in Domo before learning how to use emails to deliver content on a schedule. We spent some time learning how to use the mobile app and widget, as well as publishing slideshows. Finally, we created digital wallboards before learning about the Domo Everywhere product family product options for embedding a multi-tenant distribution.

The main takeaway from this chapter is that there are many ways to distribute content to our audiences. The no-code mobile app for iOS and Android is pretty amazing as well. And if time-to-market, world-class analytic features on a PaaS model are needed in your custom apps' external portals or the intranet, then Domo Everywhere is a great choice. You can even use Domo Everywhere to orchestrate content syndication.

In the next chapter, we will discuss the platform's capabilities around alerting.

Further reading

To learn more about the topics that were covered in this chapter, take a look at the following resources:

- Using mobile widgets: `https://domohelp.domo.com/hc/en-us/articles/360044406813-Mobile-Widgets`

- Domo Embed overview: `https://developer.domo.com/docs/overview/embedded-analytics`

- Using Domo Embed to distribute cards and dashboards outside of Domo: `https://domohelp.domo.com/hc/en-us/articles/360043437993-Sharing-Cards-and-Dashboards-Outside-of-Domo-Using-Domo-Embed`

12
Alerting

We spend so much time as data consumers working to make sense of data. Increasingly, technology is allowing us to turn that dynamic around and make the data work for us. A major way of putting data to work is through alerting technology. Alerts allow continuous monitoring of data, looking for conditions to be met, crossed, removed, and created or have failed. A well-thought-through alerting strategy can be more impactful on business performance than a great dashboard. In fact, an argument can be made that the main reason to have a monitoring dashboard is to automate alerts. People intuitively know what to alert on but lack the technology to execute it. The Domo ecosystem provides the data capture and alerting technology to enable us to implement automated alerting and discover what others are alerting on.

In this chapter we will specifically cover the following:

- Defining an alerting strategy
- Understanding the Alert Center
- Setting alerts on a card
- Setting alerts on a dataset
- Discovering management chain alerts

Technical requirements

To follow along with this chapter, you will need the following:

- Internet access

- Your Domo instance and login

- Ability to download example files from GitHub

- Dashboard and cards created in *Chapter 6*, *Creating Dashboards*, and *Chapter 7*, *Working with Drill Pathways*

- Use of Google Sheets to access the following `Alert Strategy` worksheet: `https://docs.google.com/spreadsheets/d/1pi9rVxrhBVRHGkm1Bb vptsksC4TgTxJg2_zMev5wiho/edit?usp=sharing`

> **Important Note**
>
> If you don't have a Domo instance, get a free trial instance here: `https://www.domo.com/start/free`.

Defining an alerting strategy

Asking the question *What is our alerting strategy?* will likely result in a bunch of blank stares. We spend so much time on creating monitoring dashboards and then often fail to take the next step and activate the data through alerts. Or, we go the other direction and alert on so many things that the alerts become meaningless noise. Defining an alerting strategy is about taking a planned and governed approach to determine what the system should monitor, by what method, how frequently, and what actions are expected. It is suggested that the strategy is created outside of the tool first, and then implemented in the toolset. This discipline forces strategic thinking and allows for rationalization of the number of conditions to be monitored. Keep in mind that a strategy, by design, is a high-level blueprint and will not be specific enough to cover all implementation details. The specific elements involved in defining a good strategy are contained in the **Alert Strategy** worksheet, as seen in *Figure 12.1*:

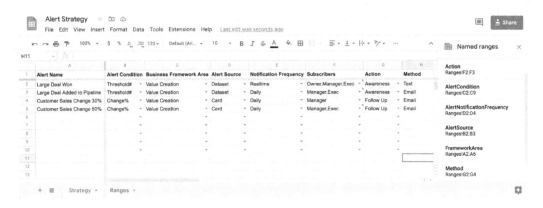

Figure 12.1 – Alert Strategy worksheet

The following are descriptions of the elements of the Alert Strategy:

- **Alert Name**: A brief label of what the alert is for.

- **Alert Condition**: The type of condition the alert uses. For example, **Threshold** is a condition comparison to an actual number, for instance, Amount > 100,000.

- **Business Framework Area**: The area of the business the alert is on; see *Chapter 10, Telling Relevant Stories*, for more information on the Business Framework.

- **Alert Source**: Optional for the strategy but, if known, can be a dataset or card. Try to set the alert on the dataset when possible, because this is more efficient than setting the alert on one of the potentially many cards on the dataset. Use card alerting when the summary number is the alert metric.

- **Notification Frequency**: Specifies how often the alert should be communicated. The options are real-time or batched in periodic communication.

- **Subscribers**: Indicates the type of people to be notified when the alert is triggered.

- **Action**: The intended action to be taken, whether simple awareness or specific follow up is intended.

- **Method**: The communication channel to be used for the alert notification.

A good practice is to go through each dashboard and fill in the Alert Strategy before creating alerts.

Next, let's learn about the Alert Center options in Domo.

Understanding the Alert Center

The **Alert Center** is the hub for viewing, editing, sharing, subscribing, reviewing trigger history, and changing alerts in Domo. It's also where you can discover insights surfaced by Mr. Roboto – the Domo **artificial intelligence** (**AI**) engine. The Alert Center is accessed from the **ALERTS** icon on the main menu bar, as seen in *Figure 12.2*:

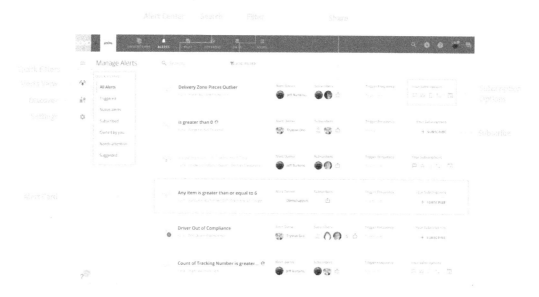

Figure 12.2 – Alert Center alerts view

The following are descriptions of the Alert Center **Alerts View** page:

- **Alert Center** is the main menu bar option to access alerts.
- **Search** is where to search alert content.
- **Filter** contains filtering options for alerts.
- **Share** is where to share alerts.
- **Subscription Options** enable the selection of notification channels (text, email, push notifications, and phone calls) and scheduling options (daily or weekly) on alerts that the user is subscribed to.
- **Subscribe** subscribes the user to the alert.
- **Quick Filters** are predefined search filters.
- **Alert Card** is where each row is a summary of the specific alert. Clicking on the card will take you to the alert card information page, as seen in *Figure 12.3*:

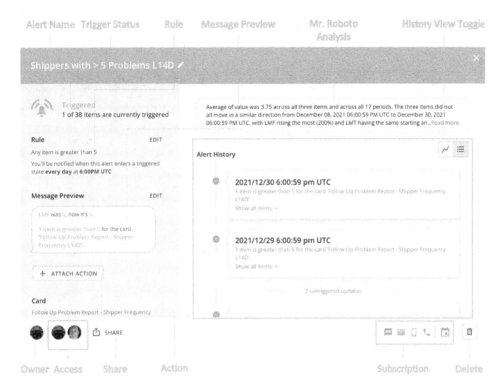

Figure 12.3 – Alert card information page

The following describes the alert card information page features:

- **Alert Name** is the title of the alert.

- **Trigger Status** is the status of the rule being triggered for the alert.

- **Rule** is the logic for triggering the alert.

- **Message Preview** shows what the alert message will look like when triggered.

- **Mr. Roboto Analysis** is the Domo AI analysis of the alert's triggered history.

- **History View Toggle** is where you can change between a chart and a list view of the alert history.

- **Owner** shows the owner of the alert.

- **Access** shows who has subscribed to the alert.

- **Share** enables sharing of the alert with others.

- **Action** takes you to the feature to set up webhooks or create tasks from an alert trigger, as seen in *Figure 12.4*:

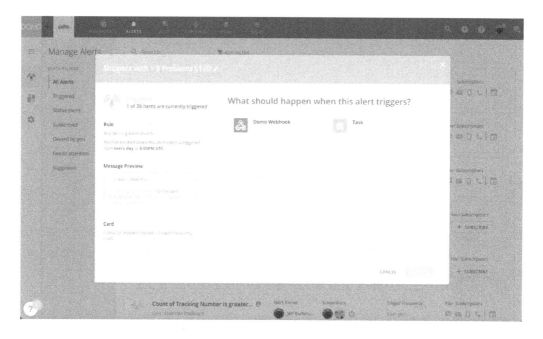

Figure 12.4 – Alert action configuration page

- **Subscription** displays subscription options when subscribed, or a subscribe button when not subscribed to the alert.

- **Delete** removes the alert permanently.

- **Alerts View** is the sidebar option to view the **Manage Alerts** page, as seen in *Figure 12.2*.

- **Discover** is the sidebar option to view the **Discover** page, as seen in *Figure 12.5*.

- **Settings** is the sidebar option to view the alert settings for instant, daily, and weekly alerts, as seen in *Figure 12.7*.

Next, let's review the features on the Alert Center **Discover** page, as seen in *Figure 12.5*:

Figure 12.5 – Alert Center Discover page

The following are descriptions of the Alert Center **Discover** page:

- **Topics** are predefined and categorized system alerts and trending cards; for example, suggested alerts to subscribe to, new people in the Domo instance, followed alerts, cards recently shared with you, suggested cards to follow, and cards with Buzz conversations. Mr. Roboto, the Domo AI, constantly scans the system and surfaces the alerts in a heat map. Hovering over a category stops the hopper rotation and highlights the heat map items for the category. Clicking on the category opens a card carousel for the category items, as seen in *Figure 12.6*:

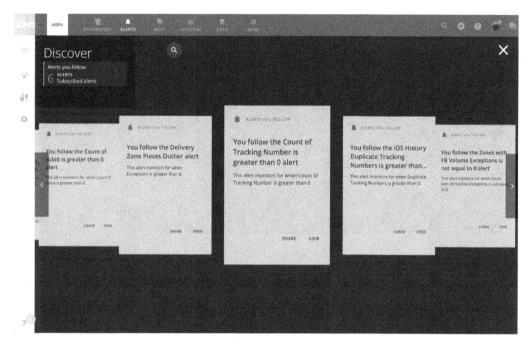

Figure 12.6 – Discover alert card carousel

The following are the feature descriptions for the carousel:

- **VIEW** opens the alert card information page.

- **SHARE** enables sharing of the alert.

- **Search** enables searching the system for discovered alerts.

- **Heat Map** visualizes all the categorized alerts in a real-time heat map. Each rectangle represents an alert.

- **Hopper** shows, in real time, alert cards being surfaced by Mr. Roboto.

Next, let's review the features on the Alert Center **Settings** page, starting with the **INSTANT ALERTS** tab, as seen in *Figure 12.7*:

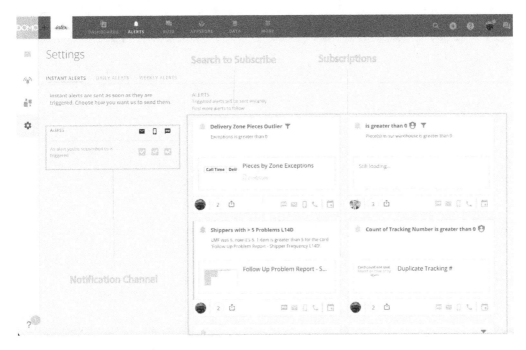

Figure 12.7 – INSTANT ALERTS settings

The following are descriptions of the **INSTANT ALERTS** settings:

- **Search to Subscribe** enables searching the unsubscribed alerts to subscribe to.
- **Subscriptions** shows the subscribed alerts.
- **Notification Channel** controls the communications method for alerts designated as instant alerts. The default alert setting is **Instant**.

Now, let's review the features on the **DAILY ALERTS** tab on the Alert Center **Settings** page, as seen in *Figure 12.8*:

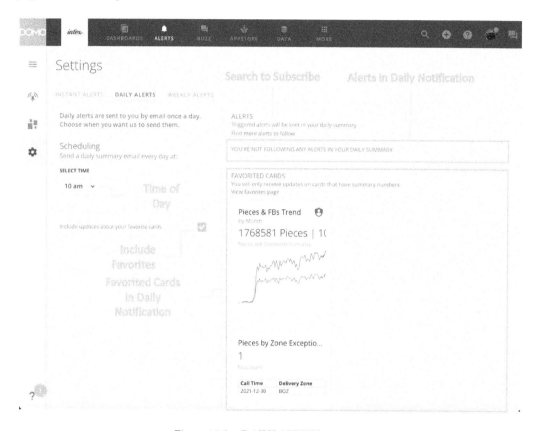

Figure 12.8 – DAILY ALERTS settings

The following are descriptions of the **DAILY ALERTS** settings:

- **Search to Subscribe** enables searching the unsubscribed alerts to subscribe to. Unfortunately, this adds the alert to **INSTANT ALERTS**, and we have to go there, then to the specific alert's options, then click the calendar option, and then select **Daily**.

- **Alerts in Daily Notification** shows a list of all the alerts in the daily message.

- **Time of Day** configures what time the daily message is sent.

- **Include Favorites** sets whether to include favorited cards with summary numbers in the daily message.

- **Favorited Cards in Daily Notification** is a list of all the favorited cards in the daily message. Notice the list here is read-only. Click the **View favorites page** link to manage favorites.

Next, let's review the features on the **WEEKLY ALERTS** tab on the Alert Center **Settings** page, as seen in *Figure 12.9*:

Figure 12.9 – WEEKLY ALERTS settings

The **WEEKLY ALERTS** settings are nearly the same as the **DAILY ALERTS** settings, except they add the ability to specify on what day of the week the message is sent, and don't have the option to include favorited card updates.

That concludes the overview of the alerting features. Now that we know how to manage alerts, let's set an alert.

Setting alerts on a card

Referring to our Alert Strategy, one of the alerts needed is for when a customer's trailing 12-month sales fall below 50% for the prior 12 months. That way, the system watches and alerts us when a customer's sales dramatically drop.

> **Important Note**
> When creating alerts, it is good to create a card specifically supporting each alert. This allows the summary number to be tailored with beast modes for the alert condition. It also reduces the chances that someone may change the card summary number and inadvertently disable the alert.

The following are the steps to set this alert:

1. On the **Sales** dashboard, click on the **Change In Sales by Account** card gear dropdown and select **Save as**.

2. Enter `Alert – Change in Sales by Account` as the card title, check the **Take me to the new card when I'm done** option if it isn't already, then click **SAVE** in the **Save as** dialog.

3. Click the Analyzer icon, click and drag the **Sales % Change** beast mode into the **FILTERS** area, choose the **is less than or equal to** option, enter `-50` as the threshold, then click **Apply**.

4. In the **Y AXIS** area, format the **Sales % Change** field as **Percentage** and verify the **Multiply by 100** option is not checked.

5. Click the **CHART TYPES** drop-down menu, select **Tables and Textboxes**, and click **Mega Table**.

6. Click **ADD CALCULATED FIELD**, add the `Alert – Sales % Change <= -.5` title, and fill it in with the following formula:

```
-- on real data use CURRENT_DATE()
-- L12M sales
case when
(case when `CloseDate`<= date('12/31/2020') and
`CloseDate`> DATE_SUB(date('12/31/2020'),interval 12
Month) then `Amount` else 0 end
/
-- Year ago sales
case when `CloseDate` <= DATE_SUB(date('12/31/2020'),
interval 12 Month) and `CloseDate` > DATE_
SUB(date('12/31/2020'),interval 24 Month) then `Amount`
else 0 end) -1
<= -.5 then 1 else 0 end
```

This formula counts the occurrences where the change in sales is less than or equal to -50%.

7. Click **SAVE & CLOSE** to save the beast mode.

8. Click the summary number dropdown and change the column to the **Alert - Sales % Change <= -.5** beast mode, as seen in *Figure 12.10*:

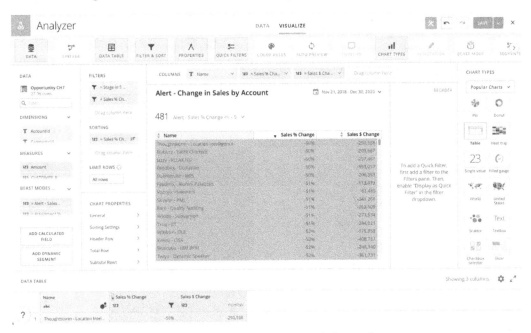

Figure 12.10 – Alert - Change in account sales configuration

9. Click the **SAVE** dropdown and click **Save and Close**.

10. Click the **See Alerts** bell icon in the card toolbar, and click **+ NEW ALERT** in the **ALERTS** panel.

11. Enter `Account YoY Sales Dropped %50 or More` over the **Add an alert name** title.

12. Set **Meets this condition** to **Changes by (Δ)** and enter 1 in **For this value**, as seen in *Figure 12.11*:

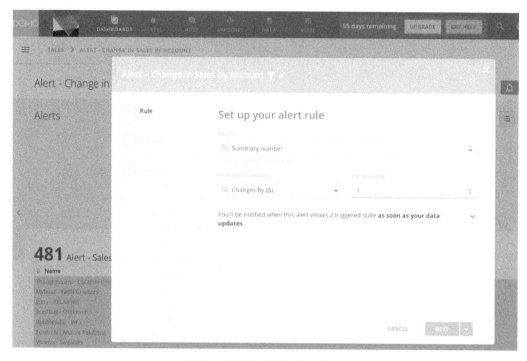

Figure 12.11 – Alert rule setup

The *Changes by 1* setting means that anytime the count moves up or down by one or more from the previous alert check, then the alert will trigger.

13. Click **NEXT**.

14. In **Compose a notification message**, let's add a timestamp to the end of the message by placing the cursor before the period and entering `last checked at`, then click **+ METRIC** and click **Last checked at**, as shown in *Figure 12.12*:

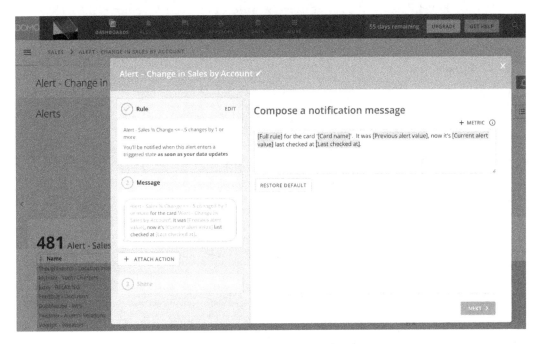

Figure 12.12 – Alert message configuration

15. Click **NEXT**, which will take us to the alert sharing dialog.

Sharing alerts

To share the alert, we have two options – with people or to a Buzz conversation:

1. On the **PEOPLE** tab, in the **Enter users, groups or emails** box, enter SALES to share the alert with the **SALES** group users.

2. Customize the message by entering Sharing this alert on Customers whose sales are down 50%+ YOY, as seen in *Figure 12.13*:

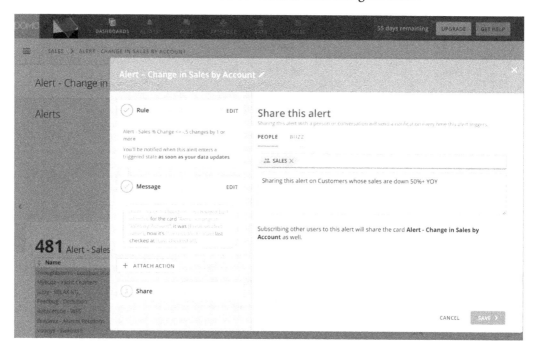

Figure 12.13 – Alert sharing dialog

3. Click **SAVE**, and then click **X** to close the alert dialog. The new alert will appear in the **Alerts** panel, as seen in *Figure 12.14*:

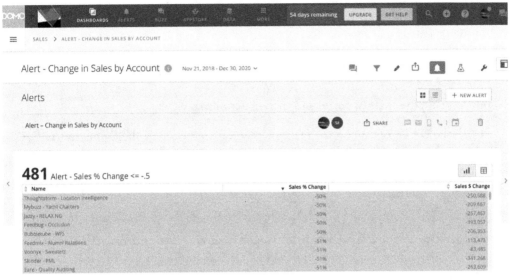

Figure 12.14 – Card alert panel

Now, whenever the data is updated for the card's dataset, the alert will check to see whether the number of accounts with a sales drop of 50% or more has changed and if it has, will send out the alert according to the alert preferences to people and to the Buzz channel, if specified.

What if we want to know not only whether the number changed but what accounts have changed too? To do that, we must create an alert using the dataset alert feature.

Setting alerts on a dataset

Setting alerts at the dataset level is more flexible in some ways than setting the alert at the card level. In fact, dataset alerting was created to address several limitations of setting alerts at the card level. Those card-level limitations included only having one criterion and only being able to include the triggered value and not additional information in the alert. Let's create an alert using the dataset alert feature and see how some of these limitations were overcome. Let's make an alert for the West Region to alert the sales manager, which triggers immediately when each account in the region drops 50% or more below the prior year's sales and creates a task for the manager to follow up.

Creating a supporting materialized view for the alert

We need to create an account aggregate view and then create a **materialized dataset** from the view to support the alert.

> **Important Note**
>
> Views, fusions, and beast modes are not supported via the dataset alert directly. These must be physical datasets, and not virtual artifacts, for the dataset alerting to work.

The following are the steps to create and then materialize a view:

1. Click the **DATA** option on the main menu toolbar, then click the **Opportunity CH7** dataset.

2. Click **OPEN WITH** and select **Views Explorer**, and rename the view to `Alert - Account Totals`.

3. Click **SELECT** and then click the trash can icon to delete all the default columns.

4. Click **+ Group** and drag **AccountId**, **StageName**, and **Region** from **AVAILABLE COLUMNS** and drop them in the **CATEGORIES** area, and click **FINISH**, as seen in *Figure 12.15*:

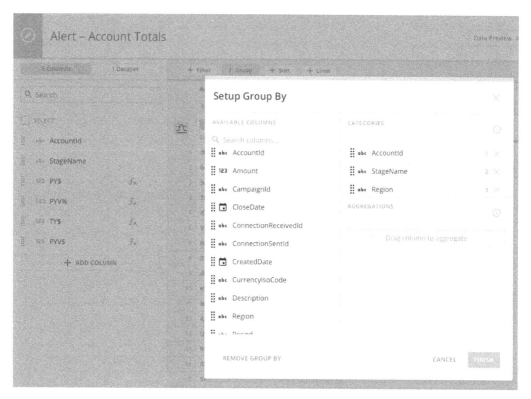

Figure 12.15 – View Setup Group By with AccountId

5. Click **+ ADD COLUMN** and scroll down and check **PY$**, **PYV%**, **TY$**, and **PYV$**, then click **ADD 4 COLUMNS**, as seen in *Figure 12.16*:

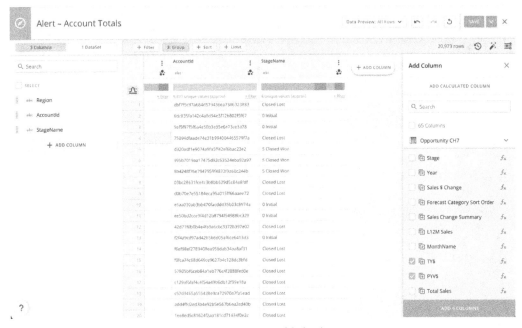

Figure 12.16 – View added columns

6. Click **SAVE AS**, then click **SAVE**.

Now, we have a view of opportunities aggregated by account, stage, and region. We will materialize the view via an **ETL** (short for **Extract, Transform, and Load**) job because the alerting feature needs a materialized source to show actual value changes:

1. Click **DATA** on the main menu and click **ETL** in the **Magic Transform** area.

2. Change **DataFlow Name** to `Materialize Account Alerts`.

3. Click and drag an **Input DataSet** tile to the canvas, then click **CHOOSE DATASET**, click **Alert – Account Totals**, and click **CHOOSE DATASET**.

4. Click and drag an **Output DataSet** tile to the canvas, then change **Output DataSet Name** to `Materialized Alert - Account Totals`.

5. Click the **Settings** icon for the job and click **Only when DataSets are updated** in **When should this DataFlow run?**, then check **Alert – Account Totals**, as seen in *Figure 12.17*:

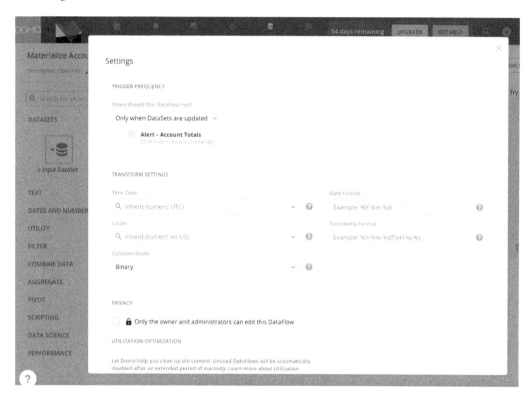

Figure 12.17 – ETL job settings

6. Scroll down and click **APPLY**, and compare the job configuration to *Figure 12.18*:

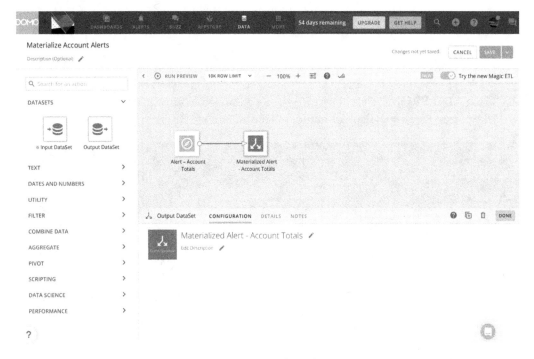

Figure 12.18 – ETL job configuration

7. Click on the **SAVE** dropdown, click **Save and Run**, and then click on **SAVE AND RUN** again.

Great! Next, let's continue our setup and steps and create a project to hold the tasks created by alert triggers.

Creating a project for the tasks

The steps for creating a project that will hold the tasks added by alert triggers are as follows:

1. Click on the **MORE** option on the main menu and then click **PROJECTS AND TASKS**.

2. Click **ADD PROJECT** and enter `West Region Accounts with a Large Drop in Sales` as **Project name**, as seen in *Figure 12.19*:

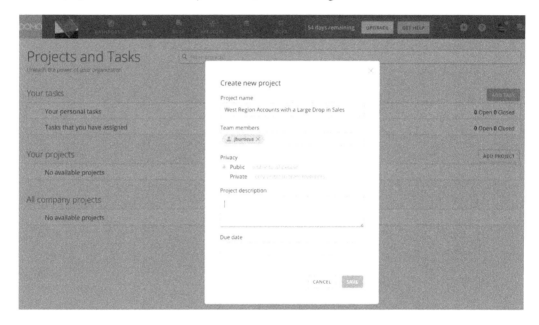

Figure 12.19 – New project

3. Select a team member that is the West Region sales manager.

4. Click **SAVE**, which lands us on the project tasks page, as seen in *Figure 12.20*:

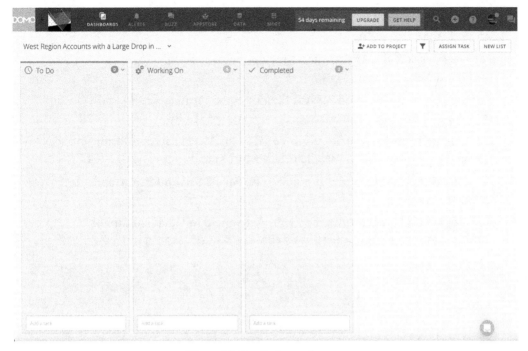

Figure 12.20 – Project tasks page

5. Tasks created by the alert triggers will appear here as tiles in the list swim lanes and allow the manager to track activities taken on accounts with declining sales.

Next, we will create the dataset alert.

Creating a dataset alert

The following are the steps to create the dataset alert:

1. Click the **DATA** option on the main menu toolbar, click the **Materialized Alert - Account Totals** dataset, then click **ALERTS** in the dataset header toolbar option, as seen in *Figure 12.21*:

Figure 12.21 – Dataset alert feature

2. Click **+ NEW ALERT**, and enter `West Region Account Sales 50% or more Decrease` as the title.

3. For **Metric**, select the **Any row is added** option, which will trigger the alert when the filter criteria are met.

4. Select **PYV%** for **Metric**, and **Is less than or equal** for **Meets this condition**, and enter `-.5` in **For this value**, then click on **+ ADD FILTER**.

5. Select **StageName** for **Metric**, and **Is equal to** for **Meets this condition**, and select **5 Closed Won** in **For this value**, then click **APPLY**, and click **+ ADD FILTER**.

6. Select **Region** for **Metric**, and **Is equal to** for **Meets this condition**, and select **West** in **For this value**, then click **APPLY**.

7. Add **Region** and **StageName** along with **AccountId** to **What column or combination of columns identifies a row as unique ?**, as seen in *Figure 12.22*:

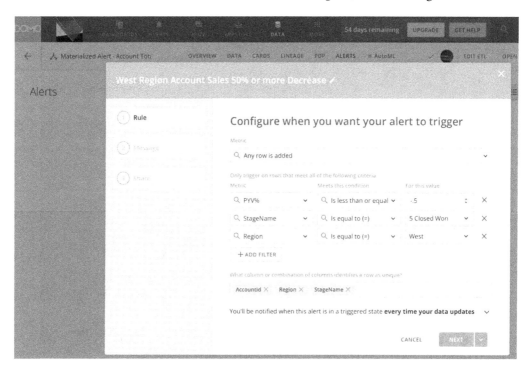

Figure 12.22 – Dataset alert rule configuration

8. Click **NEXT**.

9. Enter the following in **Body message**:

```
Region: [Region]
AccountId: [AccountId]
StageName: [StageName]
PYV%: [PYV%]
PY$: [PY$]
TY$: [TY$]
PYV$: [PYV$]
```

This can be seen in *Figure 12.23*:

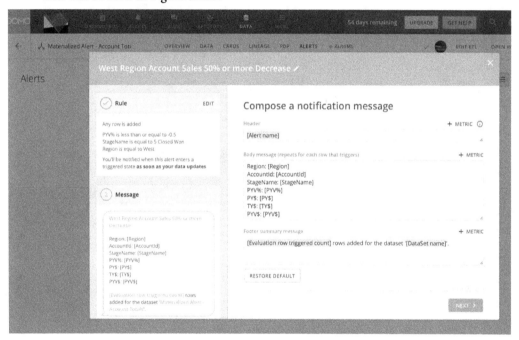

Figure 12.23 – Dataset alert message configuration

10. Click **NEXT**, share the alert with the West Region manager by entering their username in the **PEOPLE** area, click **SAVE**, and click **X**.

Fantastic! Now, every time an account in the West Region drops 50% from the prior year, the manager will be alerted with details!

But what about accountability? Yes, let's configure this alert to create a task as well.

Using alerts to invoke actions

Creating a task when an alert triggers is a great feature to create accountability on follow-up items. The following are the steps to do this:

1. On the **Materialized Alert - Account Totals** dataset detail view under the **ALERTS** tab, click the **West Region Account Sales 50% or more Decrease** alert, as seen in *Figure 12.24*:

Figure 12.24 – Editing the alert

2. In the **EDIT** dialog, scroll down in the left panel and click **+ ATTACH ACTION**, as seen in *Figure 12.25*:

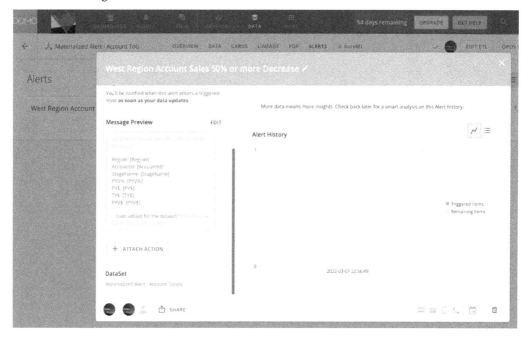

Figure 12.25 – Attach action alert

This takes us to the action dialog presenting a choice between **Domo Webhook** and **Task**, as seen in *Figure 12.26*:

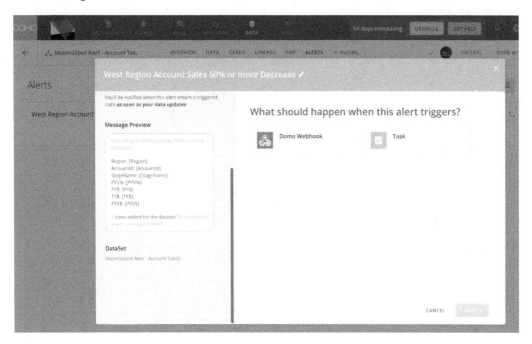

Figure 12.26 – Attach action alert options

The **Domo Webhook** option provides a mechanism to input the webhook URL, choose the HTTP API method (**POST, DELETE, PATCH, GET, and PUT**), then enter a secret authorization key, if required, for any available third-party web service endpoint, as seen in *Figure 12.27*:

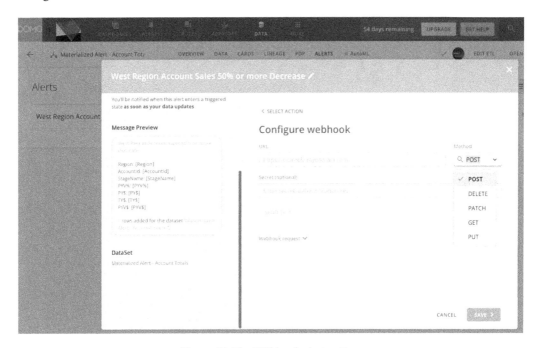

Figure 12.27 – Webhook alert option

The **Task** option provides a method to create a task in a Domo project with the alert information. The following are the steps for the **Task** option:

1. Click the **Task** option from **+ ATTACH ACTIONS** and take the defaults, as seen in *Figure 12.28*:

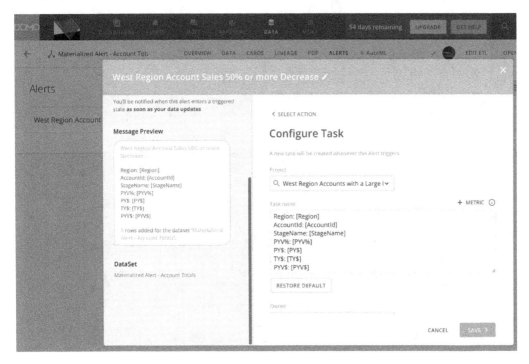

Figure 12.28 – Attach task alert configuration

2. Click **SAVE**. Now, when the alert triggers, a task will be added to the project to follow up on the account.

Now, we can set alerts at the dataset level, pass along detailed information with the alert, and, when the alert requires, combine several filter fields.

Next, let's discuss how you can use Domo alerts to see what your manager is watching.

Discovering management chain alerts

A nice communication feature in Domo is that we can see what alerts people are subscribed to. If we don't have permissions, we cannot see the alert information, but we can see the title of the alert. This can give us great insights into what is important to our management team, and if we are directly responsible for one of the alerts our management is monitoring, we will want to create an alert with a threshold that notifies us before it triggers for them!

The following are the steps to see what alerts a person is subscribed to:

1. Click on the profile picture on the main menu, which takes us to our **personal profile**, as seen in *Figure 12.29*:

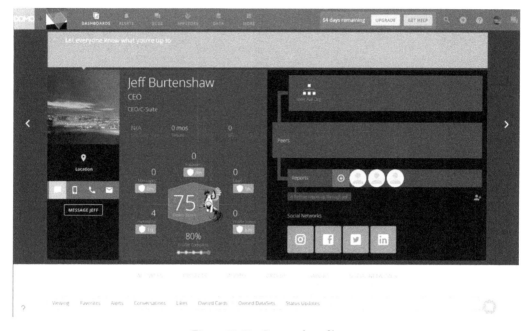

Figure 12.29 – Personal profile

2. Click on **View Full Org** to see an interactive organization chart, as seen in *Figure 12.30*:

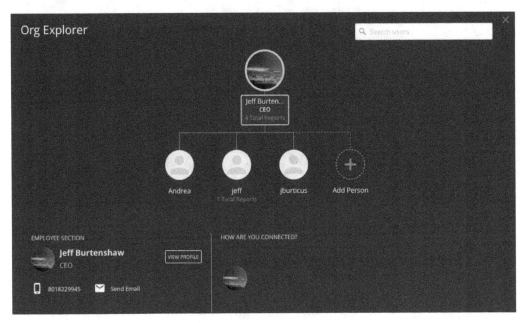

Figure 12.30 – Org Explorer

From **Org Explorer**, we can search for users or browse through the organization chart.

> **Important Note**
>
> The organization chart works only if the **Reports** and **Reports to** properties are set. This can be done via the profile page.

3. Click on a profile avatar and then, in the lower left, click **VIEW PROFILE** to see their profile including alerts they subscribe to.

4. Click on the **Alerts** tab, as seen in *Figure 12.31*:

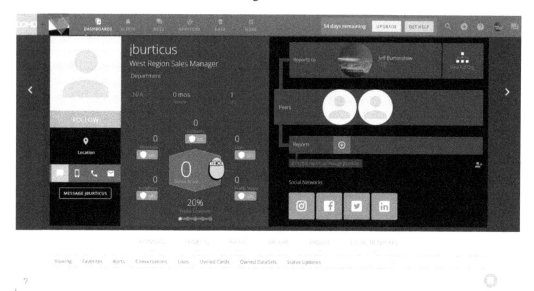

Figure 12.31 – Profile Alerts tab

5. Note the content in the **Alerts** tab under **Which alerts does {user} follow?**, as seen in *Figure 12.32*:

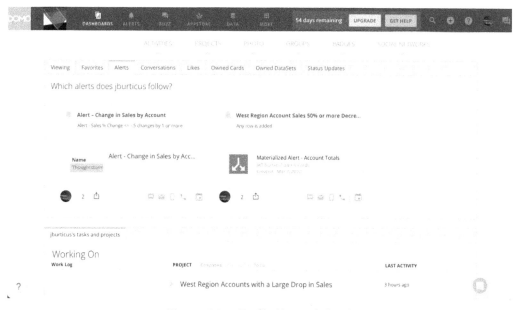

Figure 12.32 – Profile Alerts tab detail

We can quickly see that this user is the West Region sales manager and is following the **Change in Sales by Account** and **West Region Account Sales 50% or more Decrease** alerts. This feature flattens an organization and is a terrific goal alignment tool, as it provides incredible transparency as to what is important, especially when the Alert Strategy is solid.

Let's summarize what we learned in this chapter.

Summary

In this chapter, we learned that defining an alert strategy is a best practice to avoid alert spam and brings focus to what the important monitoring points and thresholds are in a business area. The alert strategy paired with the alert discovery capabilities brings unparalleled transparency and focus to an organization. We saw how to use a worksheet template to create an alert strategy and then walked through the pertinent functions in the Domo platform for monitoring and managing alerts. Truly, the platform enables us to individually create, publish, and share alerts that put the data to work for us rather than us having to work the data. We can also readily find and subscribe to alerts others have created. Alerts can bring peace of mind knowing that Mr. Roboto is always on duty surfacing triggers for things we know are important and discovering things we may not know in our data.

We learned ways to create alerts at both the card and data levels and discussed functional tradeoffs at each level. We even saw how to share alerts with others and create automated tasks for follow-up when alerts trigger. It is exciting to see how much work we can offload to the Domo alerting framework, which shifts our time allocation from monitoring to taking action.

In the next chapter, we will discuss the platform's capabilities around instant messaging.

Further reading

The following are related items of interest:

- For reference material on alerting and notifications: `https://domohelp.domo.com/hc/en-us/sections/360007334393-Notifications-and-Alerts`

- Creating an alert for a dataset: `https://domohelp.domo.com/hc/en-us/articles/360042925994-Creating-an-Alert-for-a-DataSet`

13
Buzzing

Collaboration among users is a core part of making sense of information. The ability for users to dialog around content is critical. A big gap in the efficiency and effectiveness of most messaging solutions is that users need to bring their own content into the message, supporting the discussion from outside the communication tool. Take time to notice how much time is spent on searching for files to attach or working to paste the content into a message. Now, imagine a solution that eliminates the need for attachments or cut and paste operations, one where users don't have to bring their own content to the conversation. For example, instead of having to find and attach a chart to an email or text, what if we could simply start commenting on the chart itself so the context and the messaging were always together?

Buzz is Domo's topic-based, content-integrated messaging application! It goes one step further than tools such as **Slack** (which provides self-defined, topic-threaded conversations) in that Buzz not only provides an automated contextual topic but also provides the content. **Contextually aware** means that the Buzz message topic follows users' navigation, changing message topics automatically as we navigate through pages and card content. Say goodbye to attachments and cut and paste insertions for data content; just add the message to the content and provided topic. Of course, Buzz also supports custom topic creation via a public broadcast channel, private one-on-one, or group channels. There are also page, card, and project channels. Leveraging the context and content auto-provided nature of Buzz is a major productivity booster in collaborative communication and democratizes data sharing in a new way. In fact, Domo allows unlimited **Social users** to use Buzz without a cost or licensing impact, even if the Social user does not have a Domo license.

In this chapter, we will specifically cover the following topics:

- Learning the Buzz menu options
- Using the Universal Compose feature
- Leveraging Social users
- Understanding the Buzz Navigation pane
- Touring the Buzz Conversation pane
- Discovering the Buzz Conversation Detail pane

Technical requirements

To follow along with this chapter, you will need the following:

- Internet access
- Your Domo instance and login
- The ability to download and install files
- The dashboard and cards created in *Chapter 6, Creating Dashboards*, and *Chapter 7, Working with Drill Pathways*.

> **Important Note**
>
> If you don't have a Domo instance, get a free trial instance here: `https://www.domo.com/start/free`.

Learning the Buzz menu options

The Buzz main menu options are in the Buzz pane header, as seen in *Figure 13.1*:

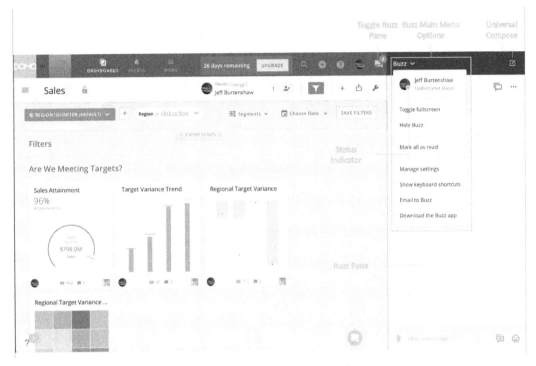

Figure 13.1 – Buzz main menu options

The following are descriptions of each option:

- Toggle **Buzz Pane**: Hide and show the Buzz pane.

- **Buzz Main Menu Options**: The drop-down menu showing the available options.

- **Universal Compose**: Access to create a new Buzz message regardless of which Buzz pane is active.

- **Update your status**: Select your public collaboration availability; status options are **Available**, **Busy**, **Away**, and **Out of office**. For the **Out of office** status, a delegate can be set up.

- **Status indicator**: A colored dot is automatically updated by the system indicating a user's activity in Domo, where green is active, yellow is inactive, and gray is logged out of Domo. A customizable status message is available to provide more information to users viewing your status.

- Toggle **Full-screen**: Switch Buzz from the right pane view to the full-screen view.

- **Hide Buzz**: Collapse the Buzz pane.

- **Mark all as read**: Set all messages to read.

- **Manage settings**: Opens the **Settings** page with a **Buzz** settings section that controls context auto-switching and the message removal period, as seen in *Figure 13.2*:

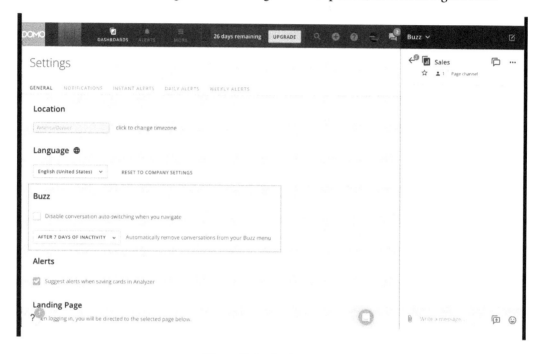

Figure 13.2 – Settings page

- **Show keyboard shortcuts**: Displays a popup of the Buzz keyboard shortcuts, as seen in *Figure 13.3*:

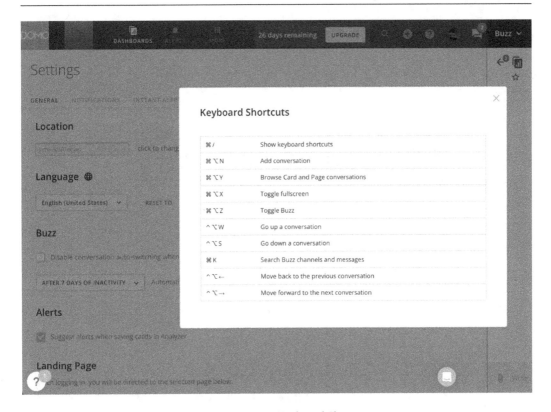

Figure 13.3 – Buzz Keyboard Shortcuts

- **Email to Buzz**: Opens the **Convert an email to a Buzz conversation** dialog.

- **Download the Buzz app**: Opens the dialog to download the desktop app for Buzz.

- **Buzz Pane**: The area where the various Buzz panes (Conversation, Navigation, and Universal Compose) appear.

Now that we understand the Buzz main menu options, next, let's create a Buzz conversation using the Universal Compose feature.

Using the Universal Compose feature

Buzz Universal Compose is the conversation creation feature of Buzz and is available from the Buzz header, as seen in *Figure 13.4*:

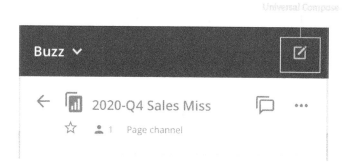

Figure 13.4 – Buzz Universal Compose access

The compose dialog has several functions, as shown in *Figure 13.5*:

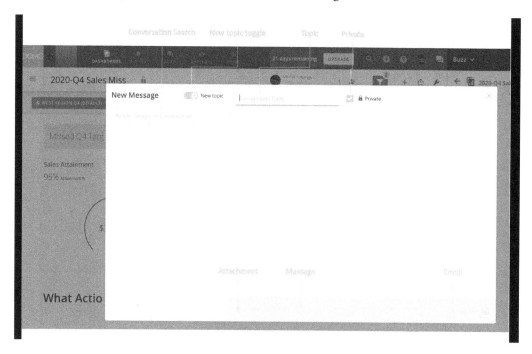

Figure 13.5 – Universal Compose dialog

The following are descriptions of the Universal Compose feature:

- **Conversation Search** provides a search of people, groups, and conversations. Search results are presented, and we can choose from the returned search hits and add to the ongoing conversation.

- The **New topic** toggle unhides the **Topic** box for entry of a new conversation topic.

- Use the **Topic** box to enter a new conversation topic title.

- **Private** indicates the conversation is a private one between participants specifically invited. If not checked, then the message is publicly viewable.

- **Message** is the conversation message entered.

- **Emoji** allows emojis to be inserted into the message as text tokens that are converted to the emoji images on sending.

Let's compose a message:

1. Click on the **Universal Compose** icon.

2. Set the **New topic** toggle to *on*, enter `My Test Topic` as **Topic**, and then check **Private**.

3. Enter `This is a great time to be alive!` as the message and click the **Emoji** icon. Click on a smiley face emoji and press the *Return* key to send the message. The new conversation will appear in the Conversation pane, as seen in *Figure 13.6*:

Figure 13.6 – New conversation in the Conversation pane

Next, let's discuss how we can involve people that are not licensed users in the Domo instance to collaborate in Buzz.

Leveraging Social users

A **Social user** is a role for unlicensed users in Domo to be able to collaborate via Buzz. Social users can leverage Social collaboration features without a Domo license, and Social users are free and unlimited in number. The Social user is limited to the Buzz functionality and Projects and Tasks features. Social users cannot see pages and cards directly but can see card images in the Conversation Details pane when the topic is a card reference. Let's invite someone as a Social user:

1. In the `My Test Topic` conversation, click the ... icon to see the Conversation Details pane.

2. Scroll down and click the **INVITE** button.

3. Enter a test email address that you have access to that is not the email you are using as your current Domo user.

4. If the email is not already a Domo user in your instance, a popup with **Invite to Domo** will appear, as seen in *Figure 13.7*:

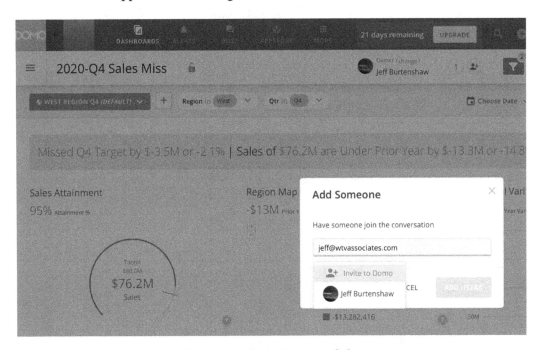

Figure 13.7 – Invite Buzz user dialog

5. Click **Invite to Domo,** and then click **Invite** to send the invite to the email address entered.

6. The person invited will see an email as seen in *Figure 13.8* in their inbox:

Figure 13.8 – Invitation to Domo

7. Next, let's mention the new user in the conversation. Enter @ and start typing in the email address of the person invited. Note, the @ mention will look ahead as you type and try to find the person and present the matches. Choose the match from the list. In my case, I entered @J, and then jeff appeared in the list, as seen in *Figure 13.9*:

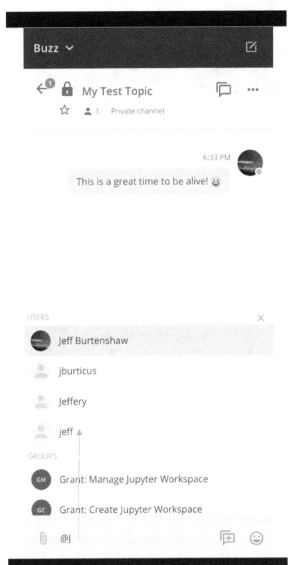

Figure 13.9 – The @mention feature

8. Click the name to mention it in the message.

9. Press the spacebar after the name mentioned in the message and enter the message Welcome to Domo Buzz messaging. Then, press the *Return* key to send the message.

10. A response highlighted in gray can be seen in *Figure 13.10*:

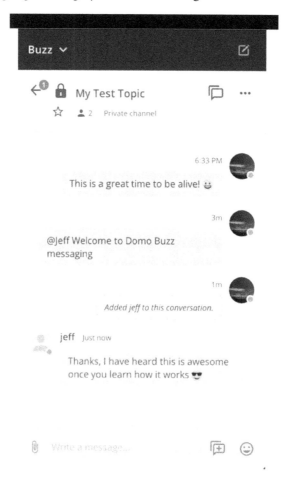

Figure 13.10 – Conversation response

Great! We have created a conversation and seen what a response looks like. Next, let's learn how to use the Buzz Navigation pane.

Understanding the Buzz Navigation pane

The **Buzz Navigation pane** is the directory for all conversation topics and messages. By clicking ← in the upper-left corner of the Conversation pane, we can familiarize ourselves with the features in the Buzz Navigation pane, as seen in *Figure 13.11*:

Figure 13.11 – Buzz Navigation pane

The following are the Navigation pane feature descriptions:

- **Universal Compose** to create a new conversation.

- **Search** to find a conversation.

- **Flag for follow-up** opens a list of the conversations that have been flagged. A number in a gray box will show the number of conversations flagged. A conversation is flagged in the Conversation pane.

- **Discover** opens a panel showing grouped lists of conversations including **TRENDING TOPICS** from MR. ROBOTO, who is **ONLINE NOW, MOST REACTED MESSAGES, RECENT CONVERSATIONS, Public channel** conversations, **1on1 conversations, Card Channel** conversations, and **Card** annotations.

- The **FAVORITES** are conversations set as favorites via the Conversation pane.

- **Reorder favorites** icons drag conversations into the desired order in the list.

- The **Current Conversation Highlight** field shades the conversation open in the Conversation pane with a gray background.

- The **CONVERSATIONS** list displays all conversations that have not been favorited.
- Select the **New Conversation** button to create a new conversation.
- **Unread** is an indicator of the number of unread messages in the conversation.
- The **Invite team member** field adds a Social user to be able to use Domo Buzz.
- The **Conversation Type** icons indicate the context type of the conversation, as seen in *Figure 13.12*:

![icon]	**Public** – all users in the Domo instance are members of public conversations.
![icon]	**Group** – a conversation limited to specified members.
![icon]	**One on One** – a private conversation between two people.
 ![icon]	**Page** – conversation is related to a specific page. Only users with access to the page can participate.
 ![icon]	**Card** – conversation related to a card. Only users with access to the card can participate.
![icon]	**Project** – conversation around a project task.

Figure 13.12 – Conversation Type indicators

Now that we understand how to create, find, and navigate conversations, let's review the features for interacting with specific conversations.

Touring the Buzz Conversation pane

The **Buzz Conversation pane** is Domo's instant messaging service for real-time digital collaboration. It is where we interact with others through threaded conversations. *Figure 13.13* highlights the primary features:

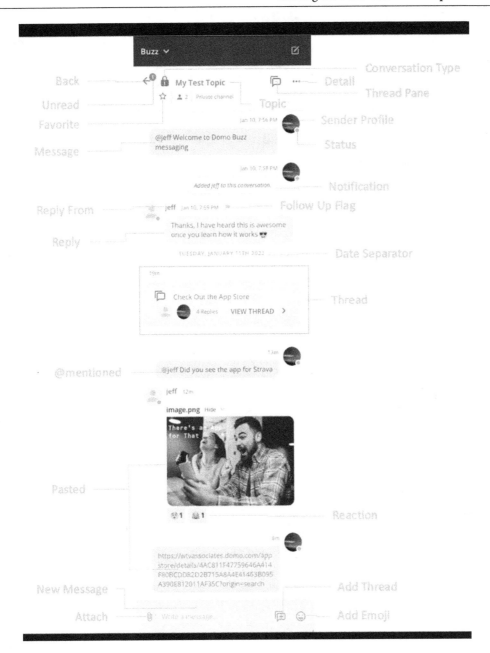

Figure 13.13 – Buzz Conversation pane features

The following are descriptions of the Buzz Conversation pane features:

- **Back** goes back to the Buzz Navigation pane.

- **Unread** indicates the number of new messages in the conversation.

- **Conversation Type** indicates the type of conversation (see *Figure 13.12* for details).

- **Detail** opens the Conversation Detail pane.

- **Thread Pane** opens the pane showing the top level of the threads in the conversation.

- **Topic** is the title of the conversation.

- **Favorite** adds the conversation to the **FAVORITES** list in the Navigation pane.

- **Sender Profile** is the personal profile image of the message sender. Clicking on the picture brings up additional profile information.

- **Status** indicates the online presence of the sender.

- **Message** is the text, photo, or URL content of the message.

- **Notification** is a message generated by the system.

- **Reply From** is the person the message reply is from.

- **Reply** is the reply message.

- **Date Separator** shows the date demarcation; messages and threads are sorted from oldest to newest.

- **Follow Up Flag** is the visual flag indicator to show the message is added to the **Flag for follow-up** list accessed from the Navigation pane.

- **Thread** is a subtopic inside the main conversation. The messages in the thread are accessed by clicking **VIEW THREAD**.

- **@mentioned** uses the @ convention to direct the message to a specific user or group, card, or page. Notice that not just people but also pages and cards can be referenced by @. A conversation on the `2020-Q4 Sales Miss` page with a message having a card reference to `@Largest Laggards` is seen in *Figure 13.14*. The card image is automatically added to the message from the @ reference.

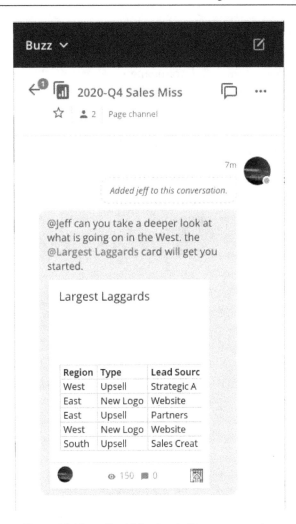

Figure 13.14 – @Card Mention in Buzz conversation

Clicking on the card name in the card image will bring up the card details.

- **Pasted** is content cut and pasted into the message directly and can be images or URLs as well as text.

- **Reaction** is an emoji reaction from replying users with a count of the number of users for each emoji reaction.

- **New Message** is the area to enter new messages. Press the *Return* key to send the message.

- **Attach** provides for attaching files to the message.

- **Add Thread** creates a new thread/subtopic in the conversation.

- **Add Emoji** inserts an emoji into the message text. Note that it comes in as a text macro but is translated into an image when you send.

Hovering on a message in the conversation will reveal the message options, as seen in *Figure 13.15*:

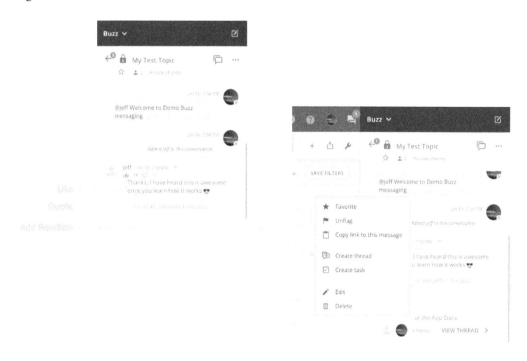

Figure 13.15 – Message options

The following are descriptions of the message options:

- **Like** enables liking or unliking the message.
- **Quote** copies the highlighted message down into the **New Message** area with double quotes surrounding it.
- **Add Reaction** provides for the selection of an emoji to add as a reaction to the message.
- **...** opens the additional options popup.
- **Favorite** adds or removes the message as favorited
- **Flag/Unflag** lets you flag or unflag the message for follow-up.
- **Copy link to this message** places a link to the message in the copy and paste buffer.
- **Create thread** opens the **Create thread** window.
- **Create task** opens the **Create task** window.
- **Edit** lets you make changes to a message that has been sent.
- **Delete** removes the message.

As you navigate through pages and cards, you may notice the conversation topic will change automatically to show the current page or card-clicked messaging.

Terrific! We have learned how to work with the Conversation pane. In the next section, we will explore the Conversation Detail features.

Discovering the Buzz Conversation Detail pane

The **Conversation Detail pane** is opened from the Conversation pane by clicking **...** in the Conversation pane header. The Conversation Detail pane has features as seen in *Figure 13.16*:

Figure 13.16 – Conversation Detail pane

The following are descriptions of the Conversation Detail pane features:

- **Detail Pane** is the panel brought up from the Conversation pane that displays the conversation's details.

- **Hide/Show Section** opens and collapses the sections in the Detail pane.

- **Share link** provides a copyable link to the conversation.

- **Import email** brings an email into the conversation as a message.

- **Follow/Unfollow** adds or removes the conversation from your Buzz Navigation pane.

- **Favorite/Unfavorite** sets or removes the conversation as a favorite.

- **Notification Settings** opens a window to control Buzz notification preferences.

- **Attachments** shows the attachments in the conversation.

- **Bots** are connections to external systems that add automated messages to the conversation.

- **Conversation Favorites** shows all messages in the conversation that have been favorited.

- **Related** displays all messages with related content detected by Domo.

- **People** shows who has access and enables inviting new users.

After going through the features and working on some examples, the overall conversation and messaging model relationships are represented in *Figure 13.17*:

Figure 13.17 – Buzz Conversation model relationships

Conversations are the same as topics and have messages and replies. Threads are subtopics within a conversation and also have messages and replies.

Next, let's review what we learned in this chapter.

Summary

In this chapter, we learned that Domo Buzz is a social messaging and collaboration tool similar to Slack, but also eliminates the need to *bring your own content* by automatically providing conversation context with Pages, Cards, and Projects in Domo. We learned that there are several panes in Buzz for navigation, conversations, and conversation details. Buzz can be run in Domo as a standalone desktop app, and on mobile via the Domo app or browser. Social users enable unlimited free access to conversations, but functionality is limited for Social users. Emails can be imported as messages into conversations, bots can run to bring in messages from other services, and the notification settings can be controlled. Domo provides conversation context automatically for pages, cards, and projects. Images and URLs can be embedded directly in a message, and file attachments are also supported. Buzz is a robust messaging communication tool.

In the next chapter, we will dive into how to extend the Domo platform via Domo apps and code scripting.

Further reading

The following are related items of interest:

- For reference material on Buzz, go to `https://domohelp.domo.com/hc/en-us/articles/360043429973-Buzz-Layout`.

- The Social user overview video can be found at `https://www.domo.com/help-center/videos/watch/LFVgZa1aIe0`.

- For help adding a bot to Buzz, go to `https://domohelp.domo.com/hc/en-us/articles/360043430273`.

Section 4: Extending

Vendor lock-in is a real thing.

Technology buyers have long been wary of vendor lock in, a situation where the technology vendor's solution is proprietary and a walled garden that the vendor can use to effectively extort continued use. Fortunately, the Domo platform is built on the philosophy that the data is the customer's data and they should be able to use their data with whatever technologies they desire or need. While the Domo platform provides the most comprehensive suite of tools and technologies for **Business Intelligence (BI)** in the industry it still is an open, community friendly platform.

In this section, you will learn how to extend the Domo platform using HTML, JavaScript, and Python to add client and server functionality to meet your bespoke app needs. You will also see how to leverage third party tools like SageMaker and Jupyter Workspaces to access integrated best of breed AI/ML technology.

After you have completed this section, you will have a basic understanding of how to create custom Domo Apps for the gaps; and how to use non-Domo industry leading AI./ML tools with Domo.

This section comprises the following chapters:

- *Chapter 14, Extending Domo with Domo Apps*
- *Chapter 15, Using Domo APIs in Python*
- *Chapter 16, Using Domo Machine Learning*

14
Extending Domo with Domo Apps

Imagine that we want to develop a new enterprise software application. We could start from scratch, or we could look for a framework to leverage. The open source community (via reusable code libraries) has both decreased the time needed to develop applications and increased the functional richness of many applications. Nevertheless, even with code libraries, it is still a tough job building a bespoke application. There are many layers of architecture, including data tiers, business logic tiers, security, user management, mobile apps, and scalable infrastructure. What if we were to discover a platform where much of what a developer needs for building an enterprise-grade application already existed and all that remained was to connect the data to the application functionality via HTML, CSS, and client scripting? That would be fantastic! We will learn what the Domo Dev Studio Framework delivers, and how to leverage it to extend the Domo platform via creating custom apps!

In this chapter, we will specifically cover the following:

- Understanding the Domo Application Development Framework
- Setting up the Domo development environment
- Creating a Domo app in Domo Dev Studio
- Publishing a Domo app

Technical requirements

To follow along with this chapter, you will need the following:

- Internet access

- Your Domo instance and login

- Ability to download and install files

- Administrative rights on your machine

- Node.js installed

- Domo Dev Studio **command-line interface** (**CLI**) installed

- Development experience:

  - Experience with web development technologies such as HTML, CSS, and JavaScript

  - Experience using development libraries and tools such as Node.js, npm, CLIs, and Git

> **Important Note**
>
> If you don't have a Domo instance, get a free trial instance here: `https://www.domo.com/start/free`.

Understanding Domo application development architecture

Before we start creating an app, it will be helpful to understand the Domo Dev Studio conceptual architecture, as seen in *Figure 14.1*:

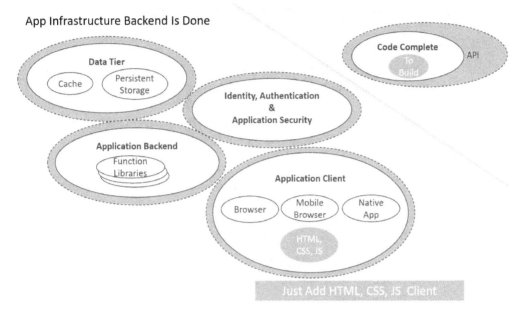

Figure 14.1 – Domo application development conceptual architecture

The white-filled circles are the architecture layers that communicate with each other through APIs. All this application infrastructure is included in the platform. The application developer can leverage this functionality through Domo APIs to custom build the UI scripting piece of the application. Certainly, having all this infrastructure and API functionality available greatly speeds up the time to market (much of the typical application code is already there). And the enterprise readiness of any app developed on the platform is already running on a proven, cloud, enterprise stack, and is deployable with a click in the Domo Appstore.

Dev Studio is a command-line utility also known as **ryuu** (separate from the Java CLI, by the way) that connects to Domo and allows applications based on HTML, CSS, and JavaScript to be published to Domo. It has app starter templates with predefined HTML pages, CSS styles, and JavaScript libraries for working with Domo APIs. It also runs a webserver on **localhost** to facilitate real-time testing of the app. The result of using Dev Studio is a custom application design that is published to the Domo **Asset Library** in a Domo instance, or worldwide in the Domo AppStore. To use the app, a Domo Card must be created using the app design. In Domo, apps are contained in, executed from, and managed as cards. When an app card is first created from the specified library asset, the application's dataset needs are mapped to live datasets via a configuration wizard.

Now, we understand the conceptual architecture and what the Dev Studio CLI does; next, let's set up our Domo development environment.

Setting up the Domo development environment

To create Domo apps, we must first set up the development environment. Dev Studio utilizes Node.js, which includes npm. Also, we can get a developer sandbox instance to house our work-in-progress app definitions and app cards. The environment will enable us to work with Domo API functions directly and use the **JavaScript** (**JS**) libraries that wrap Domo API functions. Finally, we can set up our OAuth client ID and client secrets for secure API access via a token:

1. First, check whether **Node.js** is installed by opening a terminal window and entering the `node --version` command in the shell prompt. If Node.js is installed, the version will be printed. If not, then install Node.js following the directions here: `https://nodejs.org/en/download/`.

2. Make sure that `/usr/local/bin` is in your `$PATH`. For help, see *How to add a directory to your path for Mac*, or *How to add a directory to your path for Windows*, in the *Further reading* section.

3. Update **Node Package Manager** by entering the `npm install -g npm` command. npm is installed with Node.js.

4. Enter the `npm --version` command to see the version of npm.

5. Now, let's install the Domo **Dev Studio CLI** (**DS CLI**) by entering the `npm install -g ryuu` command. Wait until completed.

6. In the terminal, enter the `domo login` command and enter your Domo domain. When logged in, you will see the screen displayed in *Figure 14.2*:

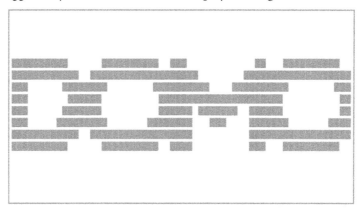

```
✓ Login to wtvassociates.domo.com successful.
Welcome, Jeff Burtenshaw.
jeffburtenshaw@Jeffs-MacBook-Pro ~ %
```

Figure 14.2 – Domo CLI successful login

7. Enter domo `--help` and press the *Return* key to see a list of DS CLI commands.

8. Enter domo `logout` to exit the DS CLI.

9. Good job on installing the CLI! Now, let's get our Domo developer account so you can publish your app, by going to https://developer.domo.com/ developer-trial-request, as seen in *Figure 14.3*:

Figure 14.3 – Domo developer account trial request

10. Complete the form and click **START DEVELOPER TRIAL**. This will send an email titled **Welcome to the Domo Developer Program!** to the registered email address, as seen in *Figure 14.4*:

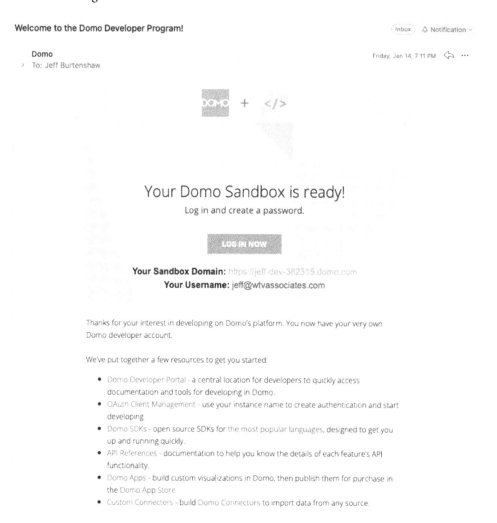

Figure 14.4 - Domo Developer Program email

11. Click **LOG IN NOW** in the email. This will prompt for a new password and then take us to our new developer sandbox instance, as seen in *Figure 14.5*. Make a note of the Domo domain name (the part of the URL before **.domo.com**) so you can log in later.

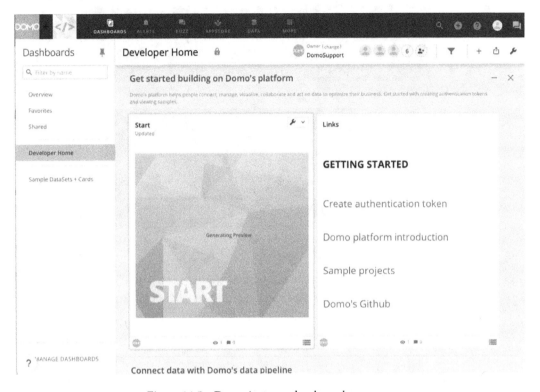

Figure 14.5 – Domo instance developer home page

12. Now, we need to get the security credentials to use the Domo APIs via the OAuth 2.0 token. Click on the **Create authentication token** link and then click **LOG IN**, as seen in *Figure 14.6*:

Figure 14.6 – Developer portal login

13. Enter the new Domo instance domain in the pop-up window and click **Continue**, which will take us to the Developer portal logged in to our account, as seen in *Figure 14.7*:

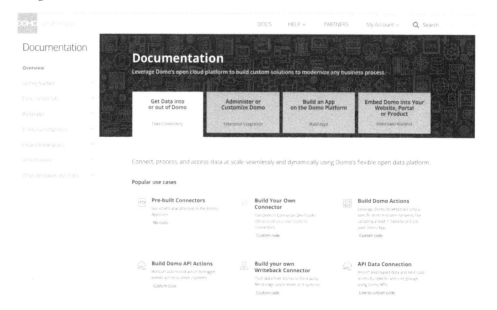

Figure 14.7 – Developer portal account

14. Click on **My Account**, and click **New Client** to go to the **Create new client** page, as seen in *Figure 14.8*:

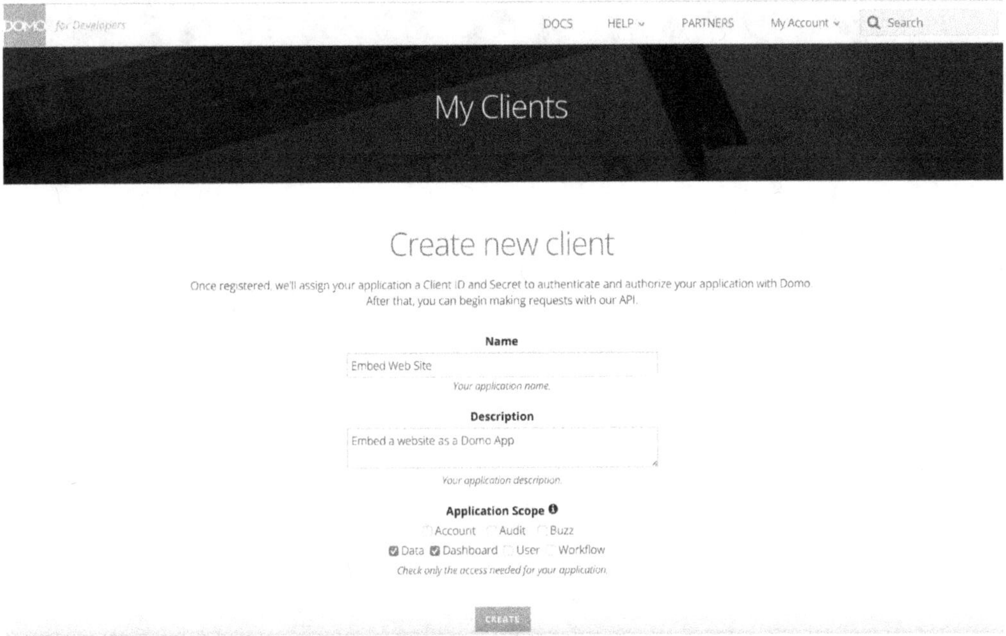

Figure 14.8 – New client setup

15. Enter Embed Web Site under **Name**, and enter Embed a website as a Domo App under **Description**.

16. **Application Scope** indicates which Domo APIs the client token will have access to. Click **Data** and **Dashboard**, as those are the APIs we will need to access for this example.

17. Click **CREATE** and we have **Client ID** and **Secret**, as seen in *Figure 14.9*:

Figure 14.9 – Client ID and Secret

18. Now, request an access token. In a terminal window, enter the following command:

```
curl -v -u {CLIENT_ID}:{CLIENT_SECRET} "https://
api.domo.com/oauth/token?grant_type=client_
credentials&scope=data%20dashboard"
```

Substitute the client ID and secret with your own and execute. You will see the results as shown in *Figure 14.10*, where the access token is highlighted:

```
* Connection state changed (MAX_CONCURRENT_STREAMS == 128)!
< HTTP/2 200
< server: nginx
< date: Mon, 17 Jan 2022 18:51:00 GMT
< content-type: application/json;charset=utf-8
< vary: Origin
< vary: Access-Control-Request-Method
< vary: Access-Control-Request-Headers
< cache-control: no-store
< pragma: no-cache
< x-content-type-options: nosniff
< x-xss-protection: 1; mode=block
< x-frame-options: DENY
< vary: Accept-Encoding, User-Agent
< strict-transport-security: max-age=31536000; includeSubDomains
<
* Connection #0 to host api.domo.com left intact
{"access_token":"ayJhbGciOiJSUzI1NiIsInR5cCI6IkpXVCJ9.eyJyb2xlIjoiQWRtaW4iLCJzY29wZSI6WyJkYXNoYm9hcmQiLCJkYXRhIl0sImRvbWFpbiI6ImplZmYtZGV2LTM4MjMxNS5
kb21vLmNvbSISImV4cCI6MTY0MjQ0OTA2MCwiZW52IjoicHJvZDYiLCJ1c2VySWQiOjgwNzEyNzkyMiwianRpIjoiVRmQ3VacWREeW93OEsweU65UEZSbDhYellBPSIsImNsaWVudF9pZCI6ImUz
YTU5ODFjLTQxOTQtNDYxYS1hODI6LWU3YWViNzJlZmNmIiIsImN1c3RvbWVyIjoibW1tbS0wMDIxLTIzMDgifQ.bSXDBEEhp2kAYcs_MD85iDL2LoJPgBAyNl5QwCfM-CM1-FVhnrQF2fzS8tX0GZ
khNlKaT0hMsxn3lcTU_jTc02s_ZKJqpmNhjn3YAsVNwnM1Vq-3xJ0K5lWBR2Dosxtqi-dQ8DSOS4srGQ8ztweDJ71aK42e3H8BnY_Cf7Ck8IA76JjeJnIiSr39SG_Pqal4SeBN3d10UiDhZzXJmvv
KPb0Qgk6hHwlfKCDEqN0NwosCTBm-nhV5C2n9Xhy6J7QL_Z_n4RNdrl4gstyvA5epggeczYdaPnIRp66cZoGPFTb3FBxHrLrfmrhUyBbsBa4_URyB8hpkouAJDDuuJB3upw","token_type":"be
arer","expires_in":3599,"scope":"dashboard data","customer":"mmmm-0021-2308","env":"prod6","userId":807127922,"role":"Admin","domain":"jeff-dev-38231
5.domo.com","jti":"YTfCuZqdDyow8KQyNyPFR18XzYA="}.        jeffburtenshaw@Jeffs-MacBook-Pro jeffburtensjeffburtenshaw@Jeffs-Mjeffburtensjeffburte
nsjeffburtensjeffburtensjeffburtensjeffburtensjjeffburtenshaw@Jeffsjeffburtenshaw@Jeffsjeffburtenshaw@Jeffs-MacBook-Pjeffburtenshaw@Jeffs-MacBook-Pje
ffburtenshaw@Jeffs-MacBook-Pjeffburtenshaw@Jeffs-MacBook-Pjeffburtenshaw@Jeffs-MacBook-Pjeffburtenshaw@Jeffsjeffburtens
haw@Jeffs-MacBook-Pjeffburtenshaw@Jeffs-MacBook-Pjeffburtenshaw@Jeffs-MacBook-Pjeffburtenjeffburtenshaw@Jeffs-MacBook-Pro ~ %
jeffburtenshaw@Jeffs-MacBook-Pro ~ %
```

Figure 14.10 – Access token response

19. Let's test the access token by using it in a direct API call. Enter the following command to list datasets in the terminal:

```
curl -v -H Authorization:'bearer {access token}
'"https://api.domo.com/v1/datasets"
```

Replace {access token} with the access token from the command results in *Step 18*. The result of the example can be seen in *Figure 14.11*:

```
* Connection #0 to host api.domo.com left intact
[{"id":"238fcef1-1756-44ff-ab23-4042b3c00416","name":"PDP Example DataSet","rows":7698,"columns":11,"owner":{"id":27,"name":"DomoSupport"},"dataCurre
ntAt":"2021-12-10T13:31:26Z","createdAt":"2021-12-10T13:31:17Z","updatedAt":"2021-12-10T13:31:27Z","pdpEnabled":false}]
jeffburtenshaw@Jeffs-MacBook-Pro ~ %
```

Figure 14.11 – List datasets command output

> **Important Note**
>
> Access tokens have an expires-in parameter that causes them to expire in about one hour.

Now we have the development environment operational, let's create a Domo app.

Creating a Domo app in Domo Dev Studio

The ability to create a custom app using **Domo Dev Studio** is amazing and makes almost any app you can imagine possible to build on the Domo platform. Let's go through some terminology to reduce the learning curve. The following are key terms used in connection with Domo Dev Studio:

- **DS CLI** (also known as **ryuu**) is a command-line utility used to connect to Domo servers, run a local web server, and manage asset resources of apps in Domo.
- **Asset Library** is the repository for app definitions in Domo.
- **App Card** is a Domo card created from assets in the Asset Library. The card is the executable container for the app.
- Domo Login is a CLI command that connects the DS CLI commands to a given Domo instance.
- Domo Init is a command in the DS CLI to copy down an app template to begin coding from a baseline app.
- Domo Publish is a DS CLI command to push the code and resource files from the development project up to the Asset Library.
- Domo Dev is a DS CLI command to start a local web server to test apps.

The following are descriptions of basic files included in an app project:

- Manifest.json is a required file that holds the metadata about our app, such as name, version, size, and dataset mappings. It provides the security context for connecting the app to Domo.

- Index.html is the HTML page that is opened by default for the app.

- App.css contains the CSS class definitions for styling the HTML.

- Thumbnail.png is required, is the brand image for the app, and is a 300x300 pixel image for the app that will display in the Asset Library and Appstore.

- Domo.js is a Domo Utility Library of JavaScript functions.

- App.js is a JavaScript file that holds our custom JavaScript for the application.

Ready to lay down some code? Let's begin by creating an app that will show a YouTube video on an app card:

1. First, we'll create a project folder to store all the files required for this app. Open the terminal program, enter the mkdir domoappprojects command, press the *Return* key, and then enter cd domoappprojects and press *Return*. This is the directory to store all your Domo app project code and file resources. Each app will have its own subdirectory.

2. In the terminal, enter domo login and press the *Return* key, then select an instance, or enter a new instance to connect to using the arrow keys and press the *Return* key. I suggest connecting to our dev instance for now. You may see a screen (as shown in *Figure 14.12*) pop up in the browser. We can close that page.

Hello Jeff Burtenshaw,

 jeff@wtvassociates.com

BETA Developer Studio has been granted access to Domo. The device will remain signed into your account until you choose to sign out.

Figure 14.12 – Dev Studio authentication to Domo confirmation

3. Enter the `domo init` command and press the *Return* key. At the **design name** prompt, enter `MyFirstApp` and press the *Return* key. In the **select a starter** prompt, use the arrow keys to highlight the **hello world** project template and press the *Return* key. This creates a new `MyFirstApp` project folder and copies the template files from the **hello world** starter template.

4. We need to add a thumbnail image to the resources for our app by entering the following URL while in the browser: `https://place-hold.it/300x300.png`. Right-click on the image and select **Save As**, enter `thumbnail.png` as the filename, and select the `MyFirstApp` directory, as shown in *Figure 14.13*:

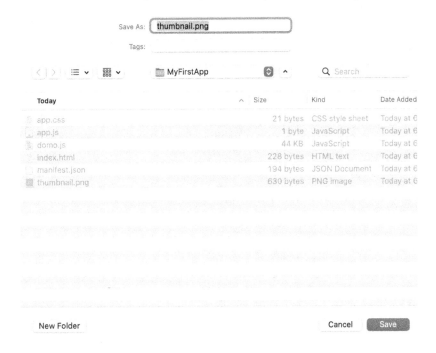

Figure 14.13 – Save thumbnail.png

5. Back in the terminal, enter the `cd MyFirstApp` command and press the *Return* key.

Important Note

The `domo publish` and `domo dev` commands must be run from the desired projects directory.

---

6. Now, to publish the MyFirstApp project design in the Asset Library of the
    instance we are logged into, enter domo publish and press the *Return* key. This
    publishes the app design from the current directory into **Asset Library**, as shown in
    Figure 14.14:

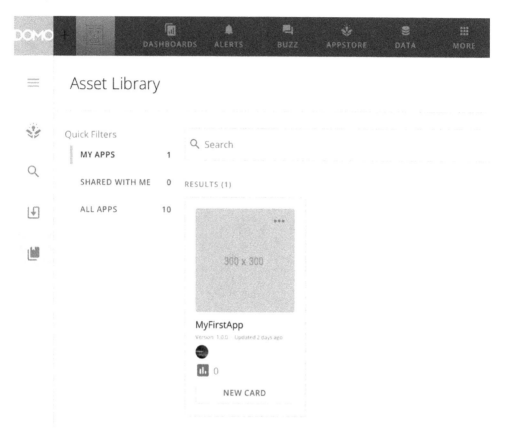

Figure 14.14 – Asset Library with MyFirstApp

7. Next, let's start the web server listening to localhost:3000 by entering the
    domo dev command and pressing the *Return* key. This command starts the Domo
    local dev web server, looks for the index.html file in the current directory, and
    renders it in the default browser with the URL localhost:3000, as shown in
    Figure 14.15:

Figure 14.15 – MyFirstApp page from localhost:3000

8.   This command sequence from the terminal can be seen in *Figure 14.16*:

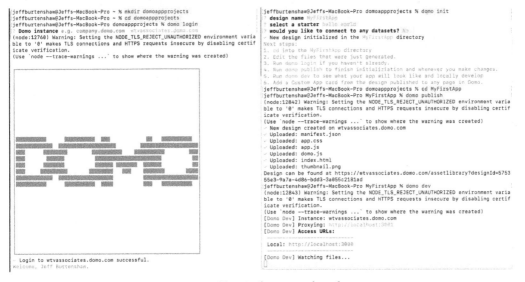

Figure 14.16 – Terminal commands and output

Important Note
To stop Domo Dev Studio, press *Ctrl + C* while in the terminal window.

9. Next, let's open our code editor/IDE; **Visual Studio Code** works well. In Visual Studio Code, click **File** > **Open**, navigate to the `domoappprojects/MyFirstApp` folder, and click **Open** to see the project files, as shown in *Figure 14.17*:

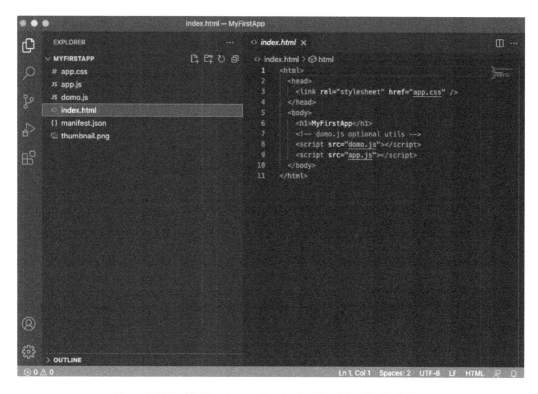

Figure 14.17 – MyFirstApp project in the Visual Studio Code IDE

10. Click on the `index.html` file on the left panel and the HTML code will open up in the right panel.

11. Let's have some fun and embed a YouTube video as our first app. Enter the following code after the closing `</h1>` tag:

```
<iframe width="560" height="315" src="https://www.
youtube.com/embed/MfHa-PGaOXw" title="YouTube video
player" frameborder="0" allow="accelerometer; autoplay;
clipboard-write; encrypted-media; gyroscope; picture-in-
picture" allowfullscreen></iframe>
```

12. Click **File** > **Save** in the IDE and look at the `localhost:3000` page in our browser; it will have automatically updated, as shown in *Figure 14.18*:

Figure 14.18 – MyFirstApp with embedded YouTube video

13. Pretty sweet! Try out the **Fullscreen** option and the video is a good overview of Dev Studio as well.

Can you believe it? We just created a cool app! Ready for something more advanced? Open the GitHub repository at `https://github.com/Burticus/Domo-App-Projects.git` and try the **CRUD Grid** project, which creates an app with an editable table to create, read, update, and delete records from the **Lead Source Mappings** dataset.

In the next section, we will learn how to deploy our app for use in our instance and in the Domo Appstore.

Publishing a Domo app

The result out of Dev Studio is an application definition that is published to the Domo Asset Library in a Domo instance or even worldwide in the Domo Appstore. To use the app, a Domo Card must be created using the app definition from the Asset Library. Let's use our **MyFirstApp** design and make a card in our instance.

Publishing an app to a single Domo instance

The following are the steps to publish and deploy a card so the app can be used in a Domo instance:

1. In the Domo instance where you published the **MyFirstApp** design, click on the **MORE** main menu option and click **Asset Library**.

2. On the **Asset Library** page, on the **MyFirstApp** tile, click on **NEW CARD**; this will bring up a page where you can name the card `YouTube Dev Studio Training` and click **SAVE & FINISH**.

3. The initial location of the app card is on the **Overview** page, as shown in *Figure 14.19*, but can be moved to any page:

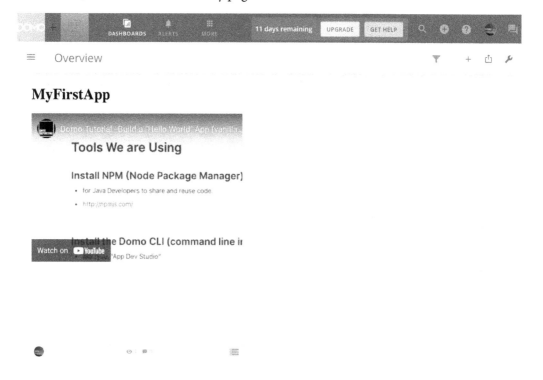

Voila! Our app is in its card home! Next, let's see how we can get information to publish an app to the Domo Appstore.

Publishing an app to the Domo Appstore

If you are interested in publishing your apps to the Domo Appstore, the requirements and guidelines can be found here: `https://developer.domo.com/docs/appstore/appstore-overview`.

Next, let's review what we learned in this chapter.

Summary

In this chapter, we learned about the Domo app development architecture and that we can leverage all that infrastructure using the Domo Dev Studio utility to create app designs in HTML, CSS, and JavaScript. Application templates are provided to accelerate the client development process with a real-time updating web service to facilitate design creation. App designs are published to the Domo Asset Library instance and deployed as app cards within an instance. We can also apply to have our custom apps published in the Domo Appstore. We created a simple app to display a YouTube video, and published and deployed it as an app card. Additional project samples using the D3 graphics library, connecting to Domo datasets, and doing CRUD operations are provided in the Git repository for those looking to get more advanced examples.

In the next chapter, we will take a foray into Domo's machine learning capabilities.

Further reading

The following are related items of interest:

- How to add a directory to your path for macOS: `https://wpbeaches.com/how-to-add-to-the-shell-path-in-macos-using-terminal/`
- How to add a directory to your path for Windows: `https://www.shellhacks.com/windows-cmd-path-variable-add-to-path-echo-path/`
- Install Node.js and npm: `https://kinsta.com/blog/how-to-install-node-js/`
- Install the Domo DS CLI: `https://developer.domo.com/docs/dev-studio/set-up`
- Overview of App Dev Studio: `https://developer.domo.com/docs/dev-studio/dev-studio-overview`
- Public assets versus authenticated assets: `https://developer.domo.com/docs/project-structure/public-assets`

- Known limitations to App Dev Studio: `https://developer.domo.com/docs/dev-studio/faq-known-limitations`

- Reference material on `domo.js`: `https://developer.domo.com/docs/dev-studio-tools/domo-js`

- DS CLI reference: `https://developer.domo.com/docs/dev-studio-tools/dev-studio-cli`

- Installing a proxy to connect to Domo APIs from a server other than Domo Dev: `https://developer.domo.com/docs/dev-studio-tools/proxy-middleware`

- Reference on `manifest.json`: `https://developer.domo.com/docs/dev-studio-project-structure/manifest`

15
Using Domo APIs in Python

Python is a high-level, interpreted programming language that trades off execution speed for development productivity when compared to something such as Java or C++. Python is very attractive for rapid application development, is simple and free to use, and is easy to distribute. All these characteristics make it a great tool for democratizing extensions to the Domo platform via Domo **application programming interfaces (APIs)**, directly or wrapped by Python libraries.

In this chapter, we'll look at the two available options for using Python with Domo. The first is through the **Extract, Transform, and Load (ETL)** scripting tile for Python. This feature is a paid subscription option from Domo as a part of the Data Science module for Magic ETL. The Python scripting tile houses a self-contained Python instance in the cloud that allows us to use the full range of Python's power in Magic ETL data pipelines. If there is not a tile to do some data transformation, no worries—you can do it in the Python tile script without having to set up and maintain a Python server. The second option is available for free by using a Python installation of our own to access Domo APIs. A common use is to automate Domo administration functions at scale. Another use for Python is as a server-side process extender for Domo using Python **REpresentational State Transfer (REST)** services called by Domo custom app clients.

In this chapter, we will specifically cover the following topics:

- Using the Magic ETL Python scripting tile
- Using the Domo Python **software development kit (SDK)**

Technical requirements

To follow along with this chapter, you will need the following:

- Internet access
- Your Domo instance and login
- Your Domo client **identifier (ID)** and client secret
- The ability to download and install files
- Administrative rights on your machine
- The latest Python 3 version installed
- Development experience with Python
- The Python scripting tile feature enabled in your Domo instance

> **Important Note**
> If you don't have a Domo instance, get a free trial instance here:
> `https://developer.domo.com/developer-trial-request`

Using the Magic ETL Python scripting tile

Access to the Python scripting tile, as seen in the following screenshot, is a paid feature for Magic ETL. Contact Domo to have the tile activated. It enables you to run Python code as part of an inline data pipeline without the need for a Python server, as that is all handled by Domo Magic:

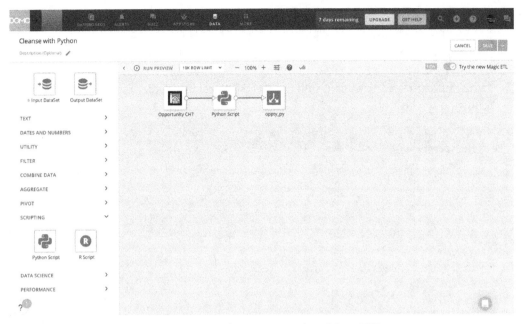

Figure 15.1 – Python scripting tile in Magic ETL

This is a powerful feature to handle edge cases that the standard Magic ETL doesn't cover. To get some experience in using this feature, let's walk through a data cleansing example that can be applied to any category attribute in a dataset, as follows:

1. Click the **DATA** main menu icon, then click on the **DataFlows** icon in the left-side toolbar, then click the **ETL** icon in the **MAGIC TRANSFORM** sub-menu.

2. Change the **Add DataFlow Name** value to Cleanse with Python.

3. Click, drag and drop an **Input Dataset** tile using the Opportunity CH7 dataset, a **Python Script** tile, and an **Output Dataset** tile (titling it oppty-py) from the **DATASETS** and **SCRIPTING** sections onto the canvas, then connect the lines between tiles, as seen in *Figure 15.1*.

4. Click on the **Python Script** tile and copy and paste the contents of the
   `PythonDataCleanse.txt` file from `https://github.com/`
   `PacktPublishing/Data-Democratization-with-Domo/`
   `blob/de5d8c3086f2eaf36ee13bca3ed5d2ff3c83cf4d/`
   `PythonDataCleanse.txt` into the **CODE** panel in the tile, as shown in the
   following screenshot:

Figure 15.2 – Python scripting tile configuration panel

Here are descriptions of the features on the tile configuration panel:

- **Python Script** is the design tile to drag and drop for use in a data flow.

- **Edit Tile Name** provides a way to rename a tile.

- **CODE** is the tab where Python code is entered.

- **SCHEMA** shows the outbound schema from the Python script and allows us to
  remove fields from the output.

- **NOTES** is an area for us to enter notes about the script.

- **Use Code Template** will replace any existing code in the **CODE** tab with a generic
  template.

- **Run Preview** runs code in the **CODE** panel populating the **CONSOLE** and **PREVIEW** tabs.

- The **CONSOLE** tab, as shown in *Figure 15.3*, appears when **Run Preview** is clicked and shows the console output.

- The **PREVIEW** tab, as shown in *Figure 15.4*, appears when **Run Preview** completes and shows a preview of the output dataset.

- **Line Out Color** sets the color of a line out of the chosen **Python Script** tile.

- **Duplicate Tile** makes a copy of a tile on the canvas.

- **Delete Tile** removes a tile from the canvas.

- **Panel Search** enables a search for items in the currently active **ENVIRONMENT**, **INPUTS**, or **PACKAGES** panels.

- The **ENVIRONMENT** panel shows the Python environment that the script is executing in.

- The **INPUTS** panel shows datasets coming in from **INPUTS** tiles. Clicking on a specific input will add a line of code into the **CODE** tab that puts the dataset data into a pandas **DataFrame**.

- The **PACKAGES** panel lists all the available Python code library packages that can be imported into the Domo-provided Python environment. Click on a package, and it will add code to import the package in the **CODE** tab.

Let's continue with our example.

5. Click **Run Preview** and wait for the process to execute, as seen in the following screenshot:

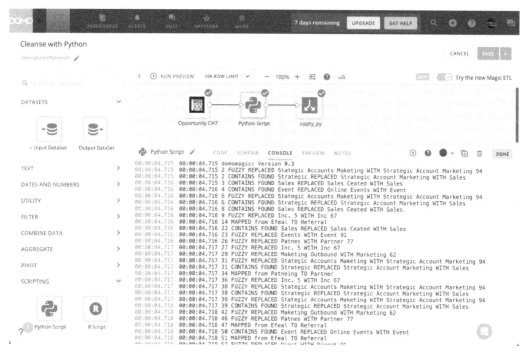

Figure 15.3 – Run Preview console output

6. The checkmarks on the tiles indicate successful execution.

7. Click on the **CONSOLE** tab to see the console output from execution. In this case, the `print()` statements in the script describe rows in the dataset where a cleansing rule was applied. The script loops through the dataset from start to finish one row at a time and looks for three types of matches on `LeadSource`: a fuzzy match (from the `fuzzywuzzy` package), a contains match, and a mapping match. Each successful match triggers a replacement of the `LeadSource` value in the DataFrame. Each time the `LeadSource` value is replaced in a row, the `LeadSourceCleaned` flag is set to `True`. The DataFrame contents are written to a new output dataset.

8. Click on the **PREVIEW** tab to see a preview of the output dataset with all the cleansing rules applied, as seen in the following screenshot:

Figure 15.4 – ETL Run Preview

9. Click on the **Save and Run** option in the **SAVE** dropdown.

10. Now, preview the data in the data center dataset list by finding the new `oppty-py` dataset created from the run, as seen in the following screenshot:

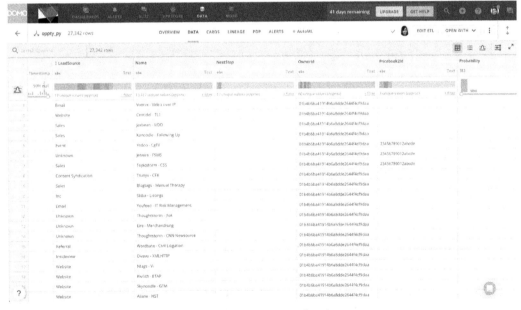

Figure 15.5 – Dataset details

That's all there is to it—very straightforward. What you can accomplish here is dependent on what you decide to do in Python. The tile dramatically speeds up the Python development process by eliminating environment setup and makes getting the dataset into a `pandas` DataFrame object easier with a simple click. The ETL wrapper also runs the process with the job automatically. It really couldn't be simpler.

Next, we will go outside the Domo Python environment into our own Python environment running locally and learn how to access Domo APIs inside our Python applications.

Using the Domo Python SDK

Domo provides an SDK for Python 3. The Python SDK is a library of Python functions that wrap the Domo APIs, so they are exposed as simple Python functions. Being able to use Python to interact with Domo APIs directly or through the SDK opens a world of possibilities for extending Domo. Let's set up our Python environment.

Setting up Python for Domo

To use Python for Domo, you will need to install Python version 3 and install the Domo SDK, as follows:

1. Download and install Python from `https://www.python.org/downloads/`.

2. Install the Python libraries required by executing each of the following commands from the command line in Windows or in Terminal for Mac:

```
pip3 install fuzzywuzzy
pip3 install pydomo
pip3 install pandas
```

Getting the client ID and client secret

We will also need to get credentials to securely access the APIs. Here's how to do this:

1. To get the client ID and client secret, go to `https://developer.domo.com` and log in, then click **My Account** and select **New Client**.

2. Enter `Python API` as the **Name** value and `Python API Access` as the **Description** value and then check **Account**, **User**, **Data**, **Audit**, and **Dashboard** for the **Application Scope** field.

3. Click **CREATE** and note the **Client ID** and **Client Secret** values for use in the code.

Next, let's create and run a Python project.

Coding and running a Python project

Let's continue with our data cleansing example, but this time, execute the project on our local Python server. Let's also make some upgrades to the project so that the dataset can replace itself after cleansing updates. Proceed as follows:

1. Download the `DimensionCleansing.py` file from GitHub here: `https://github.com/PacktPublishing/Data-Democratization-with-Domo/blob/3e080dd5acd0be797d280414c6b53fbfec2b74fe/PythonProjects/Examples/DimensionCleansing.py`.

2. Open the file with your favorite **integrated development environment** (IDE). **Visual Studio Code** (**VS Code**) is a good choice for its Python extensions and debugging ability.

3. In the same directory as `DimensionCleansing.py`, using Excel, create a CSV file called `Credentials.csv`, as seen in the following screenshot:

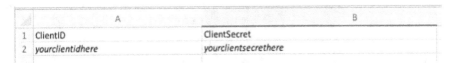

Figure 15.6 – Credentials.csv format

4. Replace `yourclientidhere` and `yourclientsecrethere` in row **2** of the worksheet with the values from the *Getting the client ID and client secret* subsection of this chapter. For the client ID and client secret, do not enclose them in quotes or braces.

5. Save the file as `Credentials.csv`. Reading in the credentials from a `.csv` file increases security as the credentials are not hardcoded into the `.py` file.

6. Update the path to `Credentials.csv` in `DimensionCleansing.py` line **8** to the correct path, as seen in the following screenshot:

```
6
7    # My Domo Client ID and Secret (https://developer.domo.com/manage-clients)
8    creddf = read_csv('./examples/Credentials.csv') # from the current directory use relative path from there
9
```

Figure 15.7 – Credentials.csv path setting

7. On line **23** in the script, we will need to replace the dataset ID of the script with the dataset ID of our `Opportunity CH7` dataset. The dataset ID is obtained by clicking on the **Opportunity CH7** row in the **DataSets** page in Domo, copying the **Uniform Resource Locator (URL)** between `datasources/ ... /details`, and pasting it into line **23** of the script, as seen in the following screenshot:

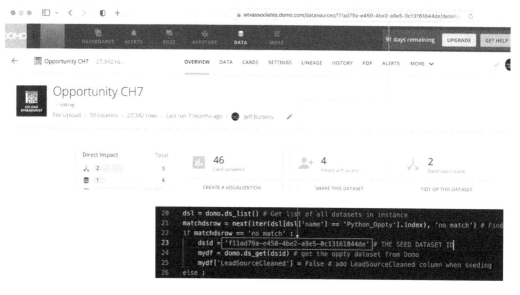

Figure 15.8 – Dataset ID location

8. Save the changes in the `DimensionCleansing.py` file.

9. To run the script in the `DimensionCleansing.py` file, use the IDE's run/debug capability, or we can execute it from Terminal by entering the following command:

```
python3 ./examples/DimensionCleansing.py
```

10. Go ahead and run the script. You will see output like this:

```
jeffburtenshaw@Jeffs-MacBook-Pro domo-python-sdk % python3 ./examples/DimensionCleansing.py
Initializing
Looping through Oppty Rows
First run created new cleansed dataset Python_Oppty with 22340 changes. View the dataset in the Domo DataS
ets page
jeffburtenshaw@Jeffs-MacBook-Pro domo-python-sdk %
```

Figure 15.9 – Terminal command output from run

Note that over 22,000 changes to clean up **LeadSources** were made!

11. Let's look at the updated data on the **DataSets** page in Domo. Make sure to refresh the page so that the new dataset appears, as seen in the following screenshot:

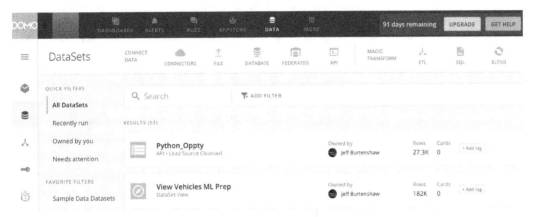

Figure 15.10 – Python_Oppty dataset

12. Click on `Python_Oppty` to go to the details page and click the **DATA** tab to preview the data, then scroll right to the **LeadSource** column, as seen in the following screenshot:

Figure 15.11 – LeadSource column data profile

We can see that the **LeadSource** column now has 9 unique values post the cleansing run, and the long-tail values are down to 5%. Not too bad.

13. Let's see if we can take one more pass and clean it even more by clicking on the **Other** category to see the details, as shown in the following screenshot:

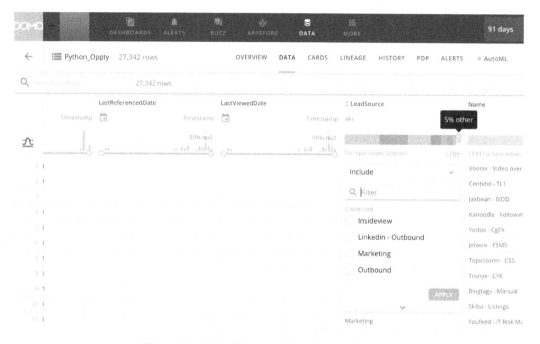

Figure 15.12 – LeadSource column long-tail values

14. Right away, we can see that **Insideview** should be exactly mapped to **Marketing**. Go back to the DimensionCleansing.py file, and starting on line **47**, let's add , 'Insideview' to 'cur': and , 'Marketing' to 'mapto':, as seen in the following screenshot:

```
46    #Mapping Array
47    mapar = {'cur':['Efeal','Patneing','Direct','Email','Chat', 'Event' , 'Inc', 'Insideview'  ],
48          'mapto':['Referral','Partner','Sales','Marketing','Marketing','Marketing','Sales' , 'Marketing'  ] }
49
```

Figure 15.13 – Mapping Insideview to Marketing

15. We can also see that **Linkedin - Outbound** should be mapped to **Marketing**. Let's use the contains approach. Back in the DimensionCleansing. py file, starting on line **42** to 'cur', add , 'Link', and to 'mapto', add , 'Marketing', as seen in the following screenshot:

```
42    # Contains Match Array
43    containar = {'cur':['Strategic','Jigsaw','Social','Event','Sales','Content','Self' , 'Link' ],
44          'mapto':['Sales','Marketing','Marketing','Event','Sales','Marketing','Sales' , 'Marketing'  ] }
45
```

Figure 15.14 – Contains search using Link to map Linkedin to Marketing

16. Let's make one more change on line **33**. We'll set the `bug` value to `True`, as seen in the following screenshot, so that we get more detail of changes in the Terminal output:

```
32   #Flags and Counters
33   bug = True # control flag for several print statements, true is for print on
34   madechanges = 0 # counter for how many times data was changed
```

Figure 15.15 – bug = True for verbose output

17. Save the changes in `DimensionCleansing.py` and run the script again by entering the following command into Terminal:

```
python3 ./examples/DimensionCleansing.py
```

18. Review the verbose Terminal output, as seen in the following screenshot:

```
jeffburtenshaw@Jeffs-MacBook-Pro domo-python-sdk % python3 ./examples/DimensionCleansing.py
Initializing
Looping through Oppty Rows
15 MAPPED from Insideview TO Marketing 57805bf6ceb84a1eb776c4f3888fed0e
33 CONTAINS FOUND Link REPLACED Linkedin - Outbound WITH Marketing d320adf1e9074a9fa5ff42ef6bac23e2
72 CONTAINS FOUND Link REPLACED Linkedin - Outbound WITH Marketing e14a4dab2ac044608b74e074c23e169f
73 CONTAINS FOUND Link REPLACED Linkedin - Outbound WITH Marketing 9b92df85e9de4bc79bfdab047a23b03e
76 CONTAINS FOUND Link REPLACED Linkedin - Outbound WITH Marketing e14a4dab2ac044608b74e074c23e169f
81 CONTAINS FOUND Link REPLACED Linkedin - Outbound WITH Marketing 0c2973c6df964b91ad36979f79dcb0ad
86 MAPPED from Insideview TO Marketing cd7d4bea12654ada9717634181c444f3
93 MAPPED from Insideview TO Marketing 7ba63ac9b5c141aeb73c56b1e600fd3a
96 CONTAINS FOUND Link REPLACED Linkedin - Outbound WITH Marketing 10222058d2364a478d47407ab79bd05a
104 MAPPED from Insideview TO Marketing a92e771a3b75421d8b8c478095c0f599

    .
    .
    .

26925 CONTAINS FOUND Link REPLACED Linkedin - Outbound WITH Marketing 378ac234f5bb43bbad3ce4e23fee93f0
26962 CONTAINS FOUND Link REPLACED Linkedin - Outbound WITH Marketing 5208c1c14b8a41f7b48f720d29373132
27070 MAPPED from Insideview TO Marketing 2438f826a68f4d99802446db0750d041
27126 CONTAINS FOUND Link REPLACED Linkedin - Outbound WITH Marketing ba14097bc66747c0a149b05534a5f125
27160 MAPPED from Insideview TO Marketing 7d63c8b2fe874a698016a23e15e73458
27223 CONTAINS FOUND Link REPLACED Linkedin - Outbound WITH Marketing a1a3d4d213db404bac222527e3c6b594
27244 MAPPED from Insideview TO Marketing 4ae4a2b32bb4439bb7fa47e9feaa1000
27268 CONTAINS FOUND Link REPLACED Linkedin - Outbound WITH Marketing da5d44a45210464184c57e68aaa3e80c
27316 CONTAINS FOUND Link REPLACED Linkedin - Outbound WITH Marketing 3e0a0a06c5584a26978dcacc4a6c1c8e
27335 CONTAINS FOUND Link REPLACED Linkedin - Outbound WITH Marketing 986b706b6f504fecb775cb4b333c9f32
27338 MAPPED from Insideview TO Marketing 78750fd4c602468286274e83950a9a00
1324 Changes Made
jeffburtenshaw@Jeffs-MacBook-Pro domo-python-sdk %
```

Figure 15.16 – Additional mapping rules: verbose Terminal output

19. Jump back to the `Python_Oppty` dataset detail page and click the browser page refresh, and then scroll over to the **LeadSource** column, as seen in the following screenshot:

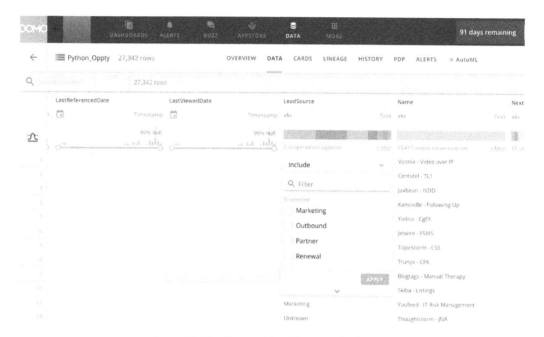

Figure 15.17 – Cleansed LeadSource values' results

All right—down to 7 **LeadSource** values; the long-tail values are gone!

Next, let's review what we learned in this chapter.

Summary

In this chapter, we went through a dimensional data cleansing exercise and learned that Python can be used to call Domo APIs directly to do **create, read, update, and delete** (**CRUD**) operations on Domo datasets. We saw that there is a Python package called `pydomo` that makes it easy to do. Then, we used standard packages such as `pandas` and `fuzzywuzzy` to do some fancy work on cleaning up the **LeadSource** dimension. We even made an iterative pass after adjusting our matching criteria based on data profiles in Domo to further reduce the long-tail values of dimension values. It doesn't take much imagination to see how this process could be generalized across multiple dimensions and run on a schedule to scan new rows that have not been cleansed to improve data quality in a dramatic fashion.

In the next chapter, we will explore one of the Domo platform's **machine learning (ML)** capabilities.

Further reading

The following list contains related items of interest:

- `fuzzywuzzy`: `https://pypi.org/project/fuzzywuzzy/`
- `pandas`: `https://pandas.pydata.org`
- `pydomo`: `https://github.com/domoinc/domo-python-sdk`
- *Django makes it easier to build better web apps more quickly and with less code*: `https://www.djangoproject.com`
- Handling different Python versions: `https://stackoverflow.com/questions/5846167/how-to-change-default-python-version`
- Running Python scripts: `https://realpython.com/run-python-scripts/`

16
Using Domo Machine Learning

Machine learning (**ML**) is everywhere these days and businesses that aren't marshaling their data with ML algorithms may not be able to compete effectively in a machine-optimized world. Democratizing ML – putting the power of ML into non-data scientists' hands – is what Domo's AutoML feature does. Domo partnered with Amazon Web Services and their **SageMaker Autopilot** product to bring this capability to Domo platform users in a turnkey way. Essentially, a Domo dataset is processed through SageMaker Autopilot with ease. This feature trains hundreds of potential ML models against a Domo dataset and determines the best model for predicting a chosen variable. The best model can be deployed in a Magic ETL dataflow using the AutoML Inference tile, which adapts to new incoming data and generates the value prediction from the model. The combination of Domo and Amazon SageMaker Autopilot helps make ML accessible to any analyst looking to apply ML insights. Domo AutoML is a paid feature available on request. Domo also has a paid option for instantiating a cloud-hosted option for **Jupyter Notebooks**.

In this chapter, we will cover the following topics:

- Understanding the AutoML process
- Training a Domo AutoML model

- Deploying a Domo AutoML model
- Supporting Jupyter Workspaces

Technical requirements

To complete this chapter, you will need the following:

- Internet access
- Ability to download a 1.5 GB file
- Your Domo instance and login details
- Ability to use the Domo File Upload Connector to import data
- Access to the Domo AutoML feature

> **Important Note**
> If you don't have a Domo instance, you can get a free trial instance here:
> `https://developer.domo.com/developer-trial-request`.

Understanding the AutoML process

The Domo AutoML process is a multi-step process, as shown in the following screenshot:

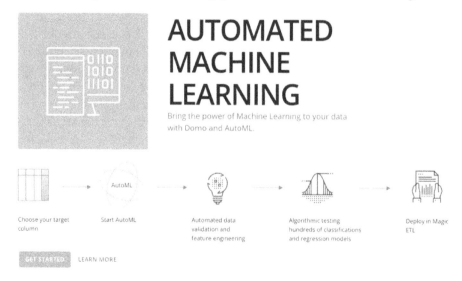

Figure 16.1 – AutoML process

Follow these steps to execute the AutoML process:

1. Prepare the data with dimensions that may influence the value to be predicted.

2. Launch an AutoML training job.

3. Evaluate model performance on the **Model Leaderboard** page.

4. Choose the model that is the best fit and deploy it via the **AutoML Inference** tile in Magic ETL.

5. Set up a model monitoring dashboard.

6. Create a story based on the insights that have been predicted.

Next, let's train a model.

Training a Domo AutoML model

To understand the utility of Domo AutoML, let's get a public dataset on the prices of used cars and use it to build a predictive model based on used car prices:

1. Download the *Used Car Dataset* from **Kaggle**: `https://www.kaggle.com/austinreese/craigslist-carstrucks-data`.

2. Use the Domo File Upload Connector to place the `vehicles.csv` data into a Domo dataset named `Vehicles`, as shown in the following screenshot. The file is large, so it may take several hours to load:

Figure 16.2 – Vehicles dataset post-import

3. After the file has been loaded in, click on **MORE**, then **AutoML**.

4. This will take us to a page that looks similar to the one shown in *Figure 16.1*. Click the **GET STARTED** button.

5. In the **AutoML** window, set the modeling parameters shown in the following screenshot. Set **price** for **Column to predict** and **Automatic (Recommended)** for **Task type**. Then, expand the **Advanced** tab and set **AUTOML CANDIDATES TO RUN** to 80:

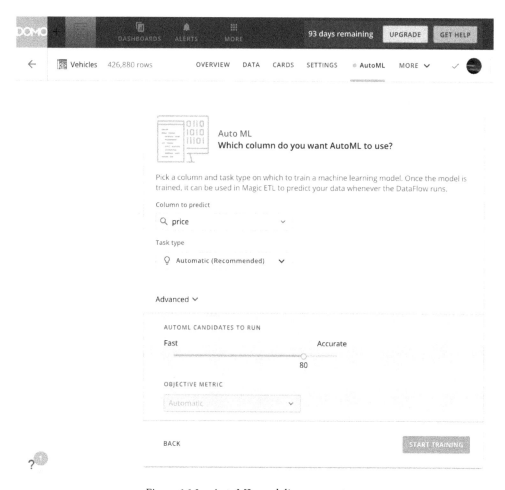

Figure 16.3 – AutoML modeling parameters

6. Click **START TRAINING** to open the model training progress page, as shown in the following screenshot:

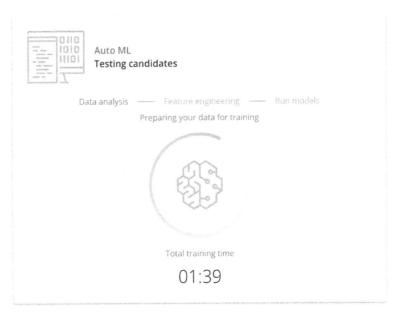

Figure 16.4 – Model training and execution

7. Here, we are running 80 model iterations on around 430k rows of data. Be patient – it is going to take a few hours for the process to figure out the best model to use. When the process completes, the results will be displayed on the **AutoML** tab, as shown in the following screenshot:

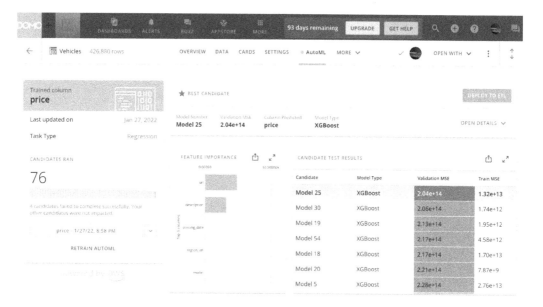

Figure 16.5 – AutoML model training results

In terms of the **BEST CANDIDATE** model for the **Regression** task type, the **FEATURE IMPORTANCE** card shows the model feature weights that the model will use in the regression equation to predict. In this case, we ran the model without doing any feature engineering – choosing and constructing features that might be better suited for inputs to predict the model. Running the model on all the features produced results that were not particularly helpful. For example, using the **URL** and **description** features is not predictive as they are too granular to be of much predictive value. So, let's back up and work on the feature list by going to the **Views Explorer** area.

8. Click **OPEN WITH** and select **Views Explorer**. Name the view `View Vehicles ML Prep`.

9. **Remove** the **url, region_url, odometer, VIN, image_url, description, lat, long**, and **posting_date** columns from the view since they are duplicative/too granular to be used effectively in a predictive model.

10. Click **ADD COLUMN** and then **ADD CALCULATED COLUMN**. Enter `mileage range` for **COLUMN NAME**.

11. Enter the following in the **CALCULATION** area:

```
case
when `odometer` < 1000 then '<1k'
when `odometer` >= 1000 and `odometer` < 5000 then
'1k-5k'
when `odometer` >= 5000 and `odometer` < 10000 then '5k -
10k'
when `odometer` >= 10000 and `odometer` < 20000 then '10k
- 20k'
when `odometer` >= 20000 and `odometer` < 30000 then '20k
- 30k'
when `odometer` >= 30000 and `odometer` < 40000 then '30k
- 40k'
when `odometer` >= 40000 and `odometer` < 50000 then '40k
- 50k'
when `odometer` >= 50000 and `odometer` < 75000 then '50k
- 75k'
when `odometer` >= 75000 and `odometer` < 100000 then
'75k - 100k'
when `odometer` >= 100000 and `odometer` < 150000 then
'100k - 150k'
when `odometer` >= 150000 and `odometer` < 200000 then
'150k - 200k'
else '>200k'
END
```

12. Click the **SAVE** dropdown and select **Save and Close**.

13. Click **ADD COLUMN** and then **ADD CALCULATED COLUMN**. Enter `posteddayofweek` for **COLUMN NAME**.

14. Enter the following in the **CALCULATION** area:

```
DAYOFWEEK(`posting_date`)
```

15. Click the **SAVE** dropdown and select **Save and Close**.

16. For the **region, year, manufacturer, model, condition, cylinders, fuel, title_status, transmission, drive, size, type, paint_color**, and **state** columns, click on the column options menu and select **Remap Nulls**. Then, enter Unknown.

17. Click **Filter** and add a filter for **Year** from 2019 to 2022.

18. Click the **SAVE** dropdown and select **Save and Close** to save the view.

19. Now, we are ready to train the model on the view. On the **View** page, click the **AutoML** tab, click **GET STARTED**, select **Price** for **Column to predict** and **Regression** for **Task type**, and click **START TRAINING**. Wait patiently for the process to run its course, as shown in the following screenshot:

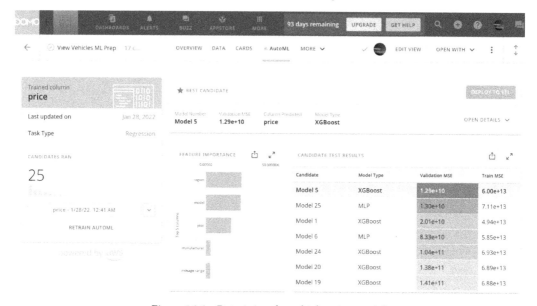

Figure 16.6 – Retraining the vehicle price model

The **FEATURE IMPORTANCE** card looks much better on this model.

Next, let's go ahead and deploy the model and see how it predicts price.

Deploying a Domo AutoML model

Now that the model has been trained, we can feed new data through the inference engine to get predictions on the new data. Let's do this for the vehicle price prediction. To make new predictions, data that is in the same format as the training model needs to be fed to the inference engine via a Magic ETL job. Let's get started::

1. Import the Get Price Prediction.csv file from this book's GitHub repository (https://github.com/PacktPublishing/ Data-Democratization-with-Domo/blob/ d6f5c34539afece0622572db0bc6109fa9dc3175/Get%20Price%20 Prediction.csv) into Domo using the File Upload Connector.

2. Click **DEPLOY TO ETL** on the **AutoML** page and choose the **View Vehicles ML Prep** dataset. Change the dataflow's name to `Vehicle Price Prediction`.

3. Drag and drop an **Input DataSet** tile onto the canvas. Then, click **CHOOSE DATASET** and pick **Get Price Prediction.csv**. Finally, click **CHOOSE DATASET**.

4. Delete the **View Vehicles ML Prep** tile from the job.

5. Connect the **Get Price Prediction.csv** tile to the **AutoML Inference** tile. Accept the default **AutoML Inference** tile settings, as shown in the following screenshot:

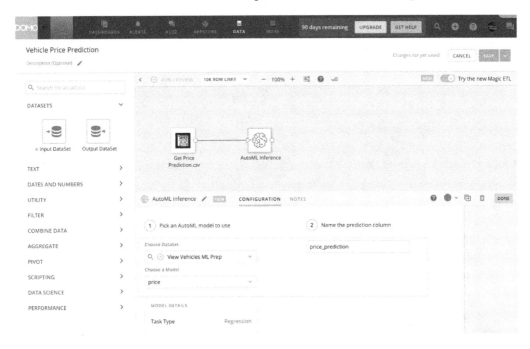

Figure 16.7 – The default AutoML Inference tile settings

Drag and drop an **Output DataSet** tile onto the canvas. Connect it to the **AutoML Inference** tile and name it **Vehicle Predicted Price**, as shown in the following screenshot:

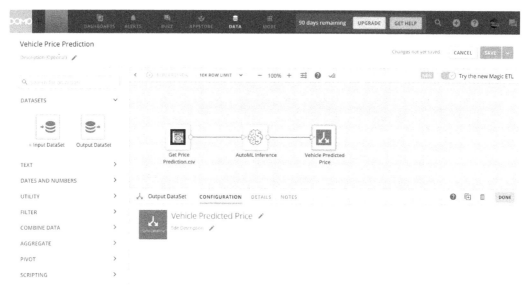

Figure 16.8 – Vehicle Predicted Price output dataset

6. Click the **SAVE** dropdown and select **Save and Run**. Click **SAVE AND RUN** on the description dialog. Wait for the process to run, as shown in the following screenshot:

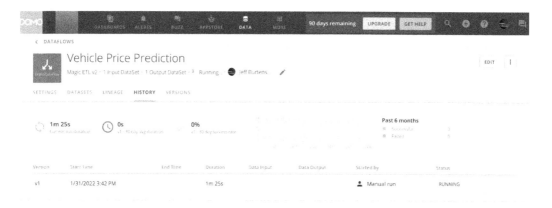

Figure 16.9 – ETL job status

7. When complete, click on the **DATASETS** tab under **OUTPUT DATASETS** and click on **Vehicle Predicted Price**.

8. On the **DataSets** page, click on the **DATA** tab and go to the last column, **price_prediction**, as shown in the following screenshot:

Figure 16.10 – Vehicle Predicted Price

Here, we can see a reasonable price estimate for that vehicle.

Next, we will discuss Jupyter Workspaces, a tool that Domo supports for interactively scripting and plotting data.

Supporting Jupyter Workspaces

Domo added the capability to run **Jupyter Workspaces** in the Domo cloud, which supports Python3 and R scripting. Setting up the Jupyter cloud environment is automatic and provides full access to the JupyterLab interface for interactive exploratory computing. This is a beta feature, so you will need to work through your Domo Account Manager to get access.

Once activated, follow these steps to access the Jupyter cloud workspace:

1. Choose the **DATA** option via the main menu. Then, click the **...** side menu option and choose **Jupyter Workspaces**, as shown in the following screenshot:

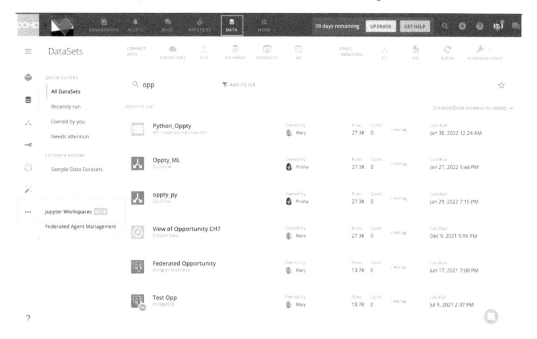

Figure 16.11 – Jupyter Workspaces access

2. Click **+ NEW WORKSPACE** and enter My Notebook as the workspace's name.

3. Click **SELECT DATASET** under **DATASETS** > **Inputs** and select **Python_Oppty**. Then, click **CHOOSE DATASET**, as shown in the following screenshot:

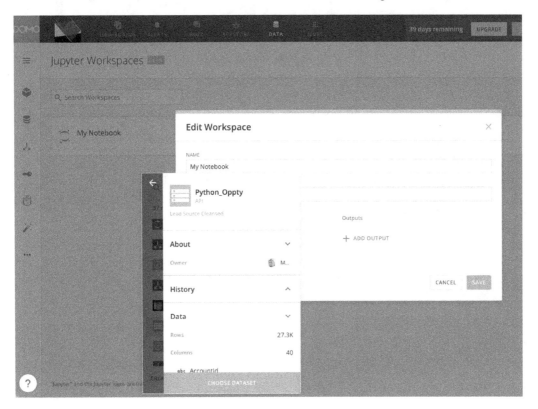

Figure 16.12 – Adding inputs to a Jupyter Workspace

4. Click **+ ADD OUPUT** and choose the default of **Output 1**, as shown in the following screenshot:

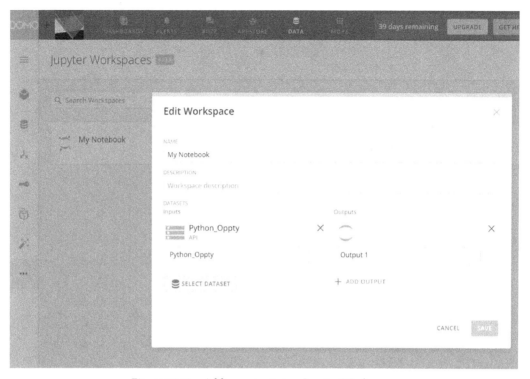

Figure 16.13 – Adding outputs to a Jupyter Workspace

5. Click **SAVE** to create a new Jupyter Workspace called **My Notebook**, as shown in the following screenshot:

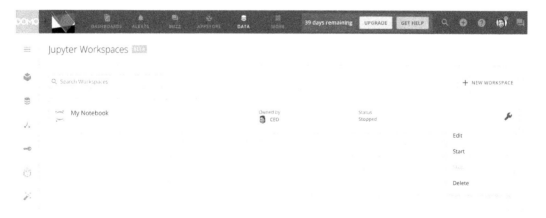

Figure 16.14 – Saved Jupyter Workspace

6. Click the wrench icon on the **My Notebook** row and click **Start**, as shown in the preceding screenshot.

7. Click **My Notebook** to open the running workspace on the **Launcher** tab, as shown in the following screenshot:

Figure 16.15 – The Launcher tab

From here, you can easily create Python and R notebooks and run console commands.

8. Click the **README.md** file to bring up some helpful information on using Jupyter Workspaces, as shown in the following screenshot:

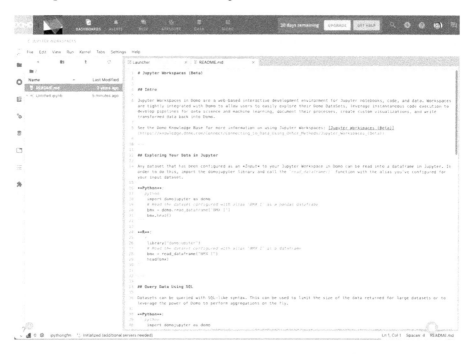

Figure 16.16 – Jupyter Workspaces README.md file

Now, let's review what we've learned in this chapter.

Summary

In this chapter, we learned that Domo provides a product that's integrated with AWS SageMaker Autopilot to automate ML modeling tasks. We learned how to prepare a dataset and train and retrain models. Then, we deployed the model via an ETL job and ran new data through it to get a predicted vehicle price from the model. Finally, we became aware of the capability to run Jupyter Workspaces in the Domo cloud. This combination of technologies makes the process complex of creating ML-based predictive modeling accessible to the typical analyst. This is another way that Domo is democratizing the analytics process.

In the next chapter, we'll look at securing assets in Domo.

Further reading

The following are related items of interest:

- AutoML model type references: https://domohelp.domo.com/hc/en-us/articles/360060598413-Machine-Learning-Concepts-to-Help-You-be-More-Successful

- Feature engineering: https://machinelearningmastery.com/discover-feature-engineering-how-to-engineer-features-and-how-to-get-good-at-it/

- *Characteristics of a Good Feature*: https://towardsdatascience.com/characteristics-of-a-good-feature-4f1ac7a90a42

- Skills to build predictive models: https://tdwi.org/articles/2017/06/26/5-skills-to-build-predictive-analytics-models.aspx

- Jupyter Workbooks in Domo: https://domohelp.domo.com/hc/en-us/articles/360047400753-Jupyter-Workspaces-Beta-

- Jupyter background: https://jupyter.org

Section 5: Governing

Governance is a broad topic relevant to the entire Domo Ecosystem; for example, Artifact Ownership, Data Freshness, Data Security, Artifact Sharing, Communications, Activity Logging, and so on.

In this section, you will gain knowledge of how to secure data and content assets. You will also walkthrough a framework for determining how to organize your BI team including roles and responsibilities and job descriptions. And finally, we will discuss the tools and methods for core standard procedures such as request management, backlog grooming, and artifact migration.

After completing this section, you will be able to govern the Domo Platform including securing the data, organizing the people, establishing necessary standard operating procedures, managing the technology usage.

This section comprises the following chapters:

- *Chapter 17, Securing Assets*
- *Chapter 18, Organizing the Team*
- *Chapter 19, Establishing Standard Procedures*

17
Securing Assets

Security is a primary concern when dealing with the content of a platform. The Domo platform has all the tools that we would expect, covering the governance of people, groups, roles, content, authentication, network security, company settings, content embedding, feature settings, and more.

However, with so many assets to manage and so much flexibility regarding how to manage usage across the platform, it is advisable to first consider a policy framework that will guide tasks in securing platform assets. Understanding organizational culture is a critical component along with the tools' capabilities in establishing an effective security policy. In this chapter, we will learn what to consider when securing assets and how to secure the assets that are aligned with a strategy with the Domo toolset.

In this chapter, we will specifically cover the following topics:

- Considering the dimensions of a security policy
- Governing people, groups, and roles
- Securing content
- Leveraging authentication standards
- Controlling network security
- Configuring the company settings

- Managing the feature settings
- Using more admin features
- Handling **Personably Identifiable Information (PII)** in Domo

Technical requirements

To follow along with this chapter, you will need the following:

- Internet access
- Your Domo instance and login credentials

> **Important Note**
> If you don't have a Domo instance, you can get a free trial instance at
> `https://developer.domo.com/developer-trial-request`.

Considering the dimensions of a security policy

The Domo platform has a robust set of tools for establishing security, as shown in *Figure 17.1*:

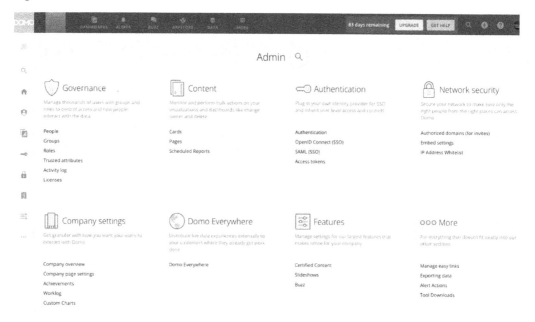

Figure 17.1 – The Domo platform's security tools

However, as good as this toolset is at implementing security, an effective security policy needs to be informed by additional dimensions, such as those shown in *Figure 17.2*:

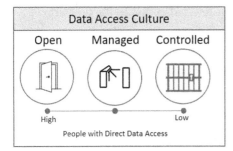
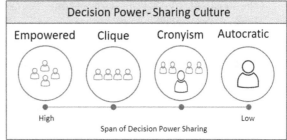
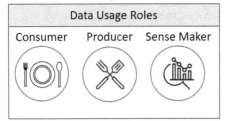
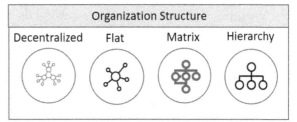

Figure 17.2 – The dimensions of a security strategy

If tasked with implementing a security policy, it is wise to understand these bigger-picture dimensional considerations. Implementing cultures with roles and structures that are not complementary will lead to a conflicted security policy.

Data access culture

Data access culture is the way an organization regards how broadly people can have access to direct source data. Some would say that having direct data access provides an unvarnished view of the business, or in other words, no spin just the data. Trying to change who has direct access to data in an organization is certainly a high-stakes, politically charged task. At worst, current stakeholders might have varnished the data to meet their needs, or they might have distorted the view, and likely, they might not appreciate alternative interpretations of the facts. At best, stakeholders have added context others might not have with only the raw data.

So, when thinking about setting up security around data access, we should consider the current culture and attitudes toward a shift to less or more direct data access.

The following list contains some theoretical markers of data access culture presented on a spectrum of the number of people having direct data access, as shown in *Figure 17.2*:

- **Open** is a culture that allows everyone free and unfettered access to the data. There are no data jailkeepers.

- **Managed** is a culture where the structure, organizational units, geographies, and groups are set up to have some degree of direct data access. Typically, this is controlled by the intersection of the structure and the decision power-sharing environment.

- **Controlled** is a culture where relatively few people have access to direct data to use creatively for discovery. In this environment, data access is a carefully considered and metered approach by functional role. People are encouraged to use the data as provided, and any attempts to go beyond the current scope are typically met with strong resistance from the controllers.

It should be noted that organizations likely have microcultures with a mix of the preceding styles. The point is to not be unaware of the dynamics.

Next, let's turn our attention to the decision power-sharing culture.

Decision power-sharing culture

The decision power-sharing culture is easy to ascertain and a favorite topic of conversation in any organization. The following list contains descriptions of the categories along a spectrum of decision-making options, as shown in *Figure 17.2*.

- **Empowered** decision-making is seen in organizations that push as much decision-making as possible down to the people who are closest to and impacted most by the decision.

- **Clique** decision-making is where affinity factions form. They work together to drive decisions in an organization to their advantage and, sometimes, at the expense of the greater good.

- **Cronyism** is a subset of the autocratic style where the top executive(s) surrounds themselves with managers that tend to go along with the prevailing winds of the executive decisions and rubber stamp the authoritative view. Any disagreement with or challenge to the party line is discouraged by way of recriminations.

- **Autocratic** is a style where a strong-willed, often charismatic, leader imposes their views on the entire organization with little or no freedom for decision-making allowed to others. In this environment, people exist to execute the vision of the autocrat.

A full discussion of the trade-offs among the spectrum of decision-making power-sharing styles is beyond the scope of this book. Nevertheless, aligning the decision support infrastructure to support or, in some cases, even challenge the current paradigm is advised.

Next, let's look at the possible data roles to consider. Several options for organizational structures, as shown in *Figure 17.2*, are discussed in the following section.

Data utility roles

The data utility roles in an organization center on how data is produced and consumed in the organization. Data usage can be grouped into the following categories, as shown in *Figure 17.2*. They are briefly discussed here:

- The **Consumer** role, as the name suggests, is a role that uses data to perform a specific task. For example, a role for accounts receivable or a raw materials purchaser needs to consume a very specific set of data to execute their responsibilities.

- The **Producer** role can be overlooked when defining a policy. As sensor feeds, machine data feeds, and IoT feeds become more prevalent producers of information for organizations, and third-party augmentation providers become more integral to creating better predictive models, policies around the people and systems that produce information and how they are utilized are becoming more essential.

- **Sense Maker** is a role that is responsible for taking data, often from disparate sources, and generating informative insights. The more sources and tools an analyst has at their command to acquire, sculpt, discover, present, and communicate, the more likely they are to produce insights that have utility.

All of these roles are couched in an organization's structure, so let's consider a framework for understanding organizational structures.

Organizational structures

An organizational structure refers to the way an organization sets up how people and resources are related to drive the mission of the organization. The following list contains brief descriptions of various types of organizational structures:

- **Decentralized** structures are getting increased attention as cryptocurrency and Web 3.0 innovators explore the boundaries of highly distributed decision-making and ownership models. History teaches us that diversity, flexibility, and openness tend to win out in technology services in comparison to tightly controlled, closed hedgehog approaches. This might yet be another point of proof of the superiority of the wisdom of crowds.

- **Flat** structures are those that have a few layers of intermediary managers between the top executives and the frontline service providers or laborers.

- **Matrix** structures have a network reporting structure where individuals have not only hierarchical line responsibilities within their department or function, but they also have dual accountability toward competency centers or balancing the interests of how they perform the work.

- **Hierarchy** structures have a tiered line reporting structure that is typical of many organizations.

The security configuration will inevitably mirror the command-and-control structures that exist in an organization.

Again, there are no one-size-fits-all or best-of-breed answers among the theoretical constructs presented here. However, an awareness of these dynamics is critical for getting a good fit for organizations using Domo. And with that, let's turn our focus on the platform tools.

Governing people, groups, and roles

To secure assets, a system must have a way in which to manage identities, groups of identities, and their associated roles. Let's start with a look at the **People** page of Domo.

Using the People page

People are the individual identities that have access to the Domo instance. The following steps are instructions on how to use the Domo **People** page features, which control what a person can and cannot do across the platform:

1. To access the **People** page, click on the **MORE** main menu option. Click on **ADMIN**, and under **Governance**, click on **People**. That takes us to the **People** page, as shown in *Figure 17.3*:

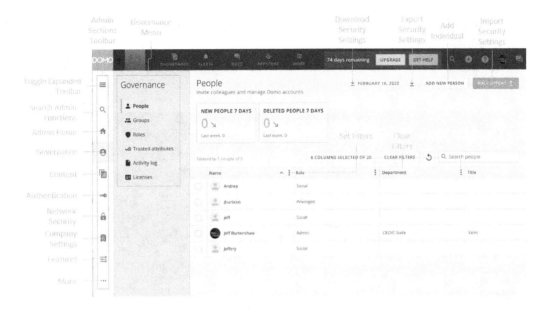

Figure 17.3 – The Governance People page

The following list contains descriptions of the primary functions that are available on the **People** page:

- **Admin Sections Toolbar** is a navigation toolbar to provide access to the various administration features.

- **Governance Menu** presents the pages supporting the governance of the **People**, **Groups**, **Roles**, **Trusted attributes**, **Activity log**, and **Licenses** features.

- **Add Individual** opens a dialog to add a person to the access list.

- **Set Filters** is active within each column and enables you to filter the rows that are displayed in the list.

- **Clear Filters** resets all of the column filters.

- **Export Security Settings** exports the current people-related security settings. This is so that they are available to be downloaded into a CSV file for use in bulk editing and the uploading of security additions and changes.

- **Download Security Settings** downloads a CSV file of the previously exported security settings.

- **Import Security Settings** updates the security settings from a CSV upload, as shown in *Figure 17.4*:

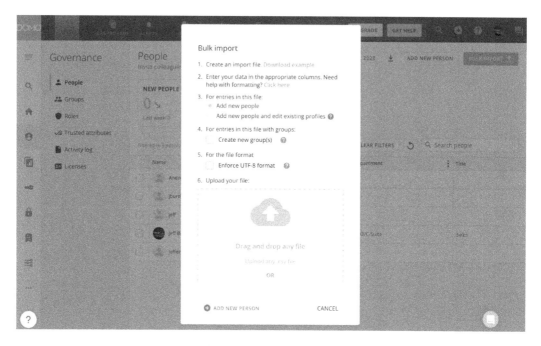

Figure 17.4 – The People page's Bulk import dialog box

2. The **People** page does have some capability to perform options on a single user or if multiple users have been selected. Click on the box next to a user to select them, and the **Action Toolbar** will appear, as shown in *Figure 17.5*:

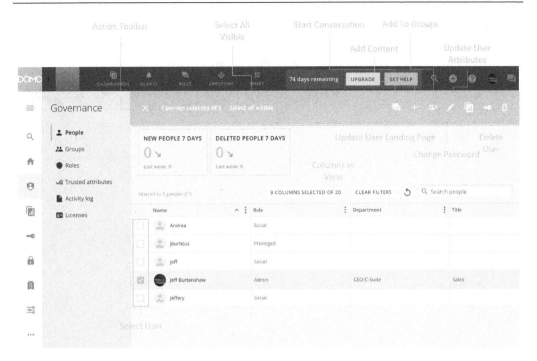

Figure 17.5 – The Governance People page's multi-select toolbar options

The following list contains descriptions of the features that are activated when one or more users are selected:

- The **Select User** area is the checkbox next to each person that indicates whether they have been selected for the operations executed on the **Action Toolbar**.

- **Action Toolbar** is the toolbar that appears when one or more users are selected in the table containing many actions relevant to the user settings.

- **Select All Visible** is a shortcut to select all the visible users in the table based on the filters applied.

- **Start Conversation** starts a new Domo Buzz conversation with the selected users.

- **Add Content** opens a search dialog to add cards and pages to the selected users' access.

- **Add to Groups** opens a search dialog to find groups and add the selected users to groups.

- **Update User Attributes** opens the user profile page if a single user has been selected. When multiple users have been selected, a page with shared attributes opens to allow you to edit all of the users' settings at once, as shown in *Figure 17.6*:

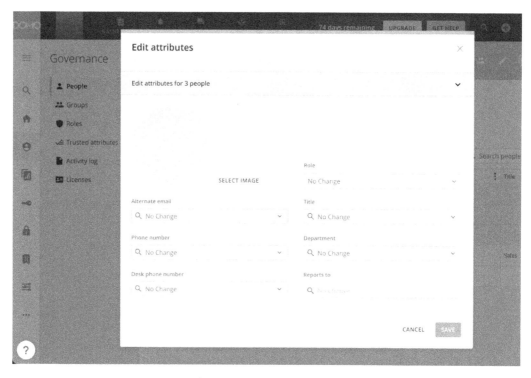

Figure 17.6 – The multi-user profile Edit attributes dialog

- **Update User Landing Page** enables you to set the desktop and mobile landing pages for all of the users selected, as shown in *Figure 17.7*:

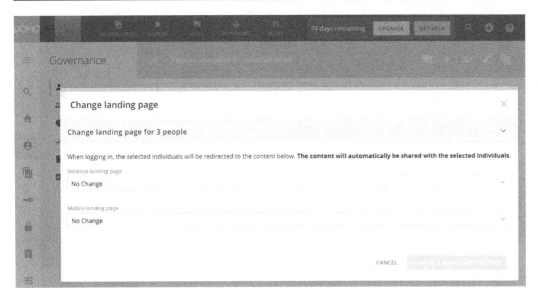

Figure 17.7 – The multi-user Change landing page dialog

- **Change Password** enables admins to change the selected user password. This feature is not multi-user enabled and only works for one selected user at a time.

- **Delete User** removes the selected users from the instance.

> **Important Note**
>
> For organizations without an identity provider, the ability to export, modify, and upload security settings in bulk is essential. Additionally, it is possible to use the Domo APIs for users, groups, and **Personal Data Privacy** (**PDP**) policies to script bulk operations, typically, on exports from other systems; for example, syncing the security profiles from an ERP system within Domo. The Domo Administration CLI is also another option for performing bulk security operations.

Now that we understand how to configure people-based security, let's discuss a related feature for setting trusted attributes to use with people.

Setting trusted attributes

Trusted attributes are attributes in the people profile that can be set to only be trusted or controlled by a trusted source. If a profile is set as trusted, typically, an HR directory or identity service is used to populate the attribute. If the attribute is trusted, then the user cannot change the attribute's value.

Let's add **Department** as a trusted attribute:

1. Click on the **Trusted attributes Governance** option, as shown in *Figure 17.1*.

2. On the **Trusted attributes** page, as shown in *Figure 17.8*, click on the checkbox for the **Department** attribute:

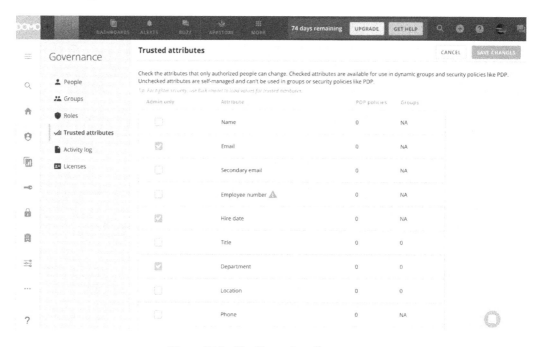

Figure 17.8 – The Trusted attributes page

3. Click on **SAVE CHANGES**.

4. This tells the system that the **Department** user profile attribute, as shown in *Figure 17.9*, cannot be edited by the user. But it has been populated by an admin or trusted source such as an HR system feed, LDAP directory, or identity service:

Figure 17.9 – The user profile page of the trusted attribute

Trusted attributes can also be used to automatically set PDP and content access policies. Next, let's cover group-based security.

Using the Groups page

Groups are logical containers of people with similar security needs or restrictions in terms of user data and user content access. Groups bring administrative efficiency so that access privileges can be given to a group once, and then any person included in the group inherits those privileges from the group. Users can participate in many groups, and privileges in Domo are optimistic. This means that if group policies conflict with each other, the most liberal access is preserved for the user.

As shown in *Figure 17.10*, the **Groups** page controls the creation and management of groups:

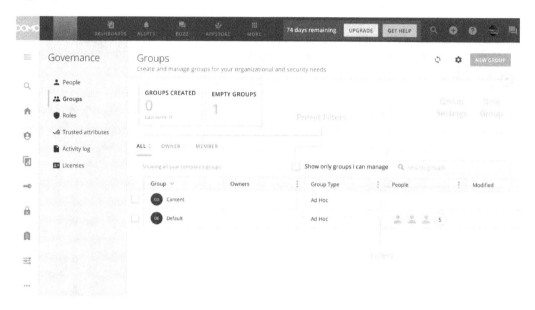

Figure 17.10 – The Governance Groups page

The following list includes brief descriptions of the major **Group** administration features:

- **Preset Filters** are predefined filters on the group table list. The **OWNER** view shows the group's owner by the current user, while **MEMBER** shows the groups that the current user is a member of. **Show only groups I can manage** filters the group list into editable groups for the current user.

- **Filters** enable access to each column to filter the group table list by the column values.

- **New Group** opens a wizard to create a new group.

- **Group Settings** opens the dialog, as shown in *Figure 17.11*:

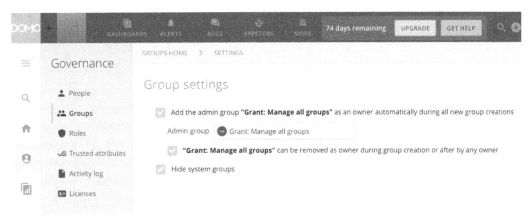

Figure 17.11 – The Group settings page

The following list contains descriptions of the **Group settings** options:

- **Add the admin group "Grant: Manage all groups" as an owner automatically during all new group creations**: When set, this means that all admins will automatically be set as co-owners of the group. If not set, then only the creator/owner of the group can manage the group.

- **"Grant: Manage all groups" can be removed as owner during group creation or after by any owner**: Put simply, if you want to have other admins be able to manage the group, then leave this checked. If unchecked, then only owners can manage the group.

- **Hide system groups**: This indicates whether internal system groups (that is, permission groups for grants) are displayed in the group list or not.

Now that we understand the **Groups** page features, let's walk through how to create a group.

Creating a new group

Creating a new group has a wizard to take us through the steps, which we'll check out here:

1. Click on the **NEW GROUP** button, as shown in *Figure 17.10*. This takes us to step **1** in the wizard to name and describe the group. Enter Sales as the **Name** value, and click on **NEXT**:

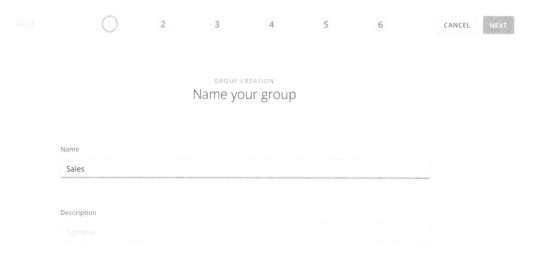

Figure 17.12 – New group wizard step 1

2. In step **2**, which is the **Select group type** page, choose the group type. Possible choices are listed in *Figure 17.13*:

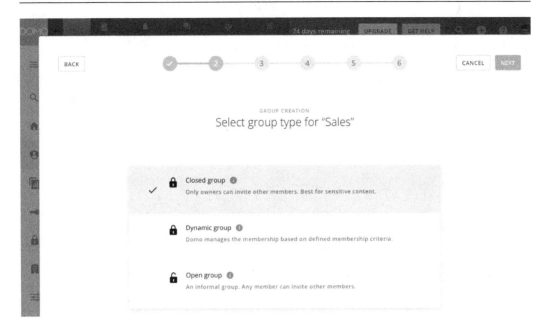

Figure 17.13 – New group wizard step 2

The following list contains additional context around the group types:

- The **Closed group** type is the most restrictive group type wherein only the group owners can invite other users into the group. This is useful for scenarios where a group wants to tightly control who can be added to the group.

- The **Dynamic group** type uses trusted attributes to automatically add people to the group. Think of an example where we are using a directory or identity service to provide the department the users are in, such as Sales. With this group type, when the department attribute is set to be trusted, then all users that have Sales as their department are automatically added to the Sales group.

- The **Open group** type is the most permissive type, allowing any member of the group to invite other users into the group.

3. Click on **Closed Group**, and click on **NEXT**.

4. Step **3**, **Select members**, as shown in *Figure 17.14.1*, is where we manually choose the members of our new group. Select a few users and click on **NEXT**:

Figure 17.14.1 – New Group wizard step 3, closed group

Alternatively, we could have chosen **Dynamic group**, as shown in *Figure 17.14.2*, to set which users are in the group, in this case, using their **Department** details:

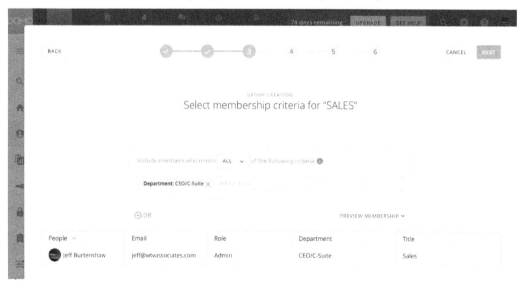

Figure 17.14.2 – New group wizard step 3, Dynamic group

Including **ALL** members means an exact match of the criteria only, whereas **ANY** will include people who match any of the criteria as members. When the field is selected by the admin, the values of the field to match for inclusion are manually chosen, too.

5. Step **4** allows us to choose an avatar for the group. Click on **NEXT**.

6. Step **5** allows us to set the owners of the group, as shown in *Figure 17.15*:

Figure 17.15 – New group wizard step 5

7. Click on **NEXT**.

Important Note

The **Grant: Manage all groups** option is automatically set as an owner.

---

8. Step **6** is **Review and save** the new group. Click on **SAVE**.

9. In *Figure 17.16*, we can see the new **SALES** group:

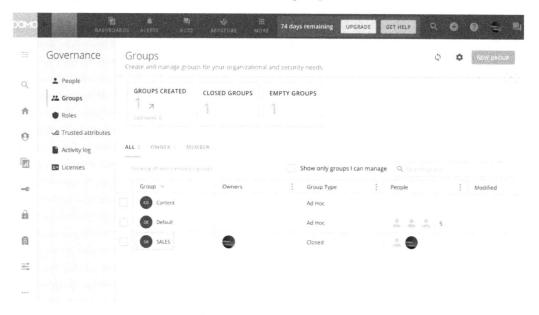

Figure 17.16 – The Groups page with the new SALES group

Important Note

Every Domo instance is set up with two user groups. The **Default** group
has all users automatically added to it when a user account is created. For
organizations that don't want this group or the ability to grant access to
everyone, it can be deleted. The **Content** group is there for convenience, as an
example group, and can also be deleted.

---

10. Click on **SALES** to open the group details, as shown in *Figure 17.17*:

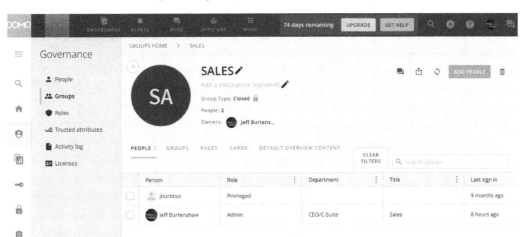

Figure 17.17 – The group details page

From the group details page, we can see **PEOPLE** and **GROUPS** who are in this group. Additionally, we can see **PAGES** and **CARDS** the group has access to and can add or remove content. We can even set up cards that provide background/training for the group participants in the **DEFAULT OVERVIEW CONTENT** tab.

Next, let's take a deeper look at the role features within Domo.

Using the Roles page

Roles are similar to groups in terms of being a logical grouping of security privileges to make permission administration scalable. However, a group's purpose is to control access to user content, whereas a role's intent is to control system permissions using grants. Users can only participate in one role.

By default, Domo has five roles, as shown in *Figure 17.18*:

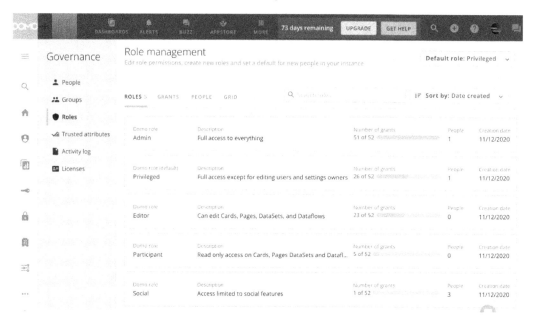

Figure 17.18 – The Role management page

To view the role details, click on the role name and it will bring up the role detail page for the **Admin** role, as shown in *Figure 17.19*:

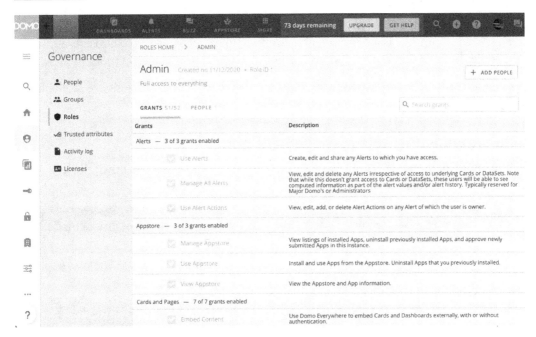

Figure 17.19 – The Admin role details page

Role **Grants** are managed in the table using checkboxes and are categorized. The **PEOPLE** tab shows who is assigned the role, and by checking the box next to the person, we can reassign their role, as shown in *Figure 17.20*:

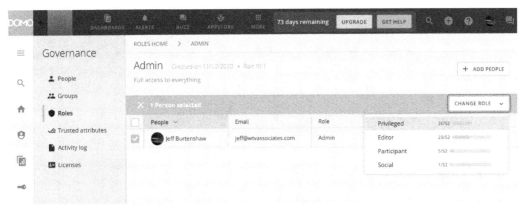

Figure 17.20 – Changing the user role

It is possible to create additional custom roles with the Enterprise version of Domo. An example of a custom role is a departmental admin who has privileges to edit users, add new people, and edit group permissions but not manage all groups permission.

Let's review the **Role management** grid to get an idea of which permissions are available to be granted to roles.

Understanding the Role management grid

The **Role management GRID** tab presents all the permission categories down the left-hand side and the **Roles** details along the top of the grid, as shown in *Figure 17.21*:

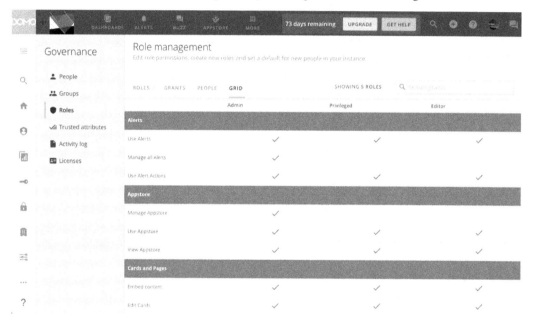

Figure 17.21 – The Role management GRID tab

Next, let's review the **GRANTS** tab.

Understanding the GRANTS tab

The **GRANTS** tab on the **Role management** page lists all the grantable permissions, as shown in *Figure 17.22*:

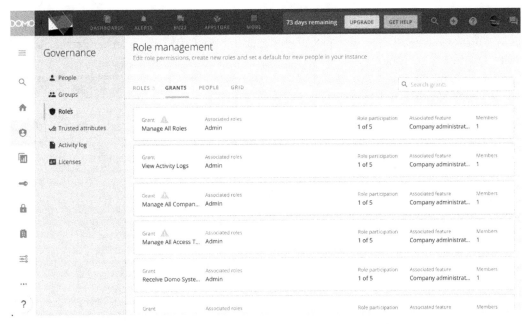

Figure 17.22 – The Role management GRANTS tab

Click on a grant name to navigate to the grant details page in **Manage All Roles**, as shown in *Figure 17.23*:

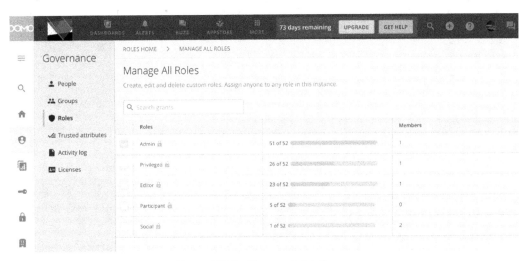

Figure 17.23 – The grant details page

The roles that are checked have this permission granted to the role.

Next, let's look at the **People** tab.

Understanding the PEOPLE tab

The **PEOPLE** tab lists all the user's metadata including **Role**. To change a user's role, click on the box next to their name and select a new role, as shown in *Figure 17.24*:

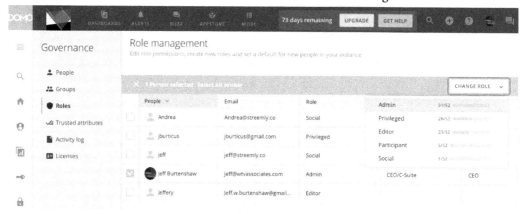

Figure 17.24 – The Role management PEOPLE tab

That completes our review of the role features. Next, let's review the activity log.

Using the Activity log page

The **Activity log** feature presents a filterable and exportable list of activities in Domo. An activity is defined by the **Time**, **Person**, **Event**, and **Object** options, as shown in *Figure 17.25*:

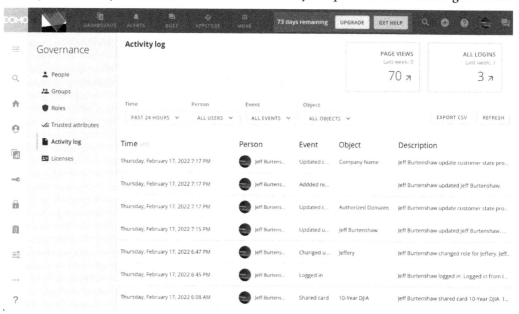

Figure 17.25 – Activity log

The **Activity log** page is useful for auditing and troubleshooting tasks.

Next, let's look at license management in Domo.

Using the Licenses page

Managing the user count in Domo is part of the governance process. To make this easy, Domo provides the **Licenses** page, as shown in *Figure 17.26*. This summarizes license use and the ability to purchase additional licenses with a click:

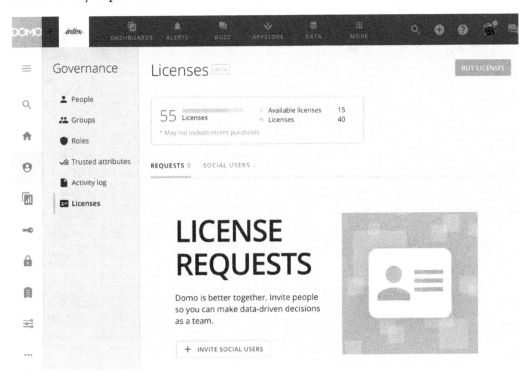

Figure 17.26 – The Licenses page

That concludes our review of the governance features. Next, we will cover how to secure content.

Securing content

The Domo administration includes features for managing content access for cards and pages and administering scheduled reports and publication groups. Let's go through each of the pages.

Understanding the Cards page

The **Cards** page lists all the cards and allows you to perform **EDIT** actions on the selected cards, as shown in *Figure 17.27*:

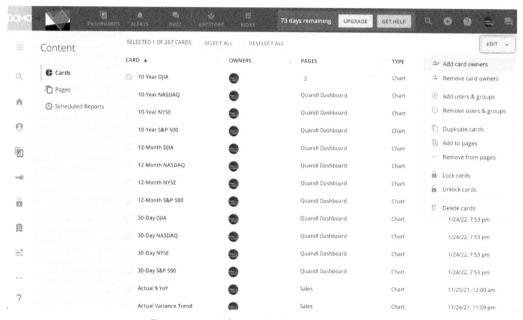

Figure 17.27 – The Cards admin content page

Individual or bulk operations are accessed using the **EDIT** drop-down menu and selecting the desired action.

Next, let's look at the **Pages** content admin page.

Understanding the Pages page

As expected, the **Pages** page lists all the pages in the Domo instance and enables you to execute administrative actions on the selected pages, as shown in *Figure 12.28*:

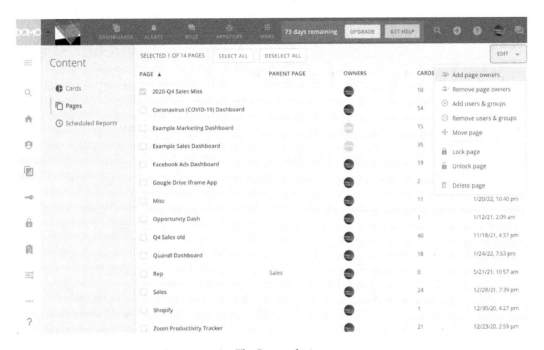

Figure 12.28 – The Pages admin content page

Select the cards to administer and then click on **EDIT**. Choose the desired function from the drop-down menu.

Next, let's review the administrative features for scheduled reports.

Understanding the Scheduled Reports page

The **Scheduled Reports** admin page lists all the scheduled reports and enables us to edit the report, edit the schedule, disable the schedule, or delete the schedule. Additionally, we can see the distribution history and manage the system-wide settings for report schedules.

Let's look at the **REPORTS** tab, as shown in *Figure 17.29*:

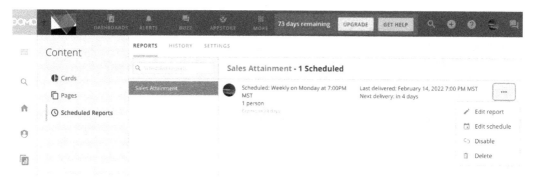

Figure 17.29 – The Scheduled Reports REPORTS tab

Clicking on the ... button opens the actions to be performed on the report schedule.

Next, let's touch on the importance of using PDP to secure content.

Using PDP

Recall that PDP and its how-to were covered, in detail, in the *Using the PDP tab* section of *Chapter 3*, *Storing Data*. However, it is important to remember that the application of PDP policies is a critical tool for controlling access to data and thereby scaling content deployment. PDP enables setting policies on what attributes to filter data on in a dataset. Trusted attributes or user-defined attributes can be used in the policy's filter rules. This policy-based row-level data filtering solution supercharges content scalability by allowing content to automatically filter the dataset rows based on the policy rules. This is so that a card or dashboard only displays the data rows that a user has access to from the PDP policy. Before PDP, dashboards and cards had to be created and filtered manually, causing mass duplication of cards and dashboards. A simple way to think of PDP is this: regardless of what a card displays, we want to control the data a user can see in a dataset based on a predefined filter policy. For example, if the user is a member of the sales department, then only show rows where the department is sales. Alternatively, if a production manager should only see data for their specific plant, then create a PDP policy for that on the dataset.

Next, let's review the features available for managing authentication.

Leveraging the authentication standards

The admin authentication section has features in which to manage password requirements, configure OpenID and **SAML Single Sign-On (SSO)**, and manage access tokens. Azure Active Directory, Google, Okta, PingIdentity, Salesforce.com, and authentication services are all supported.

Let's review the pages in this section.

Reviewing the Authentication page

The **Authentication** page controls password settings, as shown in *Figure 17.30*:

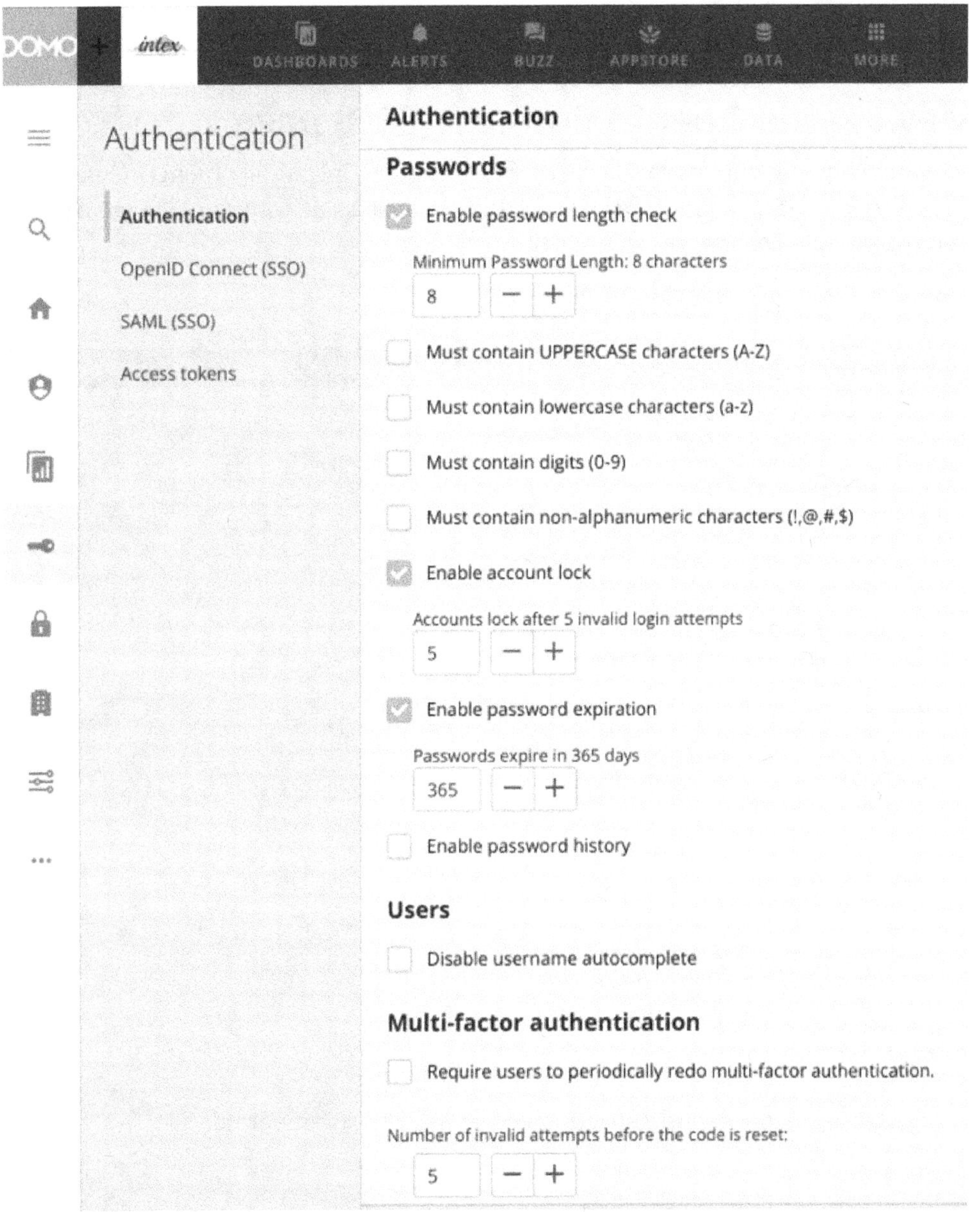

Figure 17.30 – Authentication settings

You can select options, as appropriate, for your organization's password policy.

> **Important Note**
>
> Domo supports multi-factor authentication on request as a system-wide setting, which on login, will text or email an access code to be entered to authenticate.

Reviewing the OpenID Connect (SSO) page

Domo supports the use of external authentication engines, including **OpenID Connect** (**OIDC**). To set up **SSO**, we must be in the **Admin** role or a role with the **Manage All Company Settings** grant active.

We can see the configuration page in *Figure 17.31*:

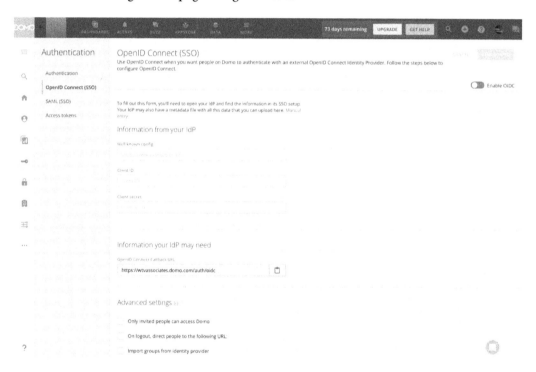

Figure 17.31 – The OpenID Connect SSO settings

Note that SAML-compliant authentication providers are also supported.

Reviewing the SAML (SSO) page

Domo supports **Security Assertion Markup Language** (**SAML**) as an authentication system. The configuration page is displayed in *Figure 17.32*:

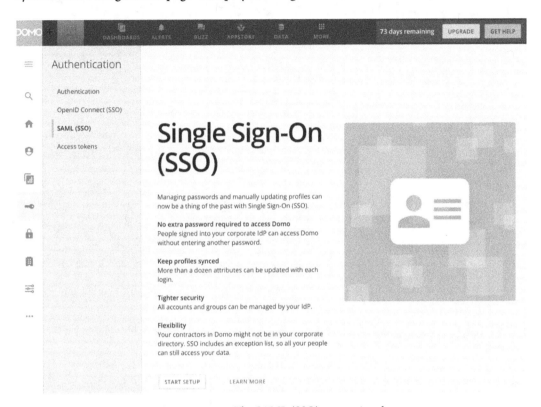

Figure 17.32 – The SAML (SSO) setup wizard

The SAML configuration will take us through a setup wizard. Click on **START SETUP** to begin. The use of SSO is recommended where feasible.

Now that we understand our SSO options, let's discuss access tokens.

Reviewing the Access tokens page

Access tokens are a way for users to enable connectors and apps using tokens to connect to Domo without having to log in, but rather using the user's security credentials via the token. To generate tokens, a user must be in a role that has the **Manage All Access Tokens** grant enabled. Tokens should not be shared as they effectively impersonate the user's access to Domo.

To create an access token, click on the **+ Generate access token** setting on the **Access tokens** page, as shown in *Figure 17.33*:

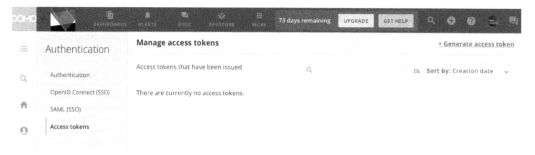

Figure 17.33 – Access tokens

> **Important Note**
> Tokens are not stored once they are generated. They should be stored by the user in their own system (such as in a password-protected file). Once you generate a token and navigate from the screen, you won't be able to see the token again.

Next, let's gain an understanding of the network security controls that are available.

Controlling network security

Domo has two options for network control: **Authorized Domains** and **IP Address Whitelist**. Network control refers to how Domo defines managing what domains and IP addresses are allowed to connect to the Domo services over the internet.

Authorized domains

Authorized domains are internet domains that are allowed to connect to the Domo instance. For example, if your company's domain is acme.com, and if you add acme.com to the **Domain name** list, then only users with emails ending in @acme.com can authenticate:

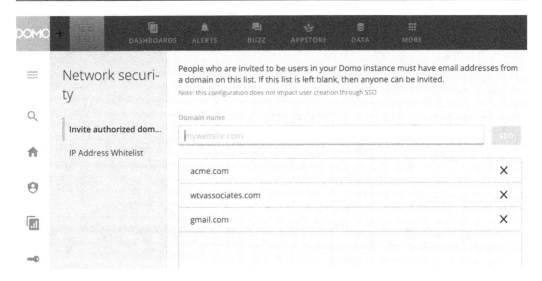

Figure 17.34 – The authorized domain list

Next, let's review the IP address whitelist feature.

IP address whitelisting

The user access control **IP Address Whitelist** feature enumerates IP addresses and ranges of IP addresses that are allowed to connect to our Domo instance. This is simply a list of the IP addresses that you want people to be allowed to log in from. If no IP addresses are listed, then all IP addresses are allowed. If there is a list, then logins are restricted to only those originating from addresses on the list, as shown in *Figure 17.35*:

Figure 17.35 – The user access control IP address whitelisting feature

> **Important Note**
>
> The user access control whitelisting feature is for user login purposes. Be aware that there is another possible need for whitelisting that sometimes gets confused with this feature. The other whitelisting purpose is for Domo connectors and federated adapters, which require customers to whitelist the Domo server IP addresses so that they can communicate with customer servers.

It is good to know that we have fine-grained control of the network locations from which users can connect.

Next, let's review the company settings of a Domo instance.

Configuring the company settings

The **Company settings** page allows us to set configurable instance settings. From the **Company overview** page, we can set the **Company name**, **Language**, and **Time zone** properties, identify the CEO of the org chart, and set other profile, security, and chart behaviors, as shown in *Figure 17.36*:

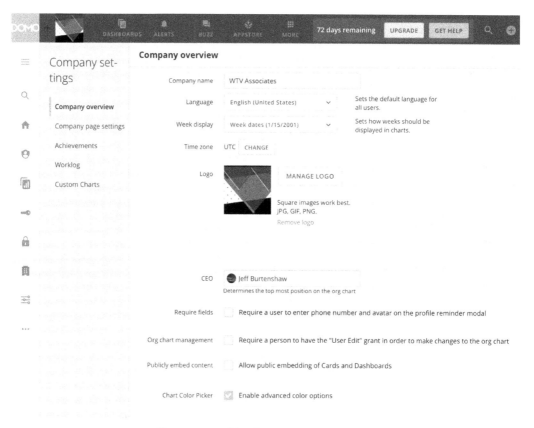

Figure 17.36 – The Company overview page

As shown in *Figure 17.37*, the **Company page settings** page enables us to set pages that will appear at the top of every user's **Dashboard** navigation panel. Additionally, we can set the default page that users land on when logging in:

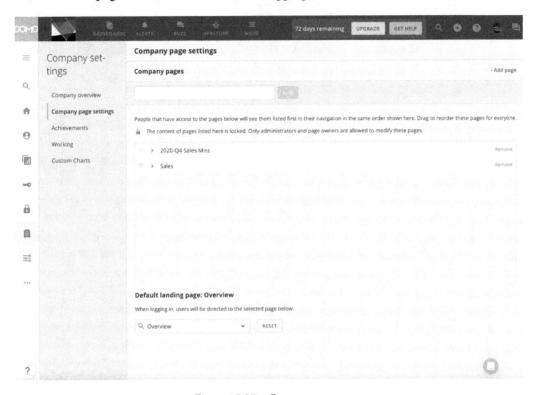

Figure 17.37 – Company pages

Click on **Add page** to search for and **add** a page as a company-wide page. Then, select an appropriate default landing page. Pages that have been added to the company-wide pages are listed at the top of the **Dashboards** navigation, as shown in *Figure 17.38*:

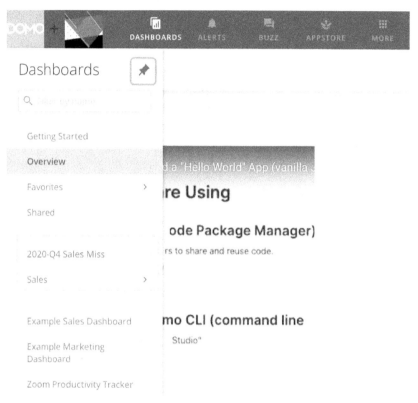

Figure 17.38 – The location of the company pages

Setting up these company settings is a good way to improve the user experience.

Next, let's set up the feature settings administration.

Managing the feature settings

The feature settings control the properties around Domo Everywhere, content certification, slideshows, and Buzz.

Configuring Domo Everywhere

The **Domo Everywhere** settings page enables us to browse through all of the embedded content links and their types. Using the **EDIT** feature, we can change the **EMBED TYPE** value, as shown in *Figure 17.39*:

Figure 17.39 – The Domo Everywhere settings

Now that we understand the settings, let's go through the steps to embed a dashboard.

Embedding the dashboards

Here are the steps to embed a Domo dashboard into a web page:

1. Navigate to the dashboard to be embedded.

> **Important Note**
> Access to **Embed Dashboard** is behind a feature switch, so you will need to reach out to Domo Sales to have this menu option activated.

2. Click on the share/export icon that is displayed on the page menu. Then, click on **Embed Dashboard**, as shown in *Figure 17.40*:

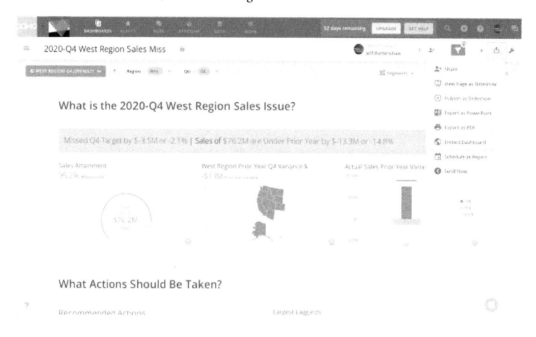

Figure 17.40 – The **Embed Dashboard** menu option

3. Set the **Embed options** drop-down menu to **Private**, as shown in *Figure 17.41*:

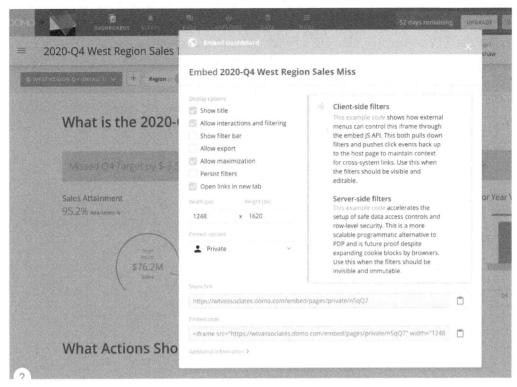

Figure 17.41 – The dashboard's embed settings

Here is a brief description of the **Embed options** settings:

- **Private** requires the user to be authenticated to a Domo instance to view the content.

- **Public** allows non-authenticated users to access the dashboard.

4. Copy the **Share link** details and paste them into a browser to view the dashboard. Anyone with this link who can authenticate to the instance will be able to access the dashboard.

5. If we want to embed the dashboard as an **iframe** on another website, copy the **Embed code** details into the HTML for the page.

Embedding dashboards is a powerful tool, allowing us to extend the reach of our content beyond the standard Domo user interface. Additionally, it can be used to accelerate the time to market for content delivery when deployed in internal or public-facing applications.

Certifying content

The **Certified Content** page allows us to establish certification processes for cards and datasets at a company and departmental level, as shown in *Figure 17.42*:

Figure 17.42 – The content certification administration page

Setting up a certification process is an important aspect of building trust in the content, as it visually differentiates the content that has been formally vetted.

Administering slideshows

The **Slideshows** administration page presents a list of slideshows that have been created and enables us to change the ownership and content of the presentation, as shown in *Figure 17.43*:

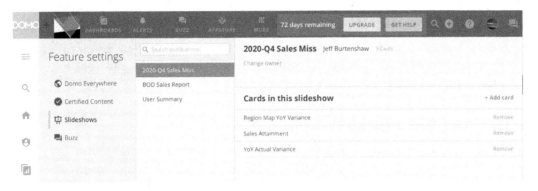

Figure 17.43 – The Slideshows administration page

Administering Buzz

The Buzz administration page enables us to control the available Buzz settings for a general company channel and whether social users can invite other social users. Additionally, it provides some statistics on the Buzz activity, as shown in *Figure 17.44*:

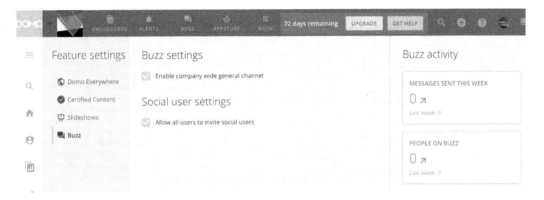

Figure 17.44 – The Buzz administration page

It is important to review these settings and make decisions on their configurations that fit the organization.

Next, let's review the final administration section.

Using more admin features

The **More administration features** section is a catch-all for miscellaneous administrative features. Currently, **Manage easy links**, **Exporting data**, **Alert Actions**, and **Tool Downloads** are located here.

Managing easy links

Easy Links are links that you send using the invite user feature to bring new team members into Domo. This page shows all the links and their statuses. Additionally, it allows us to turn the feature on or off and even **DEACTIVATE** specific links, as shown in *Figure 17.45*:

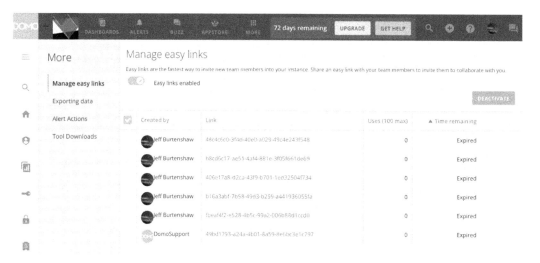

Figure 17.45 – The easy links administration page

Here is a list of some of the other options from the **More** section:

- The **Exporting data** setting has a single setting to disable Excel formulas on export for security purposes.

- **Alert Actions** has a single setting to enable webhook or API calls on alert triggers.

- **Tool Downloads** is the location where you can download Workbench, ODBC Driver, the PowerPoint, Excel, and Word plugins, the Domo CLI, the Google Sheets add-on, and Buzz Desktop.

Now that we have perused the administration features, let's discuss how to handle PII in Domo.

Handling PII in Domo

The ability to properly identify and handle the storage of PII is a necessity for preserving privacy, especially when considering the **HIPPA** and **GDPR** types of regulations. Domo's policy for PII is buyer-be-aware. This means that they are neither for nor against PII, but strongly suggest that raw PII is not stored in **Vault** without first **masking** or **redacting**. Domo has provided methods for redacting or masking data. Additionally, there are specific HIPPA-compliant server configurations available upon request and regional server configurations where solutions for jurisdictional regulations requiring data to be physically stored in the jurisdiction are available. Data is always encrypted while at rest in **Vault**. The **Bring Your Own Key** (**BYOK**) option lets the customer control their encryption keys and includes a rolling generation of data keys and a built-in kill switch user interface. Under **BYOK**, not even people from Domo's internal operations can get at the data because they don't have access to the decryption keys.

Understanding the Anonomatic partnership with Domo for PII

Domo partnered with **Anonomatic** to anonymize, mask, or redact sensitive PII data with full international data privacy compliance. **Poly-Anonymization**™ is used and takes PII data such as the name, gender, address, and government ID and replaces it with an anonymous ID in masked mode or with null in redacting mode. The mask is unique, unpredictable, and not hashed. After the data has been anonymized, it can be safely shared according to our data policies.

A Domo Workbench plugin for Anonomatic is used to apply the Poly-Anonymization methods while bringing the data into Domo through a Workbench job.

It is good to know that we have options around PII, HIPPA, and GDPR compliance. Let's review what we learned in this chapter.

Summary

In this chapter, we learned that there are a rich set of administrative features for securing the instance along with the cultural and organizational considerations that impact how to establish a security policy. Under governance, we discussed how to manage people, groups, roles, trusted attributes, activity logs, and licenses. For the content section, we reviewed the cards, pages, and scheduled reports administration features. Regarding authentication, we learned about Domo's authentication and how to use third-party authentication services for SSO. Additionally, we discussed multi-factor authentication and access tokens. The network security section covered domain and IP address whitelisting. And we went through many features around company settings, Domo Everywhere, tool downloads, and more. All of these features enable us to tailor our security policy implementation to fit our specific cultures and organizations. A number of options for safely and securely handling PII, HIPPA, GDPR, and encryption are available.

In the next chapter, we will look at organizational structures and the roles and responsibilities of the people involved with Domo in your organization.

Further reading

The following list contains some related items of interest:

- *Understanding and Configuring Domo Single Sign-On Using SAML*: `https://domohelp.domo.com/hc/en-us/articles/360042934374`

- Whitelisting for access control in Domo: `https://domohelp.domo.com/hc/en-us/articles/360043439173`

- *Whitelisting IP Addresses for Connectors & Federated Adapters*: `https://domohelp.domo.com/hc/en-us/articles/360043630093-Whitelisting-IP-Addresses-for-Connectors`

- *Enabling SSO with Azure Active Directory*: `https://domohelp.domo.com/hc/en-us/articles/360043438033-Enabling-SSO-with-Azure-Active-Directory`

- *Enabling SSO with OpenID Connect*: `https://domohelp.domo.com/hc/en-us/articles/360043438153-Enabling-SSO-with-OpenID-Connect`

- Dashboard embedding and using page filters: `https://domohelp.domo.com/hc/en-us/articles/4403508913943-Page-Filters-in-Dashboard-Embed`

- *PDP Policy Autocreation*: `https://domohelp.domo.com/hc/en-us/articles/360043439353-PDP-Policy-Autocreation`

- *Encrypting Data with Domo's Bring Your Own Key (BYOK)*: `https://domohelp.domo.com/hc/en-us/articles/360043427593-Encrypting-Data-with-Domo-s-Bring-Your-Own-Key-BYOK-`

- *What Constitutes PII?*: `https://domohelp.domo.com/hc/en-us/articles/360042934554-What-Constitutes-PII-`

- EU GDPR: `https://gdpr.eu/eu-gdpr-personal-data/`

- Using Anonomatic with Workbench for PII compliance for redacting or masking: `https://domohelp.domo.com/hc/en-us/articles/4409199658263-PII-Vault-Protection-and-Compliance`

18

Organizing the Team

Team organization is more important than technology to succeed in building analytics competency. The importance of the decisions we make about how to staff and divide responsibilities cannot be underestimated. Fortunately, there are best practices backed by years of implementation and organizational rollouts that can help us get the organizational structure optimized for success. Three key organizational roles exist to support the effective application of Domo in an organization: Executive Sponsor, MajorDomo, and Data Specialist. These roles should be formally defined and assigned. The roles can be held by a single person in a small organization or divided among many people in large organizations. In this chapter, we will define these roles and discuss trade-offs over a centralized versus a decentralized spectrum of organization structures for Domo teams.

In this chapter, we will specifically cover the following topics:

- Understanding best practices for a Domo organization structure
- Considering the implications of size on an organization's structure
- Reviewing the organization's structure roadmap
- Hiring for Domo team roles

Understanding best practices for a Domo organization structure

Organizationally, three primary roles with corresponding responsibilities are critical for a thriving analytics program, as seen in the following diagram:

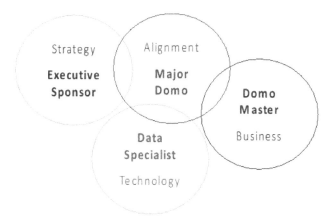

Figure 18.1 – Organization roles

Here is a description of each role and its responsibilities:

- **Executive Sponsor**: Senior executive owner of the *strategy* of where to apply Domo in the organization, including funding, initiative prioritization, success measures, staffing, and evangelization. Additional responsibilities include the following:

  - Program champion

  - Business case creation

  - Resource prioritization

  - Funding

  - Success metrics

  - Accountability

- **MajorDomo**: Leader who owns the overall success of the Domo program, ensuring value is realized. Drives tactical alignment of the business strategy and the technology. Other responsibilities include the following:

  - Managing the program

  - Implementing strategies

- Aligning initiatives and business objectives
- Communicating progress
- Establishing best practices

- **Domo Master**: Analyst who gathers business requirements and works with Data Specialists to create content. Handles user administration and project management. Their responsibilities include the following:

  - Defining requirements
  - Validating data
  - Creating content
  - User administration
  - Training
  - Adoption

- **Data Specialist**: Technical contributor role having data integration skills and access to source data. Additional responsibilities include the following:

  - Data acquisition
  - Data security
  - Data sculpting
  - Data reliability
  - Data access
  - Troubleshooting

Now that we understand the roles and responsibilities, let's consider how the size of the program impacts the distribution of roles and responsibilities.

Considering the implications of size on an organization's structure

The best-fit structure for distributing responsibilities of roles to people in an organization is largely a function of the size of the program in the organization, as seen in the following screenshot:

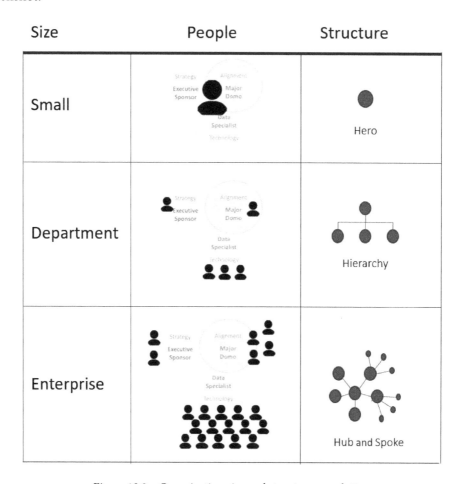

Figure 18.2 – Organization size and structure correlation

From the preceding screenshot, we can see that the following applies:

- **Small-sized programs**, which often involve a team of one individual, certainly can only be structured as a hero/heroine as the individual is simultaneously the Executive Sponsor, MajorDomo, and Data Specialist.

- **Department-sized programs** have more people involved and likely have the **Vice President** (**VP**) of the department as the sponsor, a manager as the MajorDomo, and a few **business analysts** (**BAs**) as Data Specialists.

- **Enterprise-sized programs** support many departments/regions and have many people involved. At this size, it is a good fit to use a hub-and-spoke structure.

Next, let's review how to choose where we are in the growth cycle and which organization structure is a good fit.

Reviewing the organization's structure roadmap

Organizations rarely stay static, and this is true for analytics teams supporting an organization. Consequently, we should expect the best-fit organizational structure for our analytic teams to change over time, as seen in the following diagram:

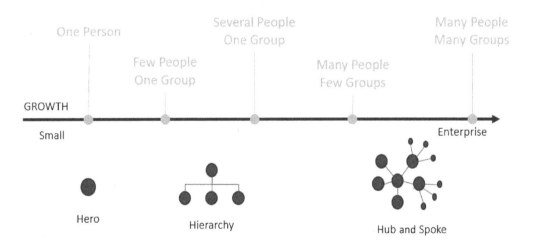

Figure 18.3 – Organization structure evolution roadmap

Fortunately, we can use *Figure 18.3* as a quick way to determine whether our current program's size and scope match the right organizational evolution for the program's current state. For example, if the program has several people supporting one group such as a marketing department or sales team, then a reporting structure of a single MajorDomo with several technical specialists would be a good fit.

However, as the team grows to support many groups and has many people, then it makes sense to have multiple MajorDomos for regions or product lines, each with its own Data Specialists organized in a hub-and-spoke fashion. The hub is a competency center or MajorDomo council, with the spokes being regional or product line-based support teams.

One of the fastest career paths to VP I have seen is via the MajorDomo role for individuals who grow a program from the Hero or Hierarchy stages to the Hub and Spoke stage. Quite often, the executive exposure of the MajorDomo's business skills during this growth can lead to General Manager-type roles as they know both the business metrics and operations and tend to have a broad understanding and good strategy-implementation experience.

> **Important Note**
>
> There are natural affiliations in organizations such as BAs familiar with their region or product lines and hence a strong affinity to have those resources reporting to the **business unit** (BU) leadership with a matrix responsibility to the competency center or council. There is also quite a natural affinity for more technical skillsets to be better served as a central pool of resources to be allocated, but as a group, they share technical standards and best practices.

Next, let's review a hiring resource for those of us who will be running analytical programs on Domo.

Hiring for Domo team roles

Having a good job description is critical in our hiring activity and, as mentioned, is the most critical decision in developing a high-performance team. Here are basic job descriptions for the MajorDomo and Data Specialist roles.

MajorDomo job description

The MajorDomo is a manager-level role directing the activity of Domo Masters and Data Specialists. The MajorDomo works with the Executive Sponsor to define and execute business strategies to be instrumented, tracked, and optimized by the Domo team. Budgeting and staffing experience is required, as well as experience in motivating teams. The right individual will have a **Master of Business Administration** (**MBA**) or equivalent business experience in driving profit and loss and controlling expense and risk. Certification for administration and governance of the Domo platform is desired as the security of the data is paramount. Establishing data policies and managing data assets are essential duties. This person will be the primary evangelist for Domo in the organization and will implement metrics to understand the uptake and value realized.

Domo Master job description

The Domo Master is an analyst who gathers business requirements and translates them into dashboards and stories. As an expert in their business area, they validate data and determine who has access to source systems data and Domo content. They work closely with Data Specialists to acquire and secure data. User administration and security setup are their responsibility. They are comfortable creating and distributing content with a focus on improving business performance.

Data Specialist job description

The Data Specialist role will help define success metrics and then acquire information assets and establish operational data pipelines to support accountability in performance. The right person will have experience with data visualization and storytelling, as well as familiarity with data pipelines, **Extract, Transform, and Load** (**ETL**) tools, **Structured Query Language** (**SQL**) statements, and Excel formulas. Experience in predictive modeling is a plus. Part BA and part data integration specialist, the person will be able to gain access to the required data sources and extract and sculpt the data using Domo connectors and Magic ETL toolsets. Beast-mode proficiency is needed, and basic experience in JavaScript, **Cascading Style Sheets** (**CSS**), and **HyperText Markup Language** (**HTML**) is desired for the creation of customized applications. The role is expected to partner with and understand the business impact of projects in which the organization is engaged and collaborate with stakeholders to produce better outcomes.

Next, let's examine a sample organization chart.

Sample organization chart

To help solidify these concepts, an example organization chart is provided for a mid-sized enterprise in the following screenshot:

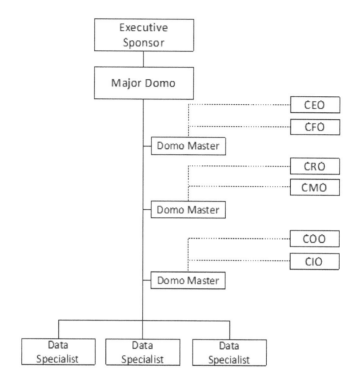

Figure 18.4 – Sample organization chart

In this sample organization, the Executive Sponsor could be any C-level individual. Whichever C-level individual is the Executive Sponsor, this will bias the direction of the program. For example, if the Executive Sponsor is a **Chief Marketing Officer (CMO)**, then customer and campaign data will be the focus; if the Executive Sponsor is a **Chief Revenue Officer (CRO)**, then the sales funnel will dominate; if the Executive Sponsor is the **Chief Executive Officer (CEO)** or **Chief Financial Officer (CFO)**, then financials will take the forefront; if the Executive Sponsor is the **Chief Operating Officer (COO)**, then operations will be paramount; and if the Executive Sponsor is the **Chief Information Officer (CIO)**, then data acquisition, structure, and security will tend to be the top priority. It is beneficial for organizations that are large enough to have Master Domos with deep domain expertise, and Data Specialists are typically a pooled resource allocated on priorities.

Let's review what we learned in this chapter.

Summary

In this chapter, we learned that organizational structure and personnel are key components that are more important than technology to run a successful analytics program. We reviewed the three key roles for a program: the Executive Sponsor, MajorDomo, and Data Specialists. The roles cover strategy creation, implementation, and technical operations. Organizational structures of Hero, Hierarchy, and Hub and Spoke were discussed. We stated that the best-fit organization structure will evolve over time and is highly correlated to the size of the program. A roadmap was reviewed, showing how to select the best-fit organization structure for the program. Hiring aids in the form of job descriptions were also provided, and a sample organization chart was presented for reference.

In the next chapter, we will look at the core processes needed for a successful program.

Further reading

Have a look at the following related item of interest:

- *Domo Program Roles*: `https://domohelp.domo.com/hc/en-us/articles/360042935174-Domo-Program-Roles`.

19
Establishing Standard Procedures

A good way to think about standard procedures is using an analogy of a traffic light. We have all waited at a busy intersection where a traffic light regulates the flow of traffic, but when that traffic light malfunctions, the wait can get much longer. Conversely, we may have been stopped at a traffic light with no other cars in sight, left wondering why we are just sitting here waiting for the light to turn green. The point is obvious: having the right amount of procedures is helpful, while having too many procedures is a waste.

When implementing Domo, there are a few core procedures that we need to establish. Here, we will show processes for indicating artifact ownership, accepting new requests, prioritizing and assigning request implementations, migrating changes from development to production, and certifying content accuracy. When these core processes are in place, trust is high, adoption increases, and performance improves.

In this chapter, we will specifically cover the following:

- Establishing ownership

- Implementing certification

- Accepting new content requests

- Managing content backlog

- Migrating artifact changes to production

Technical requirements

To follow along with this chapter, you will need the following:

- Internet access

- Your Domo instance and login

> **Important Note**
> If you don't have a Domo instance, get a free trial instance here:
> `https://www.domo.com/start/free`.

Establishing ownership

The most fundamental procedure to establish in any Domo implementation is that of ownership. More specifically, ownership of pages, cards, and datasets. Having ownership clearly defined facilitates quicker communication and problem resolution. When done correctly, proper ownership definition directs traffic to the right people as efficiently as possible. In fact, the platform has incorporated artifact ownership as an integral part of the platform, as shown in *Figure 19.1*. What is needed are clear policies and procedures to leverage these features:

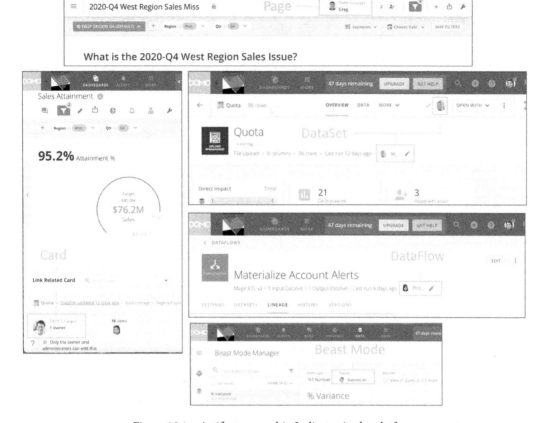

Figure 19.1 – Artifact ownership Indicator in the platform

The following table indicates the suggested relationships for artifact owners:

	ROLE				
ARTIFACT	Executive	Business Analyst	Major Domo	Data Specialist	Certification
Page	O				
Card		O			O
DataSet			O	O	O
DataFlow			O	O	
Beast Mode		O			

Figure 19.2 – Artifact ownership matrix

> **Important Note**
>
> For the highest accountability and most efficient communications, artifact ownership should always have a single person as the owner. For out-of-the-office situations, a group can also be designated as an artifact co-owner.

Let's look at more in-depth considerations for ownership policies in the following sections.

Determining page ownership

Ownership at the page level should be used to signal which business executive is responsible for the content. It is true that people other than the executive will have created the content, but it is important to signal to the organization that the business ultimately is accountable for the content. The reason for this is two-fold – trust and accuracy. If the ownership of the page is left to the designer or builder, that sends a strong signal that the business is not vested in the content. Also, the designer and creator of the page likely are not as familiar with the content as people in the executive's department. Assigning ownership to the executive forces the engagement of resources that have the context to immediately know whether the data is accurate.

If your organization's norm is for data specialists to be held accountable for the trust and accuracy of the data, then a cultural shift will be needed. Executives and analysts may not initially be comfortable taking this responsibility. We may need to explain that the data specialists will not be as familiar with the business context as the analyst or functional leader and that the analysts (because of context) are in a better position to validate data accuracy. Ideally, the dashboard ownership is held by the executive, and having their avatar as the dashboard owner sends a clear message that the content is approved by the leader. To get executive ownership buy-in, a simple approach is to establish a policy that the appropriate executive is accountable for the dashboard content and should take steps with analysts to validate that the content is accurate while holding major-domos and data specialists accountable for accessibility, timeliness, and structure of the data.

Next, let's discuss considerations for card ownership.

Determining card ownership

Each page consists of cards created by a potentially wide variety of people. At the card level, ownership should be held by the business analyst or person who is most familiar with the content and its use in the business. If this person is not the card creator, then ownership should be changed on the transition from the creator to the user or content **curator**. Often, the best owner of a card will be someone in the page owner's direct management line. Being disciplined about card ownership greatly reduces communication time around issues with the content, as the analyst owner will be the most familiar with the information presented and its status.

Next, let's discuss beast mode artifact ownership.

Determining beast mode ownership

Beast mode ownership closely follows the reasoning for card ownership and should reside with a business analyst in the executive owner's management chain. Who better to own business rule definitions contained in beast modes than the analyst who is defining the metric for the business? Note, a major-domo may assist in the creation, testing, and problem resolution activity, but should not be the designated owner. Joint or group ownership is discouraged, as it reduces accountability and confuses communication lines around issue resolution. This does not prevent the designated owner from involving committees or groups to get organizational buy-in, but that should not replace the primary ownership of a single individual.

Next, let's discuss ownership of datasets.

Determining Dataset ownership

Datasets are often created by a major-domo or data specialist, especially in all but the smallest of organizations. In the smaller organization, it may be a business analyst wearing the data specialist hat to intake data. The data specialist is the logical choice as the owner. The data specialist will be the person with access to, and familiarity with, the source data from a technical definition and interfacing perspective. Having a clear owner on the dataset facilitates rapid and direct escalation of issues around the schema and data freshness. The major-domo, or data specialist, is the **operations** owner for the dataset.

Next, let's discuss dataflow artifact ownership.

Determining DataFlow ownership

The ownership of dataflows is a clear-cut decision. The creator of the dataflow, where possible, should also be the owner of the dataflow. The creator understands the logic and process required to create the pipeline and is the best person to respond to questions and issues.

Now that we've established ownership and covered the different considerations, let's examine how certification procedures fit in.

Implementing certification

Certification differentiates content visually with a certificate icon and signals that the certified artifact has been through a rigorous vetting process that can be trusted.

Following the ownership framework in the platform enables a clear path for certification approvals. The platform supports content certification processes on cards and datasets. Approval processes to support company-wide and department levels are also supported.

Implementing the card certification process

It is easy to set up the certification process for cards and one of the most important ways to gain trust in the content from users.

Configuring company cards certification

The following are the steps to configure the company-level approval process for cards:

1. Starting on the main menu bar, click **MORE** | **ADMIN** | **Feature Settings** | **Certified Content**, and then **Certified Cards**, as shown in the following screenshot:

Figure 19.3 – Certification configuration page

2. Click on the **Company** tile under **Company certified**.

3. Click **Enabled** and enter `Please review and approve or deny certification of the requested content` in the **Instructions** box.

4. Enter the people in **Approval Chain**; in this case, we chose the people with the **MAJOR DOMO** and **CEO** roles as the approvers in **Approval Chain** for company-level cards, as seen in *Figure 19.4*:

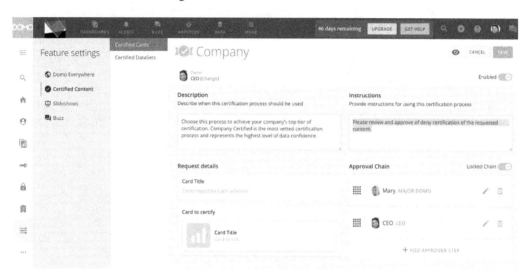

Figure 19.4 – Company cards certification configuration

5. Click **SAVE**.

Configuring department cards certification

The following are the steps to configure the department-level approval process for cards:

1. From the main menu bar, click **MORE | CERTIFICATION CENTER | Settings | Certified Cards**, then scroll down to the **Departments** section and click the + tile, as seen in *Figure 19.5*:

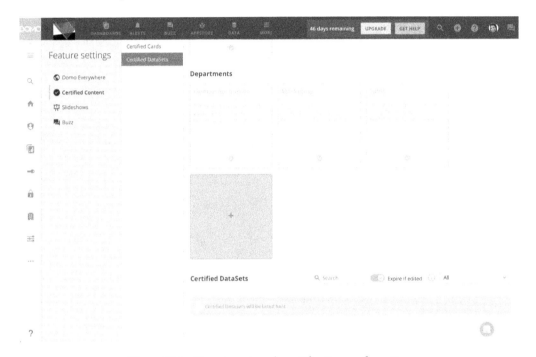

Figure 19.5 – Department cards certification configuration

2. Enter Finance in the **Department title** area.

3. Enter All cards desired to be published on the CFO page should be first vetted via this certification approval process in **Description**.

4. Enter Use the Certification Center to submit a card in **Instructions**.

5. Click **Change** under **Owner** and select the CFO user for your organization (we will need to invite the CFO and her organization into the instance first). See *Figure 19.6* for an example of the finance department organization used here:

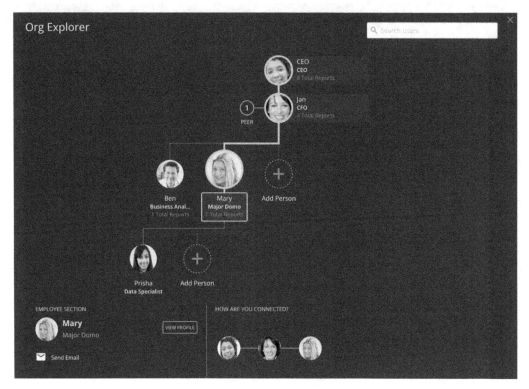

Figure 19.6 – Org Explorer

6. Click **Unlocked Chain** to allow the approvers to delegate approval to someone else on each request if they wish.

7. Click + **ADD APPROVER STEP** in the **Approval Chain** area and select the people with the **BUSINESS ANALYST** and **CFO** positions as the **Approval Chain** members for department-certified cards, as seen in *Figure 19.7*:

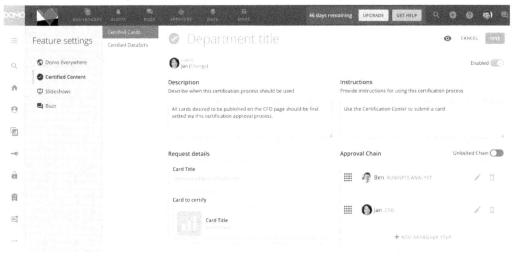

Figure 19.7 – Department card certification configuration

8. Click **SAVE**. Now, we have configured the certification approval processes for company and finance department cards, as shown in *Figure 19.8*:

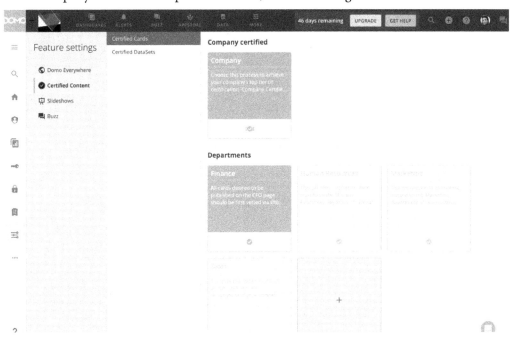

Figure 19.8 – Completed card certification approval process configuration

Next, let's discuss the dataset certification configuration.

Implementing the dataset certification process

Equal in importance to card certification is the process to certify datasets. The process to configure datasets is like the process for cards with the major difference being who participates in the approval chain. For datasets, the approval chain will likely be through a business analyst and then the major-domo. That way, a person close to the business data has approved, as well as the major-domo, who can evaluate the dataset for potential overlap and compatibility with other certified content. The certification of datasets also has distinct processes for the company and departments.

This feature is accessed by clicking the main menu option of **MORE | CERTIFICATION CENTER | Settings | Certified Datasets**, as shown in *Figure 19.9*:

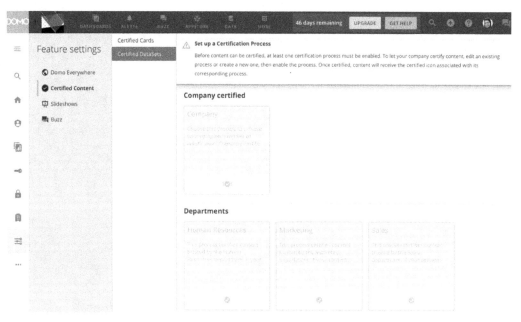

Figure 19.9 – Dataset certification configuration setup

Now that we have the certification process configured, let's see how to submit an artifact for approval.

Submitting certification requests

Submitting a card or dataset for certification approval starts from the artifact.

Submitting a card for certification

The following are the steps to submit a certification request from a card:

1. Navigate to the **Sales** dashboard and click on the **Sales Attainment** card, then click the wrench option icon, and click **Request certification**, as shown in *Figure 19.10*:

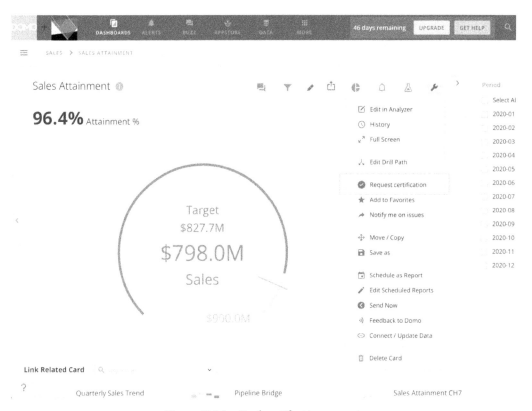

Figure 19.10 – Card certification request

2. In the **Certify Card** popup in the **Select a process** dropdown, choose **Company**, as shown in *Figure 19.11*:

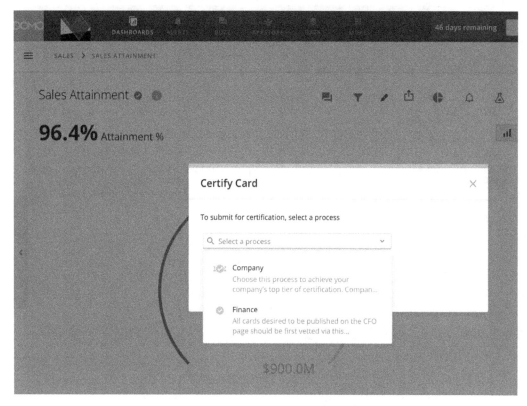

Figure 19.11 – Request company card certification

3. Review the **Company** certification request and click **SUBMIT** to initiate the approval process, as shown in *Figure 19.12*:

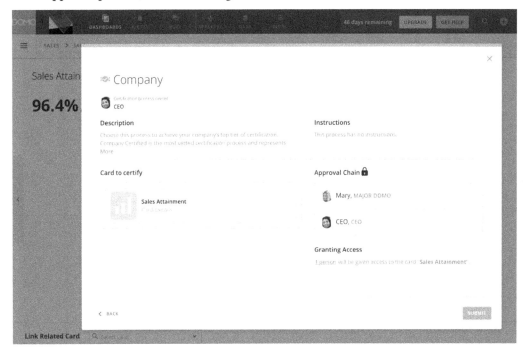

Figure 19.12 – Submit company card certification request

4. From the main menu bar, click **MORE | CERTIFICATION CENTER** and note the **SUBMITTED BY YOU** count has increased, as shown in *Figure 19.13*:

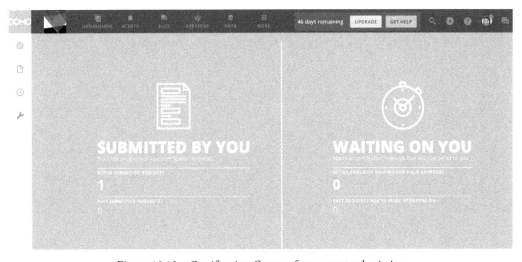

Figure 19.13 – Certification Center after request submission

5. Click on **SUBMITTED BY YOU**, then click **Sales Attainment** to see the request details, as shown in *Figure 19.14*:

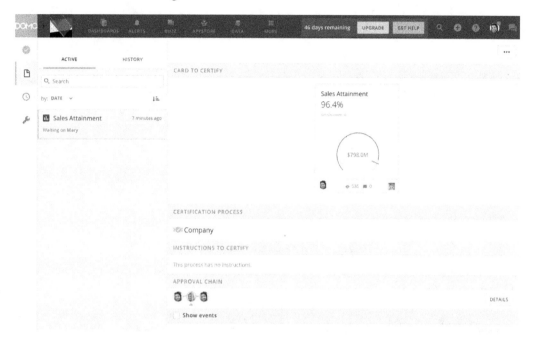

Figure 19.14 – Certification Center submitted request detail

Next, let's discuss the certification request submission for datasets.

Submitting a dataset for certification

Datasets can be individually submitted for certification.

The following are the steps to submit a certification request from a dataset:

1. From the main menu, click **DATA**, then click the **DataSets** option on the left toolbar.

2. Search for `opportunity ch7` and click the **Opportunity CH7** dataset, as shown in *Figure 19.15*:

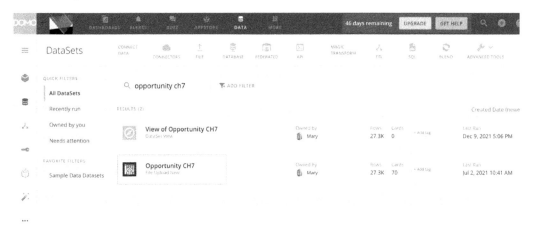

Figure 19.15 – Dataset selection

3. Click the DataSet options dropdown and click **Request Certification**, as shown in *Figure 19.16*:

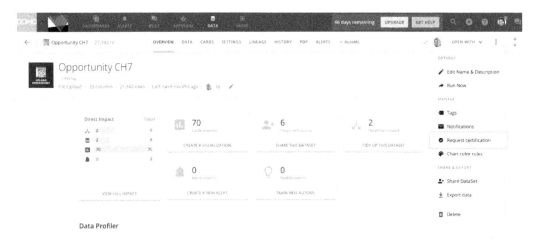

Figure 19.16 – Dataset Request certification

4. If you are not the dataset owner, it will be routed to the owner to request certification, as shown in *Figure 19.17*:

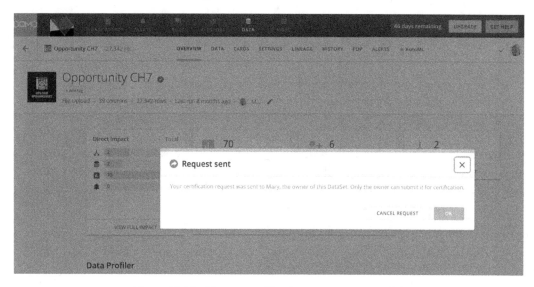

Figure 19.17 – Dataset certification request owner relay

5. Click **OK**.

6. Click **MORE** in the main menu, and click on **CERTIFICATION CENTER** to see the updated request count, as shown in *Figure 19.18*:

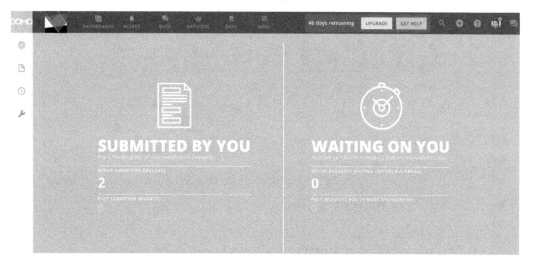

Figure 19.18 – Updated Certification Center

Fantastic! Now that we understand content certification, let's review the process for onboarding new content.

Accepting new content requests

New content is the lifeblood of a healthy data democracy. There will always be the next question to be answered and we need an effective way to capture and prioritize all the requests. To get a feel for the magnitude of demand, in a $2 billion manufacturing enterprise, there was a backlog of over 30,000 distinct requests created in a year for information. Not surprisingly, the users stopped requesting information because, at the time, the capacity to deliver was 10 requests a month. Yes, depressing isn't it? As great as the Domo platform is, there is a lot of pent-up demand in organizations for information. It is wise to have a strong request, review, prioritize, and assign or reject procedure in place to focus the energy applied to answering relevant questions.

Fortunately, the Domo platform includes tools to capture and manage requests. Using **Projects and Tasks**, we can set up work queues for vetting requests and migrating requests to the best follow-up queue, as shown in *Figure 19.19*:

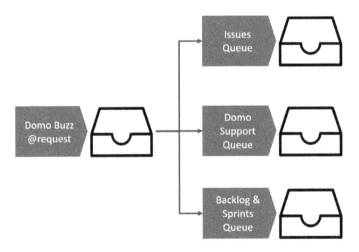

Figure 19.19 – Project and Tasks queues

The following is a description of the process:

1. Click on the main menu and click **MORE | PROJECTS AND TASKS | ADD PROJECT** to create four new projects in **Projects and Tasks**, as shown in *Figure 19.20*:

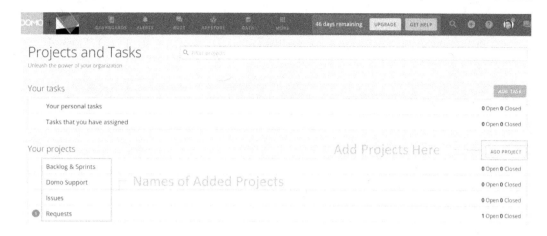

Figure 19.20 – Projects to add

2. Add the Requests project with lists, as shown in *Figure 19.21*, to queue new requests either coming from Buzz or created manually as a project task:

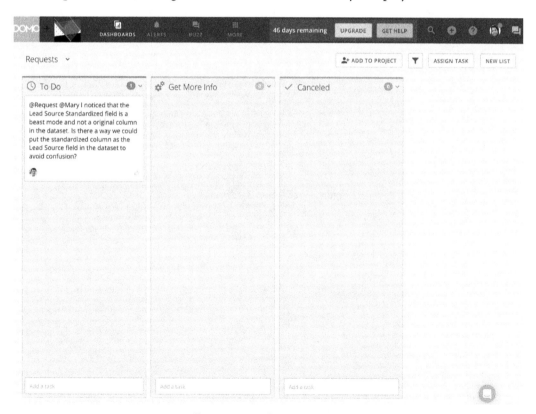

Figure 19.21 – Request project

3. Add the Issues project with lists, as shown in *Figure 19.22*, to queue new issues either coming from Buzz or created manually as a project task. Issues are typically addressed by a major-domo before being referred to Domo Support.

Figure 19.22 – Issues project

4. Add the Domo Support project with lists, as shown in *Figure 19.23*, to queue and track new support tickets either coming from Buzz or created manually as a project task and logged with Domo Support. It is advisable to have your own record of submissions to Domo Support.

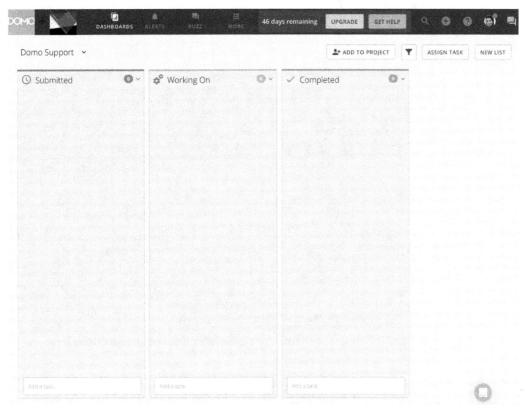

Figure 19.23 – Domo Support project

5. Add the `Backlog & Sprints` project with lists, as shown in *Figure 19.24*, to queue content additions or enhancement requests either coming from Buzz or created manually as a project task. This queue is only for development additions or changes, as the questions and support issues are handled in the other queues.

Figure 19.24 – Backlog & Sprints project

Now that the projects are set up and we have queues to organize the request flow, let's follow an issue request through the process.

Handling an issue request

Domo Buzz is a fast way to submit issue requests. Let's walk through an example:

1. First, we must do a one-time setup of a `Request` user in Domo so all requests can be sent to that operational handle.

2. Looking at the **Sales Attainment** card, we notice that it was updated 13 days ago. Then, by clicking on the **Quota** dataset, we can see who the owner is and create a Buzz message to the owner and request to see whether the dataset can be updated more frequently, as shown in *Figure 19.25*:

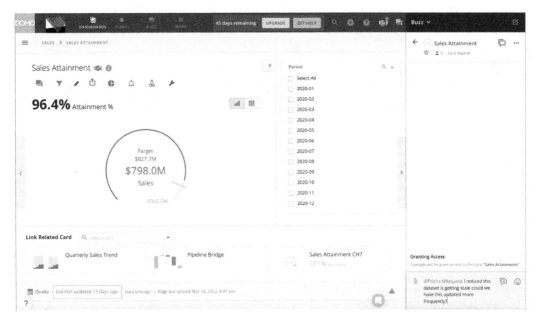

Figure 19.25 – New request

3. In the Buzz thread, we can click on the **...** message options and click **Create Task**, as shown in *Figure 19.26*:

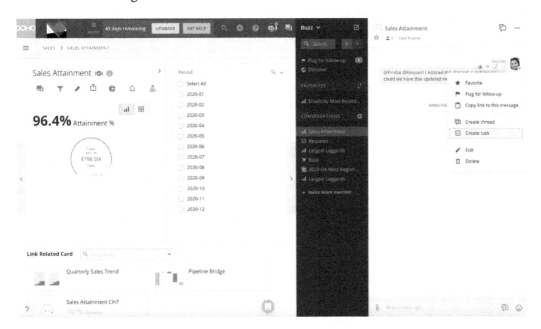

Figure 19.26 – Create task from Buzz message

4. In the task popup, click **part of a project**, choose **Requests** from the **Project** dropdown, and click **SAVE**, as shown in *Figure 19.27*:

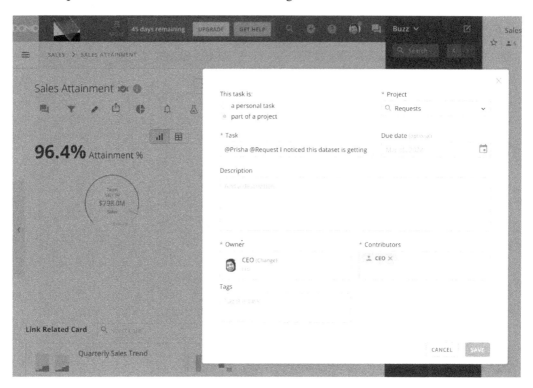

Figure 19.27 – Save project task

5. Open the Requests project in **Projects and Tasks** by clicking **Requests** under **Your Projects**, as seen in *Figure 19.20*.

6. Notice the new card on the **To Do** list, as shown in *Figure 19.28*:

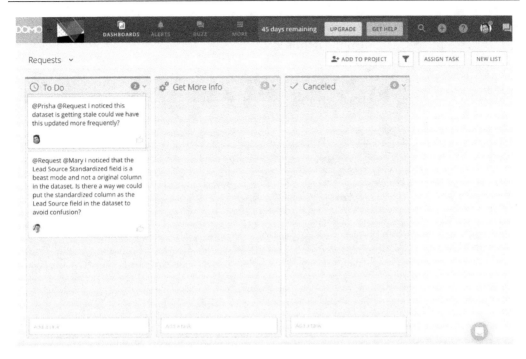

Figure 19.28 – Request To Do list

7. Click on the task card and then click **Requests** in the **in** dropdown, and select **Issues** to move the task from the Requests project to the Issues project, as shown in *Figure 19.29*:

Figure 19.29 – Task details popup

8. Click **SAVE** and then change the **Requests** dropdown to **Issues** to change to the
    Issues project, as shown in *Figure 19.30*:

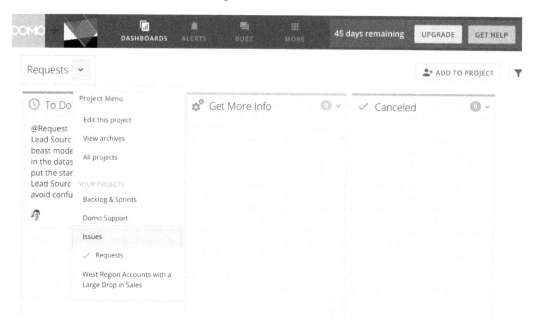

Figure 19.30 – Change project viewed

9. In the **Issues** project view, click and drag the task card from the **To Do** List and drop
    it into the **Working On** list, as shown in *Figure 19.31*:

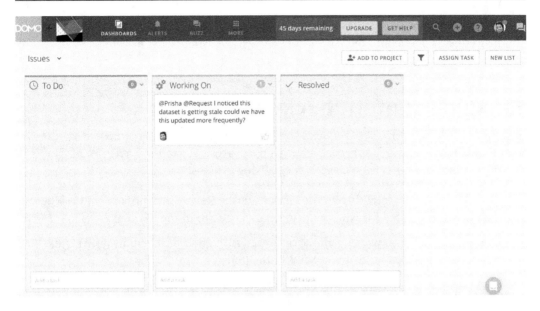

Figure 19.31 – Progress task

10. Click on the task tile and reassign a task owner, as shown in *Figure 19.32*:

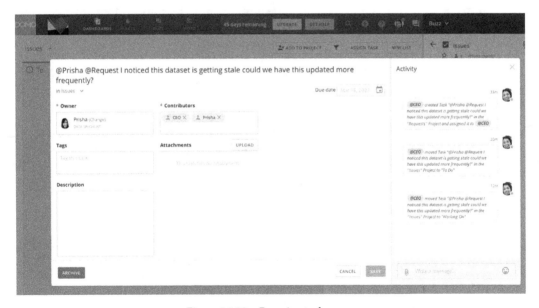

Figure 19.32 – Reassign task owner

11. Click **SAVE**.

Great! We learned how to put a task into the **Requests** queue and then how to move it from the **Requests** queue into the **Issues** queue. Next, we will discuss how to use the `Backlog & Sprints` project to manage the content backlog.

Managing content backlog

The **content backlog** is the list of requests that are candidates to be implemented via the **Sprint Planning** process. We suggest a seven-stage process, as seen in *Figure 19.24*. The following is a description of each stage in the backlog process:

- **New**: This is a list containing all the tasks from the `Requests` project, and is where each task is evaluated for what stage to be advanced to in the backlog.

- **Hold**: This is a list containing all the tasks determined to be given on-hold status. A task can be put on hold at any point in the journey.

- **Backlog**: This is a list containing all the tasks from the **New** list that have been determined to qualify for implementation at some point. This is the idea queue for potential development work. Items in the **Backlog** list can be placed in priority order from top to bottom by dragging and dropping a task card within the list.

- **Planned Sprint**: This is a list enabling the next Sprint to be planned for convenience. A common **Sprint boundary** is 2 weeks. When a new Sprint begins, the tasks from the **Planned Sprint** list are moved over to **Current Sprint**.

- **Current Sprint**: This is a list containing all the current Sprint tasks assigned. When the task is complete, it is moved to the **Done** list. It is also possible to move the task to the **Canceled** list.

- **Done**: This is a list containing all the completed tasks from the **Current Sprint** list over time. This list will eventually become large and it is suggested that tasks are archived occasionally. Tasks are archived from the task detail popup.

- **Canceled**: This is a list containing all the tasks that have been canceled and are no longer in the backlog.

Important Point

Many methods and tools exist for prioritizing the backlog, including feasibility and value matrices. The specific prioritization algorithm is less important than having a way to prioritize. **Project and Tasks** provides organizations with a tool and process if they do not already have such a process in place. Many organizations bring data from their existing request management tool of choice into Domo for management and optimization.

Migrating artifact changes to production

In code development, it is standard practice to have a formal process to migrate changes from a development environment to a production environment. However, with Business Intelligence systems that is often in the category of sitting at the traffic light in the middle of the night waiting, even though there are no other vehicles on the road. Let's discuss when the risk is high or low in the Domo platform to determine the best fit strategy for migration. The following table shows some common characteristics for this evaluation and the appropriate strategy for migration:

Characteristic	Risk	Full Stack Migration	Card Versioning	DataFlow Versioning
Street Financial Reporting	Misstating Financials Impacts on Public Stock Price	✓		
Sales Compensation	Over/Under Pay	✓		
Marketing Analytics	Misallocation of Funds		✓	✓
Operational Monitoring	Temporarily Compromised Diagnostics		✓	✓
Department KPIs	Temporarily Compromised Analysis		✓	✓
Changing Column Names in a Key Dataset	Artifact Breakage that may be Time-Consuming to fix	✓		

Figure 19.33 – Artifact migration strategy considerations

The following are broad definitions of the strategies for migration:

- **Full-stack migration**: This strategy requires the for-fee Sandbox feature to maintain a development environment separate from the production environment where all artifacts can be developed and tested before being migrated to the production environment. This strategy is best applied when there are significant and permanent consequences for producing erroneous data or not having the data available when needed (for example, public financial reporting or sales compensation). Additionally, when a key data source schema changes, then this strategy is recommended to find artifact dependencies such as card, beast mode, and SQL dataflow dependencies.

- **Card versioning**: This strategy is appropriate for most card development as it is simple to duplicate cards and make changes, then move the original card to an archive page for versioning. Cards themselves cannot change the underlying data. Employing a full-stack strategy would be overkill in many card development scenarios. For example, cards dealing with department KPIs or operational monitoring are typically easy to recover from when the mistake is discovered.

- **Dataflow versioning**: This strategy should be applied carefully and less frequently than the card versioning option. But, since every version of a dataflow is captured in Domo's version control, it is a simple operation to reset back to any prior version of the dataflow. This enables recovery to be quick and relatively painless and the speed of development is often more beneficial than the risk of a temporary outage.

The bottom line is, for the conservative among us with cash to spend on the Sandbox, use full-stack migration. For the rest of us, we can likely take the risk and be fine with the card and dataflow versioning approaches.

> **Important Note**
>
> A reminder that datasets are automatically versioned in the vault and the CLI allows accessing data versions directly. There is a limit determined by Domo operations to how many data versions are kept for storage cost efficiencies in Domo. There is no guarantee the versioned data will be available, but I have been able to use this to recover datasets in a pinch. If you want guaranteed recovery, create a scheduled ETL job to make a copy of a dataset.

Next, let's review what we learned in this chapter.

Summary

In this chapter, we learned that there is a balance between having not enough or having too much process. We discussed key procedures for artifact ownership and how ownership facilitates trust, accuracy, and communication speed. We reviewed how to configure and execute a content certification process. We defined a workflow and set up projects for collecting and managing request queues via Domo Projects and Tasks and Buzz. We walked through the phases and steps for managing the content backlog. And finally, we addressed the strategy options available for dealing with artifact development and migration. As a result, we are very prepared to put in place just the right amount of process to regulate the traffic in our data democracy.

Democracies derive power from their people where each voice matters. Democratizing data is extending the boundaries of who has access to and consumes data to meet and optimize their responsibilities. The Domo platform brings revolutionary capabilities for acquiring, sculpting, storytelling, distributing, and managing information at internet scale and speed. Organizations that embrace data democratization will thrive as better-informed decisions are made, and higher satisfaction levels are attained at all levels. Organizations that continue to live in totalitarian data regimes will not be able to compete over the long term.

Further reading

The following are related items of interest:

- Process Ownership Vital Role Six Sigma Success: `https://www.isixsigma.com/implementation/change-management-implementation/process-ownership-vital-role-six-sigma-success/`

- Backlog Grooming: `https://www.productplan.com/glossary/backlog-grooming/`

Index

C

E

Packt.com

Subscribe to our online digital library for full access to over 7,000 books and videos, as well as industry leading tools to help you plan your personal development and advance your career. For more information, please visit our website.

Why subscribe?

- Spend less time learning and more time coding with practical eBooks and Videos from over 4,000 industry professionals

- Improve your learning with Skill Plans built especially for you

- Get a free eBook or video every month

- Fully searchable for easy access to vital information

- Copy and paste, print, and bookmark content

Did you know that Packt offers eBook versions of every book published, with PDF and ePub files available? You can upgrade to the eBook version at packt.com and as a print book customer, you are entitled to a discount on the eBook copy. Get in touch with us at customercare@packtpub.com for more details.

At www.packt.com, you can also read a collection of free technical articles, sign up for a range of free newsletters, and receive exclusive discounts and offers on Packt books and eBooks.

Other Books You May Enjoy

If you enjoyed this book, you may be interested in these other books by Packt:

Learn Power BI - Second Edition

Greg Deckler

ISBN: 9781801811958

- Get up and running quickly with Power BI
- Understand and plan your business intelligence projects
- Connect to and transform data using Power Query
- Create data models optimized for analysis and reporting
- Perform simple and complex DAX calculations to enhance analysis
- Discover business insights and create professional reports
- Collaborate via Power BI dashboards, apps, goals, and scorecards
- Deploy and govern Power BI, including using deployment pipelines

Creating Actionable Insights Using Tableau CRM

Mark Tossell

ISBN: 9781801074391

- Implement and configure Tableau CRM from scratch
- Build your first Tableau CRM analytics app and embed your Tableau CRM dashboards in Salesforce to enhance user adoption
- Connect Salesforce and external data with Tableau CRM and create datasets
- Create a data recipe and get familiar with the recipe UI
- Build a custom dashboard in Tableau CRM using the dashboard editor
- Use lenses to create a Tableau CRM analytics dashboard
- Configure and implement data security and governance
- Build configured record actions to automate data directly in Salesforce

Packt is searching for authors like you

If you're interested in becoming an author for Packt, please visit `authors.packtpub.com` and apply today. We have worked with thousands of developers and tech professionals, just like you, to help them share their insight with the global tech community. You can make a general application, apply for a specific hot topic that we are recruiting an author for, or submit your own idea.

Share Your Thoughts

Now you've finished *Data Democratization with Domo*, we'd love to hear your thoughts! Scan the QR code below to go straight to the Amazon review page for this book and share your feedback or leave a review on the site that you purchased it from.

`https://packt.link/r/1-800-56842-8`

Your review is important to us and the tech community and will help us make sure we're delivering excellent quality content.

www.ingramcontent.com/pod-product-compliance
Lightning Source LLC
LaVergne TN
LVHW081327050326
832903LV00024B/1054